GUIN BOOKS

SICS OF THE FUTURE

Internationa...imed physicist Dr Michio Kaku holds the Henry
Semat Chair in ...oretical Physics at the City University of New York. He
is also an international bestselling author, his books including *Hyperspace*
and *Parallel Worlds*, and a distinguished writer, having featured in *Time*,
the *Wall Street Journal*, the *Sunday Times* and the *New Scientist* to name
but a few. Dr Kaku also hosts his own radio show, 'Science Fantastic', and
recently presented the BBC's popular series 'Time'.

Physics of the Future

The Inventions That Will Transform Our Lives

MICHIO KAKU

PENGUIN BOOKS

PENGUIN BOOKS

Published by the Penguin Group
Penguin Books Ltd, 80 Strand, London WC2R ORL, England
Penguin Group (USA) Inc., 375 Hudson Street, New York, New York 10014, USA
Penguin Group (Canada), 90 Eglinton Avenue East, Suite 700, Toronto, Ontario, Canada M4P 2Y3
(a division of Pearson Penguin Canada Inc.)
Penguin Ireland, 25 St Stephen's Green, Dublin 2, Ireland
(a division of Penguin Books Ltd)
Penguin Group (Australia), 250 Camberwell Road, Camberwell, Victoria 3124, Australia
(a division of Pearson Australia Group Pty Ltd)
Penguin Books India Pvt Ltd, 11 Community Centre, Panchsheel Park, New Delhi – 110 017, India
Penguin Group (NZ), 67 Apollo Drive, Rosedale, Auckland 0632, New Zealand
(a division of Pearson New Zealand Ltd)
Penguin Books (South Africa) (Pty) Ltd, 24 Sturdee Avenue, Rosebank, Johannesburg 2196, South Africa

Penguin Books Ltd, Registered Offices: 80 Strand, London WC2R ORL, England

www.penguin.com

First published in the United States of America by Doubleday, a division of Random House, Inc. 2011
First published in Great Britain by Allen Lane 2011
Published in Penguin Books 2012
001

978-0-141-04424-8

www.greenpenguin.co.uk

Penguin Books is committed to a sustainable
future for our business, our readers and our planet.
This book is made from Forest Stewardship
Council™ certified paper.

ALWAYS LEARNING   **PEARSON**

To my loving wife, Shizue,

and my daughters, Michelle and Alyson

CONTENTS

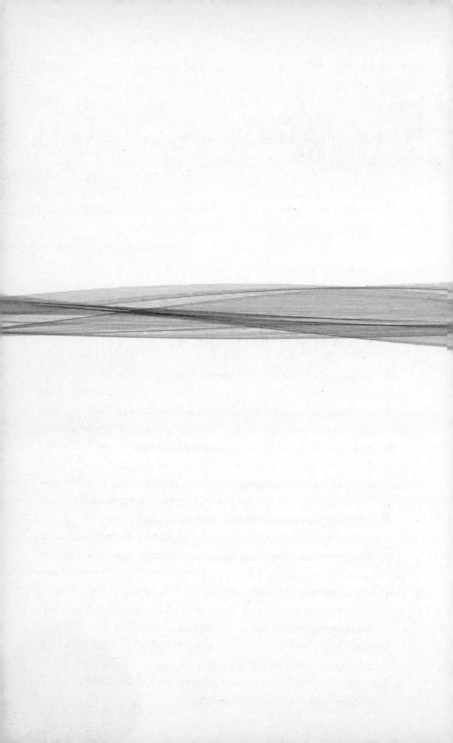

ACKNOWLEDGMENTS

I would like to thank those individuals who have worked tirelessly to make this book a success. First, I would like to thank my editors, Roger Scholl, who guided so many of my previous books and came up with the idea for a challenging book like this, and also Edward Kastenmeier, who has patiently made countless suggestions and revisions to this book that have greatly strengthened and enhanced its presentation. I would also like to thank Stuart Krichevsky, my agent for so many years, who has always encouraged me to take on newer and more exciting challenges.

And, of course, I would like to thank the more than three hundred scientists I interviewed or had discussions with concerning science. I would like to apologize for dragging a TV camera crew from BBC-TV or the Discovery and Science channels into their laboratories and thrusting a microphone and TV camera in front of their faces. This might have disrupted their research, but I hope that the final product was worth it.

I would like to thank some of these pioneers and trailblazers:

Eric Chivian, Nobel laureate, Center for Health and the Global
    Environment, Harvard Medical School

Peter Doherty, Nobel laureate, St. Jude Children's Research Hospital
Gerald Edelman, Nobel laureate, Scripps Research Institute
Murray Gell-Mann, Nobel laureate, Santa Fe Institute and Caltech
Walter Gilbert, Nobel laureate, Harvard University
David Gross, Nobel laureate, Kavli Institute for Theoretical Physics
the late Henry Kendall, Nobel laureate, MIT
Leon Lederman, Nobel laureate, Illinois Institute of Technology
Yoichiro Nambu, Nobel laureate, University of Chicago
Henry Pollack, Nobel laureate, University of Michigan
Joseph Rotblat, Nobel laureate, St. Bartholomew's Hospital
Steven Weinberg, Nobel laureate, University of Texas at Austin
Frank Wilczek, Nobel laureate, MIT

Amir Aczel, author of *Uranium Wars*
Buzz Aldrin, former NASA astronaut, second man to walk on the
    moon
Geoff Andersen, research associate, United States Air Force Academy,
    author of *The Telescope*
Jay Barbree, NBC news correspondent, coauthor of *Moon Shot*
John Barrow, physicist, University of Cambridge, author of
    Impossibility
Marcia Bartusiak, author of *Einstein's Unfinished Symphony*
Jim Bell, professor of astronomy, Cornell University
Jeffrey Bennet, author of *Beyond UFOs*
Bob Berman, astronomer, author of *Secrets of the Night Sky*
Leslie Biesecker, chief of Genetic Disease Research Branch, National
    Institutes of Health
Piers Bizony, science writer, author of *How to Build Your Own
    Spaceship*
Michael Blaese, former National Institutes of Health scientist
Alex Boese, founder of Museum of Hoaxes
Nick Bostrom, transhumanist, University of Oxford
Lt. Col. Robert Bowman, Institute for Space and Security Studies
Lawrence Brody, chief of the Genome Technology Branch, National
    Institutes of Health
Rodney Brooks, former director, MIT Artificial Intelligence Laboratory

Lester Brown, founder of Earth Policy Institute

Michael Brown, professor of astronomy, Caltech

James Canton, founder of Institute for Global Futures, author of *The Extreme Future*

Arthur Caplan, director, Center for Bioethics, University of Pennsylvania

Fritjof Capra, author of *The Science of Leonardo*

Sean Carroll, cosmologist, Caltech

Andrew Chaikin, author of *A Man on the Moon*

Leroy Chiao, former NASA astronaut

George Church, director, Center for Computational Genetics, Harvard Medical School

Thomas Cochran, physicist, Natural Resources Defense Council

Christopher Cokinos, science writer, author of *The Fallen Sky*

Francis Collins, director of the National Institutes of Health

Vicki Colvin, director of Biological and Environmental Nanotechnology, Rice University

Neil Comins, author of *The Hazards of Space Travel*

Steve Cook, director of Space Technologies, Dynetics, former NASA spokesperson

Christine Cosgrove, author of *Normal at Any Cost*

Steve Cousins, president and CEO, Willow Garage

Brian Cox, physicist, University of Manchester, BBC science host

Phillip Coyle, former assistant secretary of defense, U.S. Defense Department

Daniel Crevier, author of *AI: The Tumultuous History of the Search for Artificial Intelligence*, CEO of Coreco

Ken Croswell, astronomer, author of *Magnificent Universe*

Steven Cummer, computer science, Duke University

Mark Cutkosky, mechanical engineering, Stanford University

Paul Davies, physicist, author of *Superforce*

Aubrey de Gray, Chief Science Officer, SENS Foundation

the late Michael Dertouzos, former director, Laboratory for Computer Science, MIT

Jared Diamond, Pulitzer Prize winner, professor of geography, UCLA

Mariette DiChristina, editor in chief, *Scientific American*

Peter Dilworth, former MIT AI Lab scientist

John Donoghue, creator of BrainGate, Brown University

Ann Druyan, widow of Carl Sagan, Cosmos Studios

Freeman Dyson, emeritus professor of physics, Institute for Advanced Study, Princeton

Jonathan Ellis, physicist, CERN

Daniel Fairbanks, author of *Relics of Eden*

Timothy Ferris, emeritus professor at the University of California, Berkeley, author of *Coming of Age in the Milky Way*

Maria Finitzo, filmmaker, Peabody Award winner, *Mapping Stem Cell Research*

Robert Finkelstein, AI expert

Christopher Flavin, WorldWatch Institute

Louis Friedman, cofounder, Planetary Society

James Garvin, former NASA chief scientist, NASA Goddard Space Flight Center

Evalyn Gates, author of *Einstein's Telescope*

Jack Geiger, cofounder, Physicians for Social Responsibility

David Gelernter, professor of computer science, Yale University

Neil Gershenfeld, director, Center of Bits and Atoms, MIT

Paul Gilster, author of *Centauri Dreams*

Rebecca Goldburg, former senior scientist at Environmental Defense Fund, director of Marine Science, Pew Charitable Trust

Don Goldsmith, astronomer, author of *The Runaway Universe*

Seth Goldstein, professor of computer science, Carnegie Mellon University

David Goodstein, former assistant provost of Caltech, professor of physics

J. Richard Gott III, professor of astrophysical sciences, Princeton University, author of *Time Travel in Einstein's Universe*

the late Stephen Jay Gould, biologist, Harvard Lightbridge Corp.

Ambassador Thomas Graham, expert on spy satellites

John Grant, author of *Corrupted Science*

Eric Green, director of the National Human Genome Research Institute, National Institutes of Health

Ronald Green, author of *Babies by Design*

Brian Greene, professor of mathematics and physics, Columbia University, author of *The Elegant Universe*

Alan Guth, professor of physics, MIT, author of *The Inflationary Universe*

William Hanson, author of *The Edge of Medicine*

Leonard Hayflick, professor of anatomy, University of California at San Francisco Medical School

Donald Hillebrand, director of Center for Transportation Research, Argonne National Laboratory

Frank von Hipple, physicist, Princeton University

Jeffrey Hoffman, former NASA astronaut, professor of aeronautics and astronautics, MIT

Douglas Hofstadter, Pulitzer Prize winner, author of *Gödel, Escher, Bach*

John Horgan, Stevens Institute of Technology, author of *The End of Science*

Jamie Hyneman, host of *MythBusters*

Chris Impey, professor of astronomy, University of Arizona, author of *The Living Cosmos*

Robert Irie, former scientist at AI Lab, MIT, Massachusetts General Hospital

P. J. Jacobowitz, *PC* magazine

Jay Jaroslav, former scientist at MIT AI Lab

Donald Johanson, paleoanthropologist, discoverer of Lucy

George Johnson, science journalist, *New York Times*

Tom Jones, former NASA astronaut

Steve Kates, astronomer and radio host

Jack Kessler, professor of neurology, director of Feinberg Neuroscience Institute, Northwestern University

Robert Kirshner, astronomer, Harvard University

Kris Koenig, filmmaker and astronomer

Lawrence Krauss, Arizona State University, author of *The Physics of Star Trek*

Robert Lawrence Kuhn, filmmaker and philosopher, PBS TV series *Closer to Truth*

Ray Kurzweil, inventor, author of *The Age of Spiritual Machines*

Robert Lanza, biotechnology, Advanced Cell Technology

Roger Launius, coauthor of *Robots in Space*

Stan Lee, creator of Marvel Comics and Spider-Man

Michael Lemonick, former senior science editor, *Time* magazine, Climate Central

Arthur Lerner-Lam, geologist, volcanist, Columbia University

Simon LeVay, author of *When Science Goes Wrong*

John Lewis, astronomer, University of Arizona

Alan Lightman, MIT, author of *Einstein's Dreams*

George Linehan, author of *SpaceShipOne*

Seth Lloyd, MIT, author of *Programming the Universe*

Joseph Lykken, physicist, Fermi National Accelerator Laboratory

Pattie Maes, MIT Media Laboratory

Robert Mann, author of *Forensic Detective*

Michael Paul Mason, author of *Head Cases*

W. Patrick McCray, author of *Keep Watching the Skies!*

Glenn McGee, author of *The Perfect Baby*

James McLurkin, former scientist at MIT AI Laboratory, Rice University

Paul McMillan, director, Spacewatch, University of Arizona

Fulvio Melia, professor of physics and astronomy, University of Arizona

William Meller, author of *Evolution Rx*

Paul Meltzer, National Institutes of Health

Marvin Minsky, MIT, author of *The Society of Mind*

Hans Moravec, research professor at Carnegie Mellon University, author of *Robot*

the late Phillip Morrison, physicist, MIT

Richard Muller, astrophysicist, University of California at Berkeley

David Nahamoo, formerly with IBM Human Language Technology

Christina Neal, volcanist, Alaska Volcano Observatory, U.S. Geological Survey

Michael Novacek, curator, Fossil Mammals, American Museum of Natural History

Michael Oppenheimer, environmentalist, Princeton University

Dean Ornish, clinical professor of medicine, University of California, San Francisco

Peter Palese, professor of microbiology, Mt. Sinai School of Medicine

Charles Pellerin, former NASA official

Sidney Perkowitz, professor of physics, Emory University, author of
Hollywood Science

John Pike, director, GlobalSecurity.org

Jena Pincott, author of *Do Gentlemen Really Prefer Blondes?*

Tomaso Poggio, artificial intelligence, MIT

Correy Powell, editor in chief, *Discover* magazine

John Powell, founder, JP Aerospace

Richard Preston, author of *The Hot Zone* and *The Demon in the
Freezer*

Raman Prinja, professor of astrophysics, University College London

David Quammen, science writer, author of *The Reluctant Mr. Darwin*

Katherine Ramsland, forensic scientist

Lisa Randall, professor of theoretical physics, Harvard University,
author of *Warped Passages*

Sir Martin Rees, professor of cosmology and astrophysics, Cambridge
University, author of *Before the Beginning*

Jeremy Rifkin, founder, Foundation on Economic Trends

David Riquier, director of Corporate Outreach, MIT Media Lab

Jane Rissler, Union of Concerned Scientists

Steven Rosenberg, National Cancer Institute, National Institutes of
Health

Paul Saffo, futurist, formerly with Institute for the Future, consulting
professor at Stanford University

the late Carl Sagan, Cornell University, author of *Cosmos*

Nick Sagan, coauthor of *You Call This the Future?*

Michael Salamon, NASA's Beyond Einstein program

Adam Savage, host of *MythBusters*

Peter Schwartz, futurist, cofounder of Global Business Network,
author of *The Long View*

Michael Shermer, founder of the Skeptics Society and *Skeptic*
magazine

Donna Shirley, former manager, NASA Mars Exploration Program

Seth Shostak, SETI Institute

Neil Shubin, professor of organismal biology and anatomy, University
of Chicago, author of *Your Inner Fish*

Paul Shuch, executive director emeritus, SETI League

Peter Singer, author of *Wired for War*, Brookings Institute

Simon Singh, author of *Big Bang*

Gary Small, coauthor of *iBrain*

Paul Spudis, Planetary Geology Program of the NASA Office of Space Science, Solar System Division

Steven Squyres, professor of astronomy, Cornell University

Paul Steinhardt, professor of physics, Princeton University, coauthor of *Endless Universe*

Gregory Stock, UCLA, author of *Redesigning Humans*

Richard Stone, *The Last Great Impact on Earth*, Discover Magazine

Brian Sullivan, formerly with the Hayden Planetarium

Leonard Susskind, professor of physics, Stanford University

Daniel Tammet, autistic savant, author of *Born on a Blue Day*

Geoffrey Taylor, physicist, University of Melbourne

the late Ted Taylor, designer of U.S. nuclear warheads

Max Tegmark, physicist, MIT

Alvin Toffler, author of *The Third Wave*

Patrick Tucker, World Future Society

Admiral Stansfield M. Turner, former Director of Central Intelligence

Chris Turney, University of Exeter, UK, author of *Ice, Mud and Blood*

Neil deGrasse Tyson, director, Hayden Planetarium

Sesh Velamoor, Foundation for the Future

Robert Wallace, coauthor of *Spycraft*, former director of CIA's Office of Technical Services

Kevin Warwick, human cyborgs, University of Reading, UK

Fred Watson, astronomer, author of *Stargazer*

the late Mark Weiser, Xerox PARC

Alan Weisman, author of *The World Without Us*

Daniel Werthimer, SETI at Home, University of California at Berkeley

Mike Wessler, former scientist, MIT AI Lab

Arthur Wiggins, author of *The Joy of Physics*

Anthony Wynshaw-Boris, National Institutes of Health

Carl Zimmer, science writer, author of *Evolution*

Robert Zimmerman, author of *Leaving Earth*

Robert Zubrin, founder, Mars Society

PHYSICS OF THE FUTURE

Empires of the future will be empires of the mind.

—WINSTON CHURCHILL

INTRODUCTION   *Predicting the Next 100 Years*

When I was a child, two experiences helped to shape the person I am today and spawned two passions that have helped to define my entire life.

First, when I was eight years old, I remember all the teachers buzzing with the latest news that a great scientist had just died. That night, the newspapers printed a picture of his office, with an unfinished manuscript on his desk. The caption read that the greatest scientist of our era could not finish his greatest masterpiece. What, I asked myself, could be so difficult that such a great scientist could not finish it? What could possibly be that complicated and that important? To me, eventually this became more fascinating than any murder mystery, more intriguing than any adventure story. I had to know what was in that unfinished manuscript.

Later, I found out that the name of this scientist was Albert Einstein and the unfinished manuscript was to be his crowning achievement, his attempt to create a "theory of everything," an equation, perhaps no more than one inch wide, that would unlock the secrets of the universe and perhaps allow him to "read the mind of God."

But the other pivotal experience from my childhood was when I watched the Saturday morning TV shows, especially the *Flash Gordon* series with Buster Crabbe. Every week, my nose was glued to the TV screen. I was magically transported to a mysterious world of space aliens, starships, ray gun battles, underwater cities, and monsters. I was hooked. This was my first exposure to the world of the future. Ever since, I've felt a childlike wonder when pondering the future.

But after watching every episode of the series, I began to realize that although Flash got all the accolades, it was the scientist Dr. Zarkov who actually made the series work. He invented the rocket ship, the invisibility shield, the power source for the city in the sky, etc. Without the scientist, there is no future. The handsome and the beautiful may earn the admiration of society, but all the wondrous inventions of the future are a byproduct of the unsung, anonymous scientists.

Later, when I was in high school, I decided to follow in the footsteps of these great scientists and put some of my learning to the test. I wanted to be part of this great revolution that I knew would change the world. I decided to build an atom smasher. I asked my mother for permission to build a 2.3-million electron volt particle accelerator in the garage. She was a bit startled but gave me the okay. Then, I went to Westinghouse and Varian Associates, got 400 pounds of transformer steel, 22 miles of copper wire, and assembled a betatron accelerator in my mom's garage.

Previously, I had built a cloud chamber with a powerful magnetic field and photographed tracks of antimatter. But photographing antimatter was not enough. My goal now was to produce a beam of antimatter. The atom smasher's magnetic coils successfully produced a huge 10,000 gauss magnetic field (about 20,000 times the earth's magnetic field, which would in principle be enough to rip a hammer right out of your hand). The machine soaked up 6 kilowatts of power, draining all the electricity my house could provide. When I turned on the machine, I frequently blew out all the fuses in the house. (My poor mother must have wondered why she could not have a son who played football instead.)

So two passions have intrigued me my entire life: the desire to understand all the physical laws of the universe in a single coherent theory and the desire to see the future. Eventually, I realized that these two passions were actually complementary. The key to understanding the future is to grasp the fundamental laws of nature and then apply them to the inventions, machines, and therapies that will redefine our civilization far into the future.

There have been, I found out, numerous attempts to predict the future, many useful and insightful. However, they were mainly written by historians, sociologists, science fiction writers, and "futurists," that is, outsiders who are predicting the world of science without a firsthand knowledge of the science itself. The scientists, the insiders who are actually creating the future in their laboratories, are too busy making breakthroughs to have time to write books about the future for the public.

That is why this book is different. I hope this book will give an insider's perspective on what miraculous discoveries await us and provide the most authentic, authoritative look into the world of 2100.

Of course, it is impossible to predict the future with complete accuracy. The best one can do, I feel, is to tap into the minds of the scientists at the cutting edge of research, who are doing the yeoman's work of inventing the future. They are the ones who are creating the devices, inventions, and therapies that will revolutionize civilization. And this book is their story. I have had the opportunity to sit in the front-row seat of this great revolution, having interviewed more than 300 of the world's top scientists, thinkers, and dreamers for national TV and radio. I have also taken TV crews into their laboratories to film the prototypes of the remarkable devices that will change our future. It has been a rare honor to have hosted numerous science specials for BBC-TV, the Discovery Channel, and the Science Channel, profiling the remarkable inventions and discoveries of the visionaries who are daring to create the future. Being free to pursue my work on string theory and to eavesdrop on the cutting-edge research that will revolutionize this century, I feel I have one of the most desirable jobs in science. It is my childhood dream come true.

But this book differs from my previous ones. In books like *Beyond Einstein, Hyperspace,* and *Parallel Worlds,* I discussed the fresh, revolutionary winds sweeping through my field, theoretical physics, that are opening up new ways to understand the universe. In *Physics of the Impossible,* I discussed how the latest discoveries in physics may eventually make possible even the most imaginative schemes of science fiction.

This book most closely resembles my book *Visions,* in which I discussed how science will evolve in the coming decades. I am gratified that many of the predictions made in that book are being realized today on schedule. The accuracy of my book, to a large degree, has depended on the wisdom and foresight of the many scientists I interviewed for it.

But this book takes a much more expansive view of the future, discussing the technologies that may mature in 100 years, that will ultimately determine the fate of humanity. How we negotiate the challenges and opportunities of the next 100 years will determine the ultimate trajectory of the human race.

PREDICTING THE NEXT CENTURY

Predicting the next few years, let alone a century into the future, is a daunting task. Yet it is one that challenges us to dream about technologies we believe will one day alter the fate of humanity.

In 1863, the great novelist Jules Verne undertook perhaps his most ambitious project. He wrote a prophetic novel, called *Paris in the Twentieth Century,* in which he applied the full power of his enormous talents to forecast the coming century. Unfortunately, the manuscript was lost in the mist of time, until his great-grandson accidentally stumbled upon it lying in a safe where it had been carefully locked away for almost 130 years. Realizing what a treasure he had found, he arranged to have it published in 1994, and it became a best seller.

Back in 1863, kings and emperors still ruled ancient empires, with impoverished peasants performing backbreaking work toiling in the fields. The United States was consumed by a ruinous civil war that would almost tear the country apart, and steam power was just beginning to revolutionize the world. But Verne predicted that Paris in 1960 would have glass skyscrapers, air conditioning, TV, elevators, high-speed trains, gasoline-powered automobiles, fax machines, and even something resembling the Internet. With uncanny accuracy, Verne depicted life in modern Paris.

This was not a fluke, because just a few years later he made another spectacular prediction. In 1865, he wrote *From the Earth to the Moon,* in which he predicted the details of the mission that sent our astronauts to the moon more than 100 years later in 1969. He accurately predicted the size of the space capsule to within a few percent, the location of the launch site in Florida not far from Cape Canaveral, the number of astronauts on the mission, the length of time the voyage would last, the weightlessness that the astronauts would experience, and the final splashdown in the ocean. (The only major mistake was that he used gunpowder, rather than rocket fuel, to take his astronauts to the moon. But liquid-fueled rockets wouldn't be invented for another seventy years.)

How was Jules Verne able to predict 100 years into the future with such breathtaking accuracy? His biographers have noted that, although Verne was not a scientist himself, he constantly sought out scientists, peppering them with questions about their visions of the future. He amassed a vast archive summarizing the great scientific discoveries of his time. Verne, more than others, realized that science was the engine shaking the foundations of civilization, propelling it into a new century with unexpected marvels and miracles. The key to Verne's vision and profound insights was his grasp of the power of science to revolutionize society.

Another great prophet of technology was Leonardo da Vinci, painter, thinker, and visionary. In the late 1400s, he drew beautiful, accurate diagrams of machines that would one day fill the skies: sketches of parachutes, helicopters, hang gliders, and even airplanes. Remarkably, many of his inventions would have flown. (His flying machines, however, needed one more ingredient: at least a 1-horsepower motor, something that would not be available for another 400 years.)

What is equally astonishing is that Leonardo sketched the blueprint for a mechanical adding machine, which was perhaps 150 years ahead of its time. In 1967, a misplaced manuscript was reanalyzed, revealing his idea for an adding machine with thirteen digital wheels. If one turned a crank, the gears inside turned in sequence performing the arithmetic calculations. (The machine was built in 1968 and it worked.)

In addition, in the 1950s another manuscript was uncovered which contained a sketch for a warrior automaton, wearing German-Italian armor, that could sit up and move its arms, neck, and jaw. It, too, was subsequently built and found to work.

Like Jules Verne, Leonardo was able to get profound insights into the future by consulting a handful of forward-thinking individuals of his time. He was part of a small circle of people who were at the forefront of innovation. In addition, Leonardo was always experimenting, building, and sketching models, a key attribute of anyone who wants to translate thinking into reality.

Given the enormous, prophetic insights of Verne and Leonardo da Vinci, we ask the question: Is it possible to predict the world of 2100? In the tradition of Verne and Leonardo, this book will closely examine the work of the leading scientists who are building prototypes of the technologies that will change our future. This book is not a work of fiction, a by-product of the overheated imagination of a Hollywood scriptwriter, but rather is based on the solid science being conducted in major laboratories around the world today.

The prototypes of all these technologies already exist. As William Gibson, the author of *Neuromancer* who coined the word *cyberspace,* once said, "The future is already here. It's just unevenly distributed."

Predicting the world of 2100 is a daunting task, since we are in an era of profound scientific upheaval, in which the pace of discovery is always accelerating. More scientific knowledge has been accumulated just in the last few decades than in all human history. And by 2100, this scientific knowledge will again have doubled many times over.

But perhaps the best way to grasp the enormity of predicting 100 years into the future is to recall the world of 1900 and remember the lives our grandparents lived.

Journalist Mark Sullivan asks us to imagine someone reading a newspaper in the year 1900:

> In his newspapers of January 1, 1900, the American found no such word as radio, for that was yet twenty years in from coming; nor "movie," for that too was still mainly of the future; nor chauffeur, for the automobile was only just emerging and had been called "horseless carriage. . . ." There was no such word as aviator. . . . Farmers had not heard of tractors, nor bankers of the Federal Reserve System. Merchants had not heard of chain-stores nor "self-service"; nor seamen of oil-burning engines. . . . Ox-teams could still be seen on country roads. . . . Horses or mules for trucks were practically

universal. . . . The blacksmith beneath the spreading chestnut-tree was a reality.

To understand the difficulty of predicting the next 100 years, we have to appreciate the difficulty that the people of 1900 had in predicting the world of 2000. In 1893, as part of the World's Columbian Exposition in Chicago, seventy-four well-known individuals were asked to predict what life would be like in the next 100 years. The one problem was that they consistently underestimated the rate of progress of science. For example, many correctly predicted that we would one day have commercial transatlantic airships, but they thought that they would be balloons. Senator John J. Ingalls said, "It will be as common for the citizen to call for his dirigible balloon as it now is for his buggy or his boots." They also consistently missed the coming of the automobile. Postmaster General John Wanamaker stated that the U.S. mail would be delivered by stagecoach and horseback, even 100 years into the future.

This underestimation of science and innovation even extended to the patent office. In 1899, Charles H. Duell, commissioner of the U.S. Office of Patents, said, "Everything that can be invented has been invented."

Sometimes experts in their own field underestimated what was happening right beneath their noses. In 1927, Harry M. Warner, one of the founders of Warner Brothers, remarked during the era of silent movies, "Who the hell wants to hear actors talk?"

And Thomas Watson, chairman of IBM, said in 1943, "I think there is a world market for maybe five computers."

This underestimation of the power of scientific discovery even extended to the venerable *New York Times*. (In 1903, the *Times* declared that flying machines were a waste of time, just a week before the Wright brothers successfully flew their airplane at Kitty Hawk, North Carolina. In 1920, the *Times* criticized rocket scientist Robert Goddard, declaring his work nonsense because rockets cannot move in a vacuum. Forty-nine years later, when *Apollo 11* astronauts landed on the moon, the *Times,* to its credit, ran the retraction: "It is now definitely established that a rocket can function in a vacuum. The *Times* regrets the error.")

The lesson here is that it is very dangerous to bet against the future.

Predictions for the future, with a few exceptions, have always underestimated the pace of technological progress. History, we are told over and over again, is written by the optimists, not the pessimists. As President Dwight Eisenhower once said, "Pessimism never won a war."

We can even see how science fiction writers underestimated the pace of scientific discovery. When watching reruns of the old 1960s TV series *Star Trek,* you notice that much of this "twenty-third-century technology" is already here. Back then, TV audiences were startled to see mobile phones, portable computers, machines that could talk, and typewriters that could take dictation. Yet all these technologies exist today. Soon, we will also have versions of the universal translator, which can rapidly translate between languages as you speak, and also "tricorders," which can diagnose disease from a distance. (Excepting warp drive engines and transporters, much of this twenty-third-century science is already here.)

Given the glaring mistakes people have made in underestimating the future, how can we begin to provide a firmer scientific basis to our predictions?

UNDERSTANDING THE LAWS OF NATURE

Today, we are no longer living in the dark ages of science, when lightning bolts and plagues were thought to be the work of the gods. We have a great advantage that Verne and Leonardo da Vinci did not have: a solid understanding of the laws of nature.

Predictions will always be flawed, but one way to make them as authoritative as possible is to grasp the four fundamental forces in nature that drive the entire universe. Each time one of them was understood and described, it changed human history.

The first force to be explained was the force of gravity. Isaac Newton gave us a mechanics that could explain that objects moved via forces, rather than mystical spirits and metaphysics. This helped to pave the way for the Industrial Revolution and the introduction of steam power, especially the locomotive.

The second force to be understood was the electromagnetic force, which lights up our cities and powers our appliances. When Thomas Edison, Michael Faraday, James Clerk Maxwell, and others helped to explain electricity and magnetism, this unleashed the electronic revolution that has created a bounty of scientific wonders. We see this every time there is a power blackout, when society is suddenly wrenched back 100 years into the past.

The third and fourth forces to be understood were the two nuclear forces: the weak and strong forces. When Einstein wrote down $E = mc^2$ and when the atom was split in the 1930s, scientists for the first time began to understand the forces that light up the heavens. This revealed the secret behind the stars. Not only did this unleash the awesome power of atomic weapons, it also held out the promise that one day we would be able to harness this power on the earth.

Today, we have a fairly good grasp of these four forces. The first force, gravity, is now described through Einstein's theory of general relativity. And the other three forces are described through the quantum theory, which allows us to decode the secrets of the subatomic world.

The quantum theory, in turn, has given us the transistor, the laser, and the digital revolution that is the driving force behind modern society. Similarly, scientists were able to use the quantum theory to unlock the secret of the DNA molecule. The blinding speed of the biotechnological revolution is a direct result of computer technology, since DNA sequencing is all done by machines, robots, and computers.

As a consequence, we are better able to see the direction that science and technology will take in the coming century. There will always be totally unexpected, novel surprises that leave us speechless, but the foundation of modern physics, chemistry, and biology has largely been laid, and we do not expect any major revision of this basic knowledge, at least in the foreseeable future. As a result, the predictions we make in this book are the product not of wild speculation but are reasoned estimates of when the prototype technologies of today will finally reach maturity.

In conclusion, there are several reasons to believe that we can view the outlines of the world of 2100:

1. This book is based on interviews with more than 300 top scientists, those in the forefront of discovery.
2. Every scientific development mentioned in this book is consistent with the known laws of physics.
3. The four forces and the fundamental laws of nature are largely known; we do not expect any major new changes in these laws.
4. Prototypes of all technologies mentioned in this book already exist.

5. This book is written by an "insider" who has a firsthand look at the technologies that are on the cutting edge of research.

For countless eons we were passive observers of the dance of nature. We only gazed in wonder and fear at comets, lightning bolts, volcanic eruptions, and plagues, assuming that they were beyond our comprehension. To the ancients, the forces of nature were an eternal mystery to be feared and worshipped, so they created the gods of mythology to make sense of the world around them. The ancients hoped that by praying to these gods they would show mercy and grant them their dearest wishes.

Today, we have become choreographers of the dance of nature, able to tweak the laws of nature here and there. But by 2100, we will make the transition to being masters of nature.

2100: BECOMING THE GODS OF MYTHOLOGY

Today, if we could somehow visit our ancient ancestors and show them the bounty of modern science and technology, we would be viewed as magicians. With the wizardry of science, we could show them jet planes that can soar in the clouds, rockets that can explore the moon and planets, MRI scanners that can peer inside the living body, and cell phones that can put us in touch with anyone on the planet. If we showed them laptop computers that can send moving images and messages instantly across the continents, they would view this as sorcery.

But this is just the beginning. Science is not static. Science is exploding exponentially all around us. If you count the number of scientific articles being published, you will find that the sheer volume of science doubles every decade or so. Innovation and discovery are changing the entire economic, political, and social landscape, overturning all the old cherished beliefs and prejudices.

Now dare to imagine the world in the year 2100.

By 2100, our destiny is to become like the gods we once worshipped and feared. But our tools will not be magic wands and potions but the science of computers, nanotechnology, artificial intelligence, biotechnology, and most of all, the quantum theory, which is the foundation of the previous technologies.

By 2100, like the gods of mythology, we will be able to manipulate objects with the power of our minds. Computers, silently reading our thoughts, will be able to carry out our wishes. We will be able to move objects by thought alone, a telekinetic power usually reserved only for the gods. With the power of biotechnology, we will create perfect bodies and extend our life spans. We will also be able to create life-forms that have never walked the surface of the earth. With the power of nanotechnology, we will be able to take an object and turn it into something else, to create something seemingly almost out of nothing. We will ride not in fiery chariots but in sleek vehicles that will soar by themselves with almost no fuel, floating effortlessly in the air. With our engines, we will be able to harness the limitless energy of the stars. We will also be on the threshold of sending star ships to explore those nearby.

Although this godlike power seems unimaginably advanced, the seeds of all these technologies are being planted even as we speak. It is modern science, not chanting and incantations, that will give us this power.

I am a quantum physicist. Every day, I grapple with the equations that govern the subatomic particles out of which the universe is created. The world I live in is the universe of eleven-dimensional hyperspace, black holes, and gateways to the multiverse. But the equations of the quantum theory, used to describe exploding stars and the big bang, can also be used to decipher the outlines of our future.

But where is all this technological change leading? Where is the final destination in this long voyage into science and technology?

The culmination of all these upheavals is the formation of a planetary civilization, what physicists call a Type I civilization. This transition is perhaps the greatest transition in history, marking a sharp departure from all civilizations of the past. Every headline that dominates the news reflects, in some way, the birth pangs of this planetary civilization. Commerce, trade, culture, language, entertainment, leisure activities, and even war are all being revolutionized by the emergence of this planetary civilization. Calculating the energy output of the planet, we can estimate that we will attain Type I status within 100 years. Unless we succumb to the forces of chaos and folly, the transition to a planetary civilization is inevitable, the end product of the enormous, inexorable forces of history and technology beyond anyone's control.

WHY PREDICTIONS SOMETIMES DON'T COME TRUE

But several predictions made about the information age were spectacularly untrue. For example, many futurists predicted the "paperless office," that is, that the computer would make paper obsolete. Actually, the opposite has occurred. A glance at any office shows you that the amount of paper is actually greater than ever.

Some also envisioned the "peopleless city." Futurists predicted that teleconferencing via the Internet would make face-to-face business meetings unnecessary, so there would be no need to commute. In fact, the cities themselves would largely empty out, becoming ghost towns, as people worked in their homes rather than their offices.

Likewise, we would see the rise of "cybertourists," couch potatoes who would spend the entire day lounging on their sofas, roaming the world and watching the sights via the Internet on their computers. We would also see "cybershoppers," who would let their computer mice do the walking. Shopping malls would go bankrupt. And "cyberstudents" would take all their classes online while secretly playing video games and drinking beer. Universities would close for lack of interest.

Or consider the fate of the "picture phone." During the 1964 World's Fair, AT&T spent about $100 million perfecting a TV screen that would connect to the telephone system, so that you could see the person whom you were talking to, and vice versa. The idea never took off; AT&T sold only about 100 of them, making each unit cost about $1 million each. This was a very expensive fiasco.

And finally, it was thought that the demise of traditional media and entertainment was imminent. Some futurists claimed that the Internet was the juggernaut that would swallow live theater, the movies, radio, and TV, all of which would soon be seen only in museums.

Actually, the reverse has happened. Traffic jams are worse than ever— a permanent feature of urban life. People flock to foreign sites in record numbers, making tourism one of the fastest-growing industries on the planet. Shoppers flood the stores, in spite of economic hard times. Instead of proliferating cyberclassrooms, universities are still registering record numbers of students. To be sure, there are more people deciding to work from their homes or teleconference with their coworkers, but cities have not emptied at all. Instead, they have morphed into sprawling megacities. Today, it is easy to carry on video conversations on the Internet, but most people tend to be reluctant to be filmed, preferring face-to-face meetings. And of course, the Internet has changed the entire media landscape, as media giants puzzle over how to earn revenue on the Internet. But it is not even close to wiping out TV, radio, and live theater. The lights of Broadway still glow as brightly as before.

CAVE MAN PRINCIPLE

Why did these predictions fail to materialize? I conjecture that people largely rejected these advances because of what I call the Cave Man (or Cave Woman) Principle. Genetic and fossil evidence indicates that modern humans, who looked just like us, emerged from Africa more than 100,000 years ago, but we see no evidence that our brains and personalities have changed much since then. If you took someone from that period, he would be anatomically identical to us: if you gave him a bath and a shave, put him in a three-piece suit, and then placed him on Wall Street, he would be physically indistinguishable from everyone else. So our wants, dreams, personalities, and desires have probably not changed much in 100,000 years. We probably still think like our caveman ancestors.

The point is: whenever there is a conflict between modern technology and the desires of our primitive ancestors, these primitive desires win each time. That's the Cave Man Principle. For example, the caveman always demanded "proof of the kill." It was never enough to boast about the big one that got away. Having the fresh animal in our hands was always preferable to tales of the one that got away. Similarly, we want hard copy whenever we deal with files. We instinctively don't trust the electrons floating in our computer screen, so we print our e-mails and reports, even when it's not necessary. That's why the paperless office never came to be.

Likewise, our ancestors always liked face-to-face encounters. This helped us bond with others and to read their hidden emotions. This is why the peopleless city never came to pass. For example, a boss might want to carefully size up his employees. It's difficult to do this online, but face-to-face a boss can read body language to gain valuable unconscious information. By watching people up close, we feel a common bond and can also read their subtle body language to find out what thoughts are racing through their heads. This is because our apelike ancestors, many thousands of years before they developed speech, used body language almost exclusively to convey their thoughts and emotions.

This is the reason cybertourism never got off the ground. It's one thing to see a picture of the Taj Mahal, but it's another thing to have the bragging rights of actually seeing it in person. Similarly, listening to a CD of your favorite musician is not the same as feeling the sudden rush when actually seeing this musician in a live concert, surrounded by all the fanfare, hoopla, and noise. This means that even though we will be able to download realistic images of our favorite drama or celebrity, there is nothing like actually seeing the drama on stage or seeing the actor perform in person. Fans go to great lengths to get autographed pictures and concert tickets of their favorite celebrity, although they can download a picture from the Internet for free.

This explains why the prediction that the Internet would wipe out TV and radio never came to pass. When the movies and radio first came in, people bewailed the death of live theater. When TV came in, people predicted the demise of the movies and radio. We are living now with a mix of all these media. The lesson is that one medium never annihilates a previous one but coexists with it. It is the mix and relationship among these media that constantly change. Anyone who can accurately predict the mix of these media in the future could become very wealthy.

The reason for this is that our ancient ancestors always wanted to see something for themselves and not rely on hearsay. It was crucial for our survival in the forest to rely on actual physical evidence rather than rumors. Even a century from now, we will still have live theater and still chase celebrities, an ancient heritage of our distant past.

In addition, we are descended from predators who hunted. Hence, we love to watch others and even sit for hours in front of a TV, endlessly watching the antics of our fellow humans, but we instantly get nervous when we feel others watching us. In fact, scientists have calculated that we get nervous if we are stared at by a stranger for about four seconds. After about ten seconds, we even get irate and hostile at being stared at. This is the reason why the original picture phone was such a flop. Also, who wants to have to comb one's hair before going online? (Today, after decades of slow, painful improvement, video conferencing is finally catching on.)

And today, it is possible to take courses online. But universities are bulging with students. The one-to-one encounter with professors, who can give individual attention and answer personal questions, is still preferable to online courses. And a university degree still carries more weight than an online diploma when applying for a job.

So there is a continual competition between High Tech and High Touch, that is, sitting in a chair watching TV versus reaching out and touching things around us. In this competition, we will want both. That is why we still have live theater, rock concerts, paper, and tourism in the age of cyberspace and virtual reality. But if we are offered a free picture of our favorite celebrity musician or actual tickets to his concert, we will take the tickets, hands down.

So that is the Cave Man Principle: we prefer to have both, but if given a choice we will chose High Touch, like our cavemen ancestors.

But there is also a corollary to this principle. When scientists first created the Internet back in the 1960s, it was widely believed that it would evolve into a forum for education, science, and progress. Instead, many were horrified that it soon degenerated into the no-holds-barred Wild West that it is today. Actually, this is to be expected. The corollary to the Cave Man Principle is that if you want to predict the social interactions of humans in the future, simply imagine our social interactions 100,000 years ago and multiply by a billion. This means that there will be a premium placed on gossip, social networking, and entertainment. Rumors were essential in a tribe to rapidly communicate information, especially about the leaders and role models. Those who were out of the loop often did not survive to pass on their genes. Today, we can see this played out in grocery checkout stands, which have wall-to-wall celebrity gossip magazines, and in the rise of a celebrity-driven culture. The only difference today is that the magnitude of this tribal gossip has been multiplied enormously by mass media and can now circle the earth many times over within a fraction of a second.

The sudden proliferation of social networking Web sites, which turned young, baby-faced entrepreneurs into billionaires almost overnight, caught many analysts off guard, but it is also an example of this principle. In our evolutionary history, those who maintained large social networks could rely on them for resources, advice, and help that were vital for survival.

And last, entertainment will continue to grow explosively. We sometimes don't like to admit it, but a dominant part of our culture is based on entertainment. After the hunt, our ancestors relaxed and entertained themselves. This was important not only for bonding but also for establishing one's position within the tribe. It is no accident that dancing and singing, which are essential parts of entertainment, are also vital in the animal kingdom to demonstrate fitness to the opposite sex. When male birds sing beautiful, complex melodies or engage in bizarre mating rituals, it is mainly to show the opposite sex that they are healthy, physically fit, free of parasites, and have genes worthy enough to be passed down.

And the creation of art was not only for enjoyment but also played an important part in the evolution of our brain, which handles most information symbolically.

So unless we genetically change our basic personality, we can expect that the power of entertainment, tabloid gossip, and social networking will increase, not decrease, in the future.

SCIENCE AS A SWORD

I once saw a movie that forever changed my attitude toward the future. It was called *Forbidden Planet*, based on Shakespeare's *The Tempest*. In the movie astronauts encounter an ancient civilization that, in its glory, was millions of years ahead of us. They had attained the ultimate goal of their technology: infinite power without instrumentality, that is, the power to do almost anything via their minds. Their thoughts tapped into colossal thermonuclear power plants, buried deep inside their planet, that converted their every desire into reality. In other words, they had the power of the gods.

We will have a similar power, but we will not have to wait millions of years. We will have to wait only a century, and we can see the seeds of this future even in today's technology. But the movie was also a morality tale, since this divine power eventually overwhelmed this civilization.

Of course, science is a double-edged sword; it creates as many problems as it solves, but always on a higher level. There are two competing trends in the world today: one is to create a planetary civilization that is tolerant, scientific, and prosperous, but the other glorifies anarchy and ignorance that could rip the fabric of our society. We still have the same sectarian, fundamentalist, irrational passions of our ancestors, but the difference is that now we have nuclear, chemical, and biological weapons.

In the future, we will make the transition from being passive observers of the dance of nature, to being the choreographers of nature, to being masters of nature, and finally to being conservators of nature. So let us hope that we can wield the sword of science with wisdom and equanimity, taming the barbarism of our ancient past.

Let us now embark upon a hypothetical journey through the next 100 years of scientific innovation and discovery, as told to me by the scientists who are making it happen. It will be a wild ride through the rapid advances in computers, telecommunications, biotechnology, artificial intelligence, and nanotechnology. It will undoubtedly change nothing less than the future of civilization.

Everyone takes the limits of his own vision for the limits of the world.

—ARTHUR SCHOPENHAUER

No pessimist ever discovered the secrets of the stars or sailed to an uncharted land or opened a new heaven to the human spirit.

—HELEN KELLER

1 FUTURE OF THE COMPUTER *Mind over Matter*

I remember vividly sitting in Mark Weiser's office in Silicon Valley almost twenty years ago as he explained to me his vision of the future. Gesturing with his hands, he excitedly told me a new revolution was about to happen that would change the world. Weiser was part of the computer elite, working at Xerox PARC (Palo Alto Research Center, which was the first to pioneer the personal computer, the laser printer, and Windows-type architecture with graphical user interface), but he was a maverick, an iconoclast who was shattering conventional wisdom, and also a member of a wild rock band.

Back then (it seems like a lifetime ago), personal computers were new, just beginning to penetrate people's lives, as they slowly warmed up to the idea of buying large, bulky desktop computers in order to do spreadsheet analysis and a little bit of word processing. The Internet was still largely the isolated province of scientists like me, cranking out equations to fellow scientists in an arcane language. There were raging debates about whether this box sitting on your desk would dehumanize civilization with its cold, unforgiving stare. Even political analyst William F. Buckley had to defend the word processor against intellectuals who railed against it and refused to ever touch a computer, calling it an instrument of the philistines.

It was in this era of controversy that Weiser coined the expression "ubiquitous computing." Seeing far past the personal computer, he predicted that the chips would one day become so cheap and plentiful that they would be scattered throughout the environment—in our clothing, our furniture, the walls, even our bodies. And they would all be connected to the Internet, sharing data, making our lives more pleasant, monitoring all our wishes. Everywhere we moved, chips would be there to silently carry out our desires. The environment would be alive.

For its time, Weiser's dream was outlandish, even preposterous. Most personal computers were still expensive and not even connected to the Internet. The idea that billions of tiny chips would one day be as cheap as running water was considered lunacy.

And then I asked him why he felt so sure about this revolution. He calmly replied that computer power was growing exponentially, with no end in sight. Do the math, he implied. It was only a matter of time. (Sadly, Weiser did not live long enough to see his revolution come true, dying of cancer in 1999.)

The driving source behind Weiser's prophetic dreams is something called Moore's law, a rule of thumb that has driven the computer industry for fifty or more years, setting the pace for modern civilization like clockwork. Moore's law simply says that computer power doubles about every eighteen months. First stated in 1965 by Gordon Moore, one of the founders of the Intel Corporation, this simple law has helped to revolutionize the world economy, generated fabulous new wealth, and irreversibly altered our way of life. When you plot the plunging price of computer chips and their rapid advancements in speed, processing power, and memory, you find a remarkably straight line going back fifty years. (This is plotted on a logarithmic curve. In fact, if you extend the graph, so that it includes vacuum tube technology and even mechanical hand-crank adding machines, the line can be extended more than 100 years into the past.)

Exponential growth is often hard to grasp, since our minds think linearly. It is so gradual that you sometimes cannot experience the change at all. But over decades, it can completely alter everything around us.

According to Moore's law, every Christmas your new computer games are almost twice as powerful (in terms of the number of transistors) as those from the previous year. Furthermore, as the years pass, this incremental gain becomes monumental. For example, when you receive a birthday card in the mail, it often has a chip that sings "Happy Birthday" to you. Remarkably, that chip has more computer power than all the Allied forces of 1945. Hitler, Churchill, or Roosevelt might have killed to get that chip. But what do we do with it? After the birthday, we throw the card and chip away. Today, your cell phone has more computer power than all of NASA back in 1969, when it placed two astronauts on the moon. Video games, which consume enormous amounts of computer power to simulate 3-D situations, use more computer power than mainframe computers of the previous decade. The Sony PlayStation of today, which costs $300, has the power of a military supercomputer of 1997, which cost millions of dollars.

We can see the difference between linear and exponential growth of computer power when we analyze how people viewed the future of the computer back in 1949, when *Popular Mechanics* predicted that computers would grow linearly into the future, perhaps only doubling or tripling with time. It wrote: "Where a calculator like the ENIAC today is equipped with 18,000 vacuum tubes and weighs 30 tons, computers in the future may have only 1,000 vacuum tubes and weigh only 1½ tons."

(Mother Nature appreciates the power of the exponential. A single virus can hijack a human cell and force it to create several hundred copies of itself. Growing by a factor of 100 in each generation, one virus can generate 10 billion viruses in just five generations. No wonder a single virus can infect the human body, with trillions of healthy cells, and give you a cold in just a week or so.)

Not only has the amount of computer power increased, but the way that this power is delivered has also radically changed, with enormous implications for the economy. We can see this progression, decade by decade:

- **1950s.** Vacuum tube computers were gigantic contraptions filling entire rooms with jungles of wires, coils, and steel. Only the military was rich enough to fund these monstrosities.

- **1960s.** Transistors replaced vacuum tube computers, and mainframe computers gradually entered the commercial marketplace.
- **1970s.** Integrated circuit boards, containing hundreds of transistors, created the minicomputer, which was the size of a large desk.
- **1980s.** Chips, containing tens of millions of transistors, made possible personal computers that can fit inside a briefcase.
- **1990s.** The Internet connected hundreds of millions of computers into a single, global computer network.
- **2000s.** Ubiquitous computing freed the chip from the computer, so chips were dispersed into the environment.

So the old paradigm (a single chip inside a desktop computer or laptop connected to a computer) is being replaced by a new paradigm (thousands of chips scattered inside every artifact, such as furniture, appliances, pictures, walls, cars, and clothes, all talking to one another and connected to the Internet).

When these chips are inserted into an appliance, it is miraculously transformed. When chips were inserted into typewriters, they became word processors. When inserted into telephones, they became cell phones. When inserted into cameras, they became digital cameras. Pinball machines became video games. Phonographs became iPods. Airplanes became deadly Predator drones. Each time, an industry was revolutionized and was reborn. Eventually, almost everything around us will become intelligent. Chips will be so cheap they will even cost less than the plastic wrapper and will replace the bar code. Companies that do not make their products intelligent may find themselves driven out of business by their competitors that do.

Of course, we will still be surrounded by computer monitors, but they will resemble wallpaper, picture frames, or family photographs, rather than computers. Imagine all the pictures and photographs that decorate our homes today; now imagine each one being animated, moving, and connected to the Internet. When we walk outside, we will see pictures move, since moving pictures will cost as little as static ones.

The destiny of computers—like other mass technologies like electricity, paper, and running water—is to become invisible, that is, to disappear into the fabric of our lives, to be everywhere and nowhere, silently and seamlessly carrying out our wishes.

Today, when we enter a room, we automatically look for the light switch, since we assume that the walls are electrified. In the future, the first thing we will do on entering a room is to look for the Internet portal, because we will assume the room is intelligent. As novelist Max Frisch once said, "Technology [is] the knack of so arranging the world that we don't have to experience it."

Moore's law also allows us to predict the evolution of the computer into the near future. In the coming decade, chips will be combined with supersensitive sensors, so that they can detect diseases, accidents, and emergencies and alert us before they get out of control. They will, to a degree, recognize the human voice and face and converse in a formal language. They will be able to create entire virtual worlds that we can only dream of today. Around 2020, the price of a chip may also drop to about a penny, which is the cost of scrap paper. Then we will have millions of chips distributed everywhere in our environment, silently carrying out our orders.

Ultimately, the word *computer* itself will disappear from the English language.

In order to discuss the future progress of science and technology, I have divided each chapter into three periods: the near future (today to 2030), the midcentury (from 2030 to 2070), and finally the far future, from 2070 to 2100. These time periods are only rough approximations, but they show the time frame for the various trends profiled in this book.

The rapid rise of computer power by the year 2100 will give us power like that of the gods of mythology we once worshipped, enabling us to control the world around us by sheer thought. Like the gods of mythology, who could move objects and reshape life with a simple wave of the hand or nod of the head, we too will be able to control the world around us with the power of our minds. We will be in constant mental contact with chips scattered in our environment that will then silently carry out our commands.

I remember once watching an episode from *Star Trek* in which the crew of the starship *Enterprise* came across a planet inhabited by the Greek gods. Standing in front of them was the towering god Apollo, a giant figure who could dazzle and overwhelm the crew with godlike feats. Twenty-third-century science was powerless to spar with a god who ruled the heavens thousands of years ago in ancient Greece. But once the crew recovered from the shock of encountering the Greek gods, they soon realized that there must be a source of this power, that Apollo must simply be in mental contact with a central computer and power plant, which then executed his wishes. Once the crew located and destroyed the power supply, Apollo was reduced to an ordinary mortal.

This was just a Hollywood tale. However, by extending the radical discoveries now being made in the laboratory, scientists can envision the day when we, too, may use telepathic control over computers to give us the power of this Apollo.

NEAR FUTURE (PRESENT TO 2030)

INTERNET GLASSES AND CONTACT LENSES

Today, we can communicate with the Internet via our computers and cell phones. But in the future, the Internet will be everywhere—in wall screens, furniture, on billboards, and even in our glasses and contact lenses. When we blink, we will go online.

There are several ways we can put the Internet on a lens. The image can be flashed from our glasses directly through the lens of our eyes and onto our retinas. The image could also be projected onto the lens, which would act as a screen. Or it might be attached to the frame of the glasses, like a small jeweler's lens. As we peer into the glasses, we see the Internet, as if looking at a movie screen. We can then manipulate it with a handheld device that controls the computer via a wireless connection. We could also simply move our fingers in the air to control the image, since the computer recognizes the position of our fingers as we wave them.

For example, since 1991, scientists at the University of Washington have worked to perfect the virtual retinal display (VRD) in which red, green, and blue laser light are shone directly onto the retina. With a 120-degree field of view and a resolution of $1600 \times 1,200$ pixels, the VRD display can produce a brilliant, lifelike image that is comparable to that seen in a motion picture theater. The image can be generated using a helmet, goggles, or glasses.

Back in the 1990s, I had a chance to try out these Internet glasses. It was an early version created by the scientists at the Media Lab at MIT. It looked like an ordinary pair of glasses, except there was a cylindrical lens about ½ inch long, attached to the right-hand corner of the lens. I could look through the glasses without any problem. But if I tapped the glasses, then the tiny lens dropped in front of my eye. Peering into the lens, I could clearly make out an entire computer screen, seemingly only a bit smaller than a standard PC screen. I was surprised how clear it was, almost as if the screen were staring me in the face. Then I held a device, about the size of a cell phone, with buttons on it. By pressing the buttons, I could control the cursor on the screen and even type instructions.

In 2010, for a Science Channel special I hosted, I journeyed down to Fort Benning, Georgia, to check out the U.S. Army's latest "Internet for the battlefield," called the Land Warrior. I put on a special helmet with a miniature screen attached to its side. When I flipped the screen over my eyes, suddenly I could see a startling image: the entire battlefield with X's marking the location of friendly and enemy troops. Remarkably, the "fog of war" was lifted, with GPS sensors accurately locating the position of all troops, tanks, and buildings. By clicking a button, the image would rapidly change, putting the Internet at my disposal on the battlefield, with information concerning the weather, disposition of friendly and enemy forces, and strategy and tactics.

A much more advanced version would have the Internet flashed directly through our contact lenses by inserting a chip and LCD display into the plastic. Babak A. Parviz and his group at the University of Washington in Seattle are laying the groundwork for the Internet contact lens, designing prototypes that may eventually change the way we access the Internet.

He foresees that one immediate application of this technology might be to help diabetics regulate their glucose levels. The lens will display an immediate readout of the conditions within their body. But this is just the beginning. Eventually, Parviz envisions the day when we will be able to download any movie, song, Web site, or piece of information off the Internet into our contact lens. We will have a complete home entertainment system in our lens as we lie back and enjoy feature-length movies. We can also use it to connect directly to our office computer via our lens, then manipulate the files that flash before us. From the comfort of the beach, we will be able to teleconference to the office by blinking.

By inserting some pattern-recognition software into these Internet glasses, they will also recognize objects and even some people's faces. Already, some software programs can recognize preprogrammed faces with better than 90 percent accuracy. Not just the name, but the biography of the person you are talking to may flash before you as you speak. At a meeting this will end the embarrassment of bumping into someone you know whose name you can't remember. This may also serve an important function at a cocktail party, where there are many strangers, some of whom are very important, but you don't know who they are. In the future, you will be able to identify strangers and know their backgrounds, even as you speak to them. (This is somewhat like the world as seen through robotic eyes in *The Terminator*.)

This may alter the educational system. In the future, students taking a final exam will be able to silently scan the Internet via their contact lens for the answers to the questions, which would pose an obvious problem for teachers who often rely on rote memorization. This means that educators will have to stress thinking and reasoning ability instead.

Your glasses may also have a tiny video camera in the frame, so it can film your surroundings and then broadcast the images directly onto the Internet. People around the world may be able to share in your experiences as they happen. Whatever you are watching, thousands of others will be able to see it as well. Parents will know what their children are doing. Lovers may share experiences when separated. People at concerts will be able to communicate their excitement to fans around the world. Inspectors will visit faraway factories and then beam the live images directly to the contact lens of the boss. (Or one spouse may do the shopping, while the other makes comments about what to buy.)

Already, Parviz has been able to miniaturize a computer chip so that it can be placed inside the polymer film of a contact lens. He has successfully placed an LED (light-emitting diode) into a contact lens, and is now working on one with an 8×8 array of LEDs. His contact lens can be controlled by a wireless connection. He claims, "Those components will eventually include hundreds of LEDs, which will form images in front of the eye, such as words, charts, and photographs. Much of the hardware is semitransparent so that wearers can navigate their surroundings without crashing into them or becoming disoriented." His ultimate goal, which is still years away, is to create a contact lens with 3,600 pixels, each one no more than 10 micrometers thick.

One advantage of Internet contact lenses is that they use so little power, only a few millionths of a watt, so they are very efficient in their energy requirements and won't drain the battery. Another advantage is that the eye and optic nerve are, in some sense, a direct extension of the human brain, so we are gaining direct access to the human brain without having to implant electrodes. The eye and the optic nerve transmit information at a rate exceeding a high-speed Internet connection. So an Internet contact lens offers perhaps the most efficient and rapid access to the brain.

Shining an image onto the eye via the contact lens is a bit more complex than for the Internet glasses. An LED can produce a dot, or pixel, of light, but you have to add a microlens so that it focuses directly onto the retina. The final image would appear to float about two feet away from you. A more advanced design that Parviz is considering is to use microlasers to send a supersharp image directly onto the retina. With the same technology used in the chip industry to carve out tiny transistors, one can also etch tiny lasers of the same size, making the smallest lasers in the world. Lasers that are about 100 atoms across are in principle possible using this technology. Like transistors, you could conceivably pack millions of lasers onto a chip the size of your fingernail.

DRIVERLESS CAR

In the near future, you will also be able to safely surf the Web via your contact lens while driving a car. Commuting to work won't be such an agonizing chore because cars will drive themselves. Already, driverless cars, using GPS to locate their position within a few feet, can drive over hundreds of miles. The Pentagon's Defense Advanced Research Projects Agency (DARPA) sponsored a contest, called the DARPA Grand Challenge, in which laboratories were invited to submit driverless cars for a race across the Mojave Desert to claim a $1 million prize. DARPA was continuing its long-standing tradition of financing risky but visionary technologies.

(Some examples of Pentagon projects include the Internet, which was originally designed to connect scientists and officials during and after a nuclear war, and the GPS system, which was originally designed to guide

ICBM missiles. But both the Internet and GPS were declassified and given to the public after the end of the Cold War.)

In 2004, the contest had an embarrassing beginning, when not a single driverless car was able to travel the 150 miles of rugged terrain and cross the finish line. The robotic cars either broke down or got lost. But the next year, five cars completed an even more demanding course. They had to drive on roads that included 100 sharp turns, three narrow tunnels, and paths with sheer drop-offs on either side.

Some critics said that robotic cars might be able to travel in the desert but never in midtown traffic. So in 2007, DARPA sponsored an even more ambitious project, the Urban Challenge, in which robotic cars had to complete a grueling 60-mile course through mock-urban territory in less than six hours. The cars had to obey all traffic laws, avoid other robot cars along the course, and negotiate four-way intersections. Six teams successfully completed the Urban Challenge, with the top three claiming the $2 million, $1 million, and $500,000 prizes.

The Pentagon's goal is to make fully one-third of the U.S. ground forces autonomous by 2015. This could prove to be a lifesaving technology, since recently most U.S. casualties have been from roadside bombs. In the future, many U.S. military vehicles will have no drivers at all. But for the consumer, it might mean cars that drive themselves at the touch of a button, allowing the driver to work, relax, admire the scenery, watch a movie, or scan the Internet.

I had a chance to drive one of these cars myself for a TV special for the Discovery Channel. It was a sleek sports car, modified by the engineers at North Carolina State University so that it became fully autonomous. Its computers had the power of eight PCs. Entering the car for me was a bit of a problem, since the interior was crammed. Everywhere inside, I could see sophisticated electronic components piled on the seats and dashboard. When I grabbed the steering wheel, I noticed that it had a special rubber cable connected to a small motor. A computer, by controlling the motor, could then turn the steering wheel.

After I turned the key, stepped on the accelerator, and steered the car onto the highway, I flicked a switch that allowed the computer to take control. I took my hands off the wheel, and the car drove itself. I had full confidence in the car, whose computer was constantly making tiny adjustments via the rubber cable on the steering wheel. At first, it was a bit eerie noticing that the steering wheel and accelerator pedal were moving by themselves. It felt like there was an invisible, ghostlike driver who had taken control, but after a while I got used to it. In fact, later it became a joy to be able to relax in a car that drove itself with superhuman accuracy and skill. I could sit back and enjoy the ride.

The heart of the driverless car was the GPS system, which allowed the computer to locate its position to within a few feet. (Sometimes, the engineers told me, the GPS system could determine the car's position to within inches.) The GPS system itself is a marvel of modern technology. Each of the thirty-two GPS satellites orbiting the earth emits a specific radio wave, which is then picked up by the GPS receivers in my car. The signal from each satellite is slightly distorted because they are traveling in slightly different orbits. This distortion is called the Doppler shift. (Radio waves, for example, are compressed if the satellite is moving toward you, and are stretched if it moves away from you.) By analyzing the slight distortion of frequencies from three or four satellites, the car's computer could determine my position accurately.

The car also had radar in its fenders so that it could sense obstacles. This will be crucial in the future, as each car will automatically take emergency measures as soon as it detects an impending accident. Today, almost 40,000 people in the United States die in car accidents every year. In the future, the words *car accident* may gradually disappear from the English language.

Traffic jams may also be a thing of the past. A central computer will be able to track all the motions of every car on the road by communicating with each driverless car. It will then easily spot traffic jams and bottlenecks on the highways. In one experiment, conducted north of San Diego on Interstate 15, chips were placed in the road so that a central computer took control of the cars on the road. In case of a traffic jam, the computer will override the driver and allow traffic to flow freely.

The car of the future will also be able to sense other dangers. Thousands of people have been killed or injured in car accidents when the driver fell asleep, especially at night or on long, monotonous trips. Computers today can focus on your eyes and recognize the telltale signs of your becoming drowsy. The computer is then programmed to make a sound and wake you up. If this fails, the computer will take over the car. Computers can also recognize the presence of excessive amounts of alcohol in the car, which may reduce the thousands of alcohol-related fatalities that happen every year.

The transition to intelligent cars will not happen immediately. First, the military will deploy these vehicles and in the process work out any kinks. Then robotic cars will enter the marketplace, appearing first on long, boring stretches of interstate highways. Next, they will appear in the suburbs and large cities, but the driver will always have the ability to override the computer in case of an emergency. Eventually, we will wonder how we could have lived without them.

FOUR WALL SCREENS

Not only will computers relieve the strain of commuting and reduce car accidents, they will also help to connect us to friends and acquaintances. In the past, some people have complained that the computer revolution has dehumanized and isolated us. Actually, it has allowed us to exponentially expand our circle of friends and acquaintances. When you are lonely or in need of company, you will simply ask your wall screen to set up a bridge game with other lonely individuals anywhere in the world. When you want some assistance planning a vacation, organizing a trip, or finding a date, you will do it via your wall screen.

In the future, a friendly face might first emerge on your wall screen (a face you can change to suit your tastes). You will ask it to plan a vacation for you. It already knows your preferences and will scan the Internet and give you a list of the best possible options at the best prices.

Family gatherings may also take place via the wall screen. All four walls of your living room will have wall screens, so you will be surrounded by images of your relatives from far away. In the future, perhaps a relative may not be able to visit for an important occasion. Instead, the family may gather around the wall screen and celebrate a reunion that is part real and part virtual. Or, via your contact lens, you can see the images of all your loved ones as if they were really there, even though they are thousands of miles away. (Some commentators have remarked that the Internet was originally conceived as a "male" device by the Pentagon, that is, it was concerned with dominating an enemy in wartime. But now the Internet is mainly "female," in that it's about reaching out and touching someone.)

Teleconferencing will be replaced by telepresence—the complete 3-D images and sounds of a person will appear in your glasses or contact lens. At a meeting, for example, everyone will sit around a table, except some of the participants will appear only in your lens. Without your lens, you would see that some of the chairs around the table are empty. With your lens, you will see the image of everyone sitting in their chairs as if they were there. (This means that all participants will be videotaped by a special camera around a similar table and then their images sent over the Internet.)

In the movie *Star Wars*, audiences were amazed to see 3-D images of people appearing in the air. But using computer technology, we will be able to see these 3-D images in our contact lens, glasses, or wall screens in the future.

At first, it might seem strange talking to an empty room. But remember, when the telephone first came out, some criticized it, saying that people would be speaking to disembodied voices. They wailed that it would gradually replace direct person-to-person contact. The critics were right, but today we don't mind speaking to disembodied voices, because it has vastly increased our circle of contacts and enriched our lives.

This may also change your love life. If you are lonely, your wall screen will know your past preferences and the physical and social characteristics you want in a date, and then scan the Internet for a possible match. And since people sometimes lie in their profiles, as a security measure, your screen will automatically scan each person's history to detect falsehoods in their biography.

FLEXIBLE ELECTRONIC PAPER

The price of flat-screen TVs, once more than $10,000, has dropped by a factor of about fifty just within a decade. In the future, flat screens that cover an entire wall will also fall dramatically in price. These wall screens will be flexible and superthin, using OLEDs (organic light-emitting diodes). They are similar to ordinary light-emitting diodes, except they are based on organic compounds that can be arranged in a polymer, making them flexible. Each pixel on the flexible screen is connected to a transistor that controls the color and intensity of the light.

Already, the scientists at Arizona State University's Flexible Display Center are working with Hewlett-Packard and the U.S. Army to perfect this technology. Market forces will then drive down the cost of this technology and bring it to the public. As prices go down, the cost of these wall screens may eventually approach the price of ordinary wallpaper. So in the future, when putting up wallpaper, one might also be putting up wall screens at the same time. When we wish to change the pattern on our wallpaper, we will simply push a button. Redecorating will be so simple.

This flexible screen technology may also revolutionize how we interact with our portable computers. We will not need to lug heavy laptop computers with us. The laptop may be a simple sheet of OLEDs we then fold up and put in our wallets. A cell phone may contain a flexible screen that can be pulled out, like a scroll. Then, instead of straining to type on the tiny keyboard of your cell phone, you may be able to pull out a flexible screen as large as you want.

This technology also makes possible PC screens that are totally transparent. In the near future, we may be staring out a window, and then wave our hands, and suddenly the window becomes a PC screen. Or any image we desire. We could be staring out a window thousands of miles away.

Today, we have scrap paper that we scribble on and then throw away. In the future, we might have "scrap computers" that have no special identity of their own. We scribble on them and discard them. Today, we arrange our desk and furniture around the computer, which dominates our office. In the future, the desktop computer might disappear and the files will move with us as we go from place to place, from room to room, or from office to home. This will give us seamless information, anytime, anywhere. Today at airports you see hundreds of travelers carrying laptop computers. Once at the hotel, they have to connect to the Internet; and once they return back home, they have to download files into their desktop machines. In the future, you will never need to lug a computer around, since everywhere you turn, the walls, pictures, and furniture can connect you to the Internet, even if you are in a train or car. ("Cloud computing," where you are billed not for computers but for computer time, treating computation like a utility that is metered like water or electricity, is an early example of this.)

VIRTUAL WORLDS

The goal of ubiquitous computing is to bring the computer into our world: to put chips everywhere. The purpose of virtual reality is the opposite:

to put us into the world of the computer. Virtual reality was first introduced by the military in the 1960s as a way of training pilots and soldiers using simulations. Pilots could practice landing on the deck of an aircraft carrier by watching a computer screen and moving a joystick. In case of a nuclear war, generals and political leaders from distant locations could meet secretly in cyberspace.

Today, with computer power expanding exponentially, one can live in a simulated world, where you can control an avatar (an animated image that represents you). You can meet other avatars, explore imaginary worlds, and even fall in love and get married. You can also buy virtual items with virtual money that can then be converted to real money. One of the most popular sites, Second Life, registered 16 million accounts by 2009. That year, several people earned more than $1 million per year using Second Life. (The profit you make, however, is taxable by the U.S. government, which considers it real income.)

Virtual reality is already a staple of video games. In the future as computer power continues to expand, via your glasses or wall screen, you will also be able to visit unreal worlds. For example, if you wish to go shopping or visit an exotic place, you might first do it via virtual reality, navigating the computer screen as if you were really there. In this way, you will be able to walk on the moon, vacation on Mars, shop in distant countries, visit any museum, and decide for yourself where you want to go.

You will also, to a degree, have the ability to feel and touch objects in this cyberworld. This is called "haptic technology" and allows you to feel the presence of objects that are computer generated. It was first developed by scientists who had to handle highly radioactive materials with remote-controlled robotic arms, and by the military, which wanted its pilots to feel the resistance of a joystick in a flight simulator.

To duplicate the sense of touch scientists have created a device attached to springs and gears, so that as you push your fingers forward on the device, it pushes back, simulating the sensation of pressure. As you move your fingers across a table, for example, this device can simulate the sensation of feeling its hard wooden surface. In this way, you can feel the presence of objects that are seen in virtual reality goggles, completing the illusion that you are somewhere else.

To create the sensation of texture, another device allows your fingers to pass across a surface containing thousands of tiny pins. As your fingers move, the height of each pin is controlled by a computer, so that it can simulate the texture of hard surfaces, velvety cloth, or rough sandpaper. In the future, by putting on special gloves, it may be possible to give a realistic sensation of touch over a variety of objects and surfaces.

This will be essential for training surgeons in the future, since the surgeon has to be able to sense pressure when performing delicate surgery, and the patient might be a 3-D holographic image. It also takes us a bit closer to the holodeck of the *Star Trek* series, where you wander in a virtual world and can touch virtual objects. As you roam around an empty room, you can see fantastic objects in your goggles or contact lens. As you reach out and grab them, a haptic device rises from the floor and simulates the object you are touching.

I had a chance to witness these technologies firsthand when I visited the CAVE (cave automatic virtual environment) at Rowan University in New Jersey for the Science Channel. I entered an empty room, where I was surrounded by four walls, each wall lit up by a projector. 3-D images could be flashed onto the walls, giving the illusion of being transported to another world. In one demonstration, I was surrounded by giant, ferocious dinosaurs. By moving a joystick, I could take a ride on the back of a Tyrannosaurus rex, or even go right into its mouth. Then I visited the Aberdeen Proving Ground in Maryland, where the U.S. military has devised the most advanced version of a holodeck. Sensors were placed on my helmet and backpack, so the computer knew exactly the position of my body. I then walked on an Omnidirectional Treadmill, a sophisticated treadmill that allows you to walk in any direction while remaining in the same place. Suddenly I was on a battlefield, dodging bullets from enemy snipers. I could run in any direction, hide in any alleyway, sprint down any street, and the 3-D images on the screen changed instantly. I could even lie flat on the floor, and the screens changed accordingly. I could imagine that, in the future, you will be able to experience total immersion, e.g. engage in dogfights with alien spaceships, flee from rampaging monsters, or frolic on a deserted island, all from the comfort of your living room.

MEDICAL CARE IN THE NEAR FUTURE

A visit to the doctor's office will be completely changed. For a routine checkup, when you talk to the "doctor," it will probably be a robotic software program that appears on your wall screen and that can correctly diagnose up to 95 percent of all common ailments. Your "doctor" may look like a person, but it will actually be an animated image programmed to ask certain simple questions. Your "doctor" will also have a complete record of your genes, and will recommend a course of medical treatments that takes into account all your genetic risk factors.

To diagnose a problem, the "doctor" will ask you to pass a simple probe over your body. In the original *Star Trek* TV series, the public was amazed to see a device called the tricorder that could instantly diagnose any illness and peer inside your body. But you do not have to wait until the twenty-third century for this futuristic device. Already, MRI machines, which weigh several tons and can fill up an entire room, have been miniaturized to about a foot, and will eventually be as small as a cell phone. By passing one over your body, you will be able to see inside your organs. Computers will process these 3-D images and then give you a diagnosis. This probe will also be able to determine, within minutes, the presence of a wide variety of diseases, including cancer, years before a tumor forms. This probe will contain DNA chips, silicon chips that have millions of tiny sensors that can detect the presence of the telltale DNA of many diseases.

Of course, many people hate going to the doctor. But in the future, your health will be silently and effortlessly monitored several times a day without your being aware of it. Your toilet, bathroom mirror, and clothes will have DNA chips to silently determine if you have cancer colonies of only a few hundred cells growing in your body. You will have more sensors hidden in your bathroom and clothes than are found in a modern hospital or university today. For example, simply by blowing on a mirror, the DNA for a mutated protein called p53 can be detected, which is implicated in 50 percent of all common cancers. This means that the word *tumor* will gradually disappear from the English language.

Today, if you are in a bad car accident on a lonely road, you could easily bleed to death. But in the future, your clothes and car will automatically spring into action at the first sign of trauma, calling for an ambulance, locating your car's position, uploading your entire medical history, all while you are unconscious. In the future, it will be difficult to die alone. Your clothes will be able to sense any irregularities in your heartbeat, breathing, and even brain waves by means of tiny chips woven into the fabric. When you get dressed, you go online.

Today, it is possible to put a chip into a pill about the size of an aspirin, complete with a TV camera and radio. When you swallow it, the "smart pill" takes TV images of your gullet and intestines, and then radios the signals to a nearby receiver. (This gives new meaning to the slogan "Intel inside.") In this way, doctors may be able to take pictures of a patient's intestines and detect cancers without ever performing a colonoscopy (which involves the inconvenience of inserting a six-foot-long tube up your large intestine). Microscopic devices like these also will gradually reduce the necessity of cutting skin for surgery.

This is only a sample of how the computer revolution will affect our health. We will discuss the revolution in medicine in much more detail in Chapters 3 and 4, where we also discuss gene therapy, cloning, and altering the human life span.

LIVING IN A FAIRY TALE

Because computer intelligence will be so cheap and widespread in the environment, some futurists have commented that the future might look like something out of a fairy tale. If we have the power of the gods, then the heaven we inhabit will look like a fantasy world. The future of the Internet, for example, is to become the magic mirror of Snow White. We will say, "Mirror, mirror on the wall," and a friendly face will emerge, allowing us to access the wisdom of the planet. We will put chips in our toys, making them intelligent, like Pinocchio, the puppet who wanted to be a real boy. Like Pocahontas, we will talk to the wind and the trees, and they will talk back. We will assume that objects are intelligent and that we can talk to them.

Because computers will be able to locate many of the genes that control the aging process, we might be forever young like Peter Pan. We will be able to slow down and perhaps reverse the aging process, like the boys from Neverland who didn't want to grow up. Augmented reality will give us the illusion that, like Cinderella, we can ride to fantasy balls in a royal coach and dance gracefully with a handsome prince. (But at midnight, our augmented reality glasses turn off and we return to the real world.) Because computers are revealing the genes that control our bodies, we will be able to reengineer our bodies, replacing organs and changing our appearance, even at the genetic level, like the beast in "Beauty and the Beast."

Some futurists have even feared that this might give rise to a return to the mysticism of the Middle Ages, when most people believed that there were invisible spirits inhabiting everything around them.

MIDCENTURY (2030 TO 2070)

END OF MOORE'S LAW

We have to ask: How long can this computer revolution last? If Moore's law holds true for another fifty years, it is conceivable that computers will rapidly exceed the computational power of the human brain. By midcentury, a new dynamic occurs. As George Harrison once said, "All things must pass." Even Moore's law must end, and with it the spectacular rise of computer power that has fueled economic growth for the past half century.

Today, we take it for granted, and in fact believe it is our birthright, to have computer products of ever-increasing power and complexity. This is why we buy new computer products every year, knowing that they are almost twice as powerful as last year's model. But if Moore's law collapses—and every generation of computer products has roughly the same power and speed of the previous generation—then why bother to buy new computers?

Since chips are placed in a wide variety of products, this could have disastrous effects on the entire economy. As entire industries grind to a halt, millions could lose their jobs, and the economy could be thrown into turmoil.

Years ago, when we physicists pointed out the inevitable collapse of Moore's law, traditionally the industry pooh-poohed our claims, implying that we were crying wolf. The end of Moore's law was predicted so many times, they said, that they simply did not believe it.

But not anymore.

Two years ago, I keynoted a major conference for Microsoft at their main headquarters in Seattle, Washington. Three thousand of the top engineers at Microsoft were in the audience, waiting to hear what I had to say about the future of computers and telecommunications. Staring out at the huge crowd, I could see the faces of the young, enthusiastic engineers who would be creating the programs that will run the computers sitting on our desks and laps. I was blunt about Moore's law, and said that the industry has to prepare for this collapse. A decade earlier, I might have been met with laughter or a few snickers. But this time I only saw people nodding their heads.

So the collapse of Moore's law is a matter of international importance, with trillions of dollars at stake. But precisely how it will end, and what will replace it, depends on the laws of physics. The answers to these physics questions will eventually rock the economic structure of capitalism.

To understand this situation, it is important to realize that the remarkable success of the computer revolution rests on several principles of physics. First, computers have dazzling speed because electrical signals travel at near the speed of light, which is the ultimate speed in the universe. In one second, a light beam can travel around the world seven times or reach the moon. Electrons are also easily moved around and loosely bound to the atom (and can be scraped off just by combing your hair, walking across a carpet, or by doing your laundry—that's why we have static cling). The combination of loosely bound electrons and their enormous speed allows us to send electrical signals at a blinding pace, which has created the electric revolution of the past century.

Second, there is virtually no limit to the amount of information you can place on a laser beam. Light waves, because they vibrate much faster than sound waves, can carry vastly more information than sound. (For example, think of stretching a long piece of rope and then vibrating one end rapidly. The faster you wiggle one end, the more signals you can send along the rope. Hence, the amount of information you can cram onto a wave increases the faster you vibrate it, that is, by increasing its frequency.) Light is a wave that vibrates at roughly 10^{14} cycles per second (that is 1 with 14 zeros after it). It takes many cycles to convey one bit of information (a 1 or a 0). This means that a fiber-optic cable can carry roughly 10^{11} bits of information on a single frequency. And this number can be increased by cramming many signals into a single optical fiber and then bundling these fibers into a cable. This means that, by increasing the number of channels in a cable and then increasing the number of cables, one can transmit information almost without limit.

Third, and most important, the computer revolution is driven by miniaturizing transistors. A transistor is a gate, or switch, that controls the flow of electricity. If an electric circuit is compared to plumbing, then a transistor is like a valve controlling the flow of water. In the same way that the simple twist of a valve can control a huge volume of water, the transistor allows a tiny flow of electricity to control a much larger flow, thereby amplifying its power.

At the heart of this revolution is the computer chip, which can contain hundreds of millions of transistors on a silicon wafer the size of your fingernail. Inside your laptop there is a chip whose transistors can be seen only under a microscope. These incredibly tiny transistors are created the same way that designs on T-shirts are made.

Designs on T-shirts are mass-produced by first creating a stencil with the outline of the pattern one wishes to create. Then the stencil is placed over the cloth, and spray paint is applied. Only where there are gaps in the stencil does the paint penetrate to the cloth. Once the stencil is removed, one has a perfect copy of the pattern on the T-shirt.

Likewise, a stencil is made containing the intricate outlines of millions of transistors. This is placed over a wafer containing many layers of silicon, which is sensitive to light. Ultraviolet light is then focused on the stencil, which then penetrates through the gaps of the stencil and exposes the silicon wafer.

Then the wafer is bathed in acid, carving the outlines of the circuits and creating the intricate design of millions of transistors. Since the wafer consists of many conducting and semiconducting layers, the acid cuts into the wafer at different depths and patterns, so one can create circuits of enormous complexity.

One reason why Moore's law has relentlessly increased the power of chips is because UV light can be tuned so that its wavelength is smaller and smaller, making it possible to etch increasingly tiny transistors onto silicon wafers. Since UV light has a wavelength as small as 10 nanometers (a nanometer is a billionth of a meter), this means that the smallest transistor that you can etch is about thirty atoms across.

But this process cannot go on forever. At some point, it will be physically impossible to etch transistors in this way that are the size of atoms. You can even calculate roughly when Moore's law will finally collapse: when you finally hit transistors the size of individual atoms.

Around 2020 or soon afterward, Moore's law will gradually cease to hold true and Silicon Valley may slowly turn into a rust belt unless a replace-

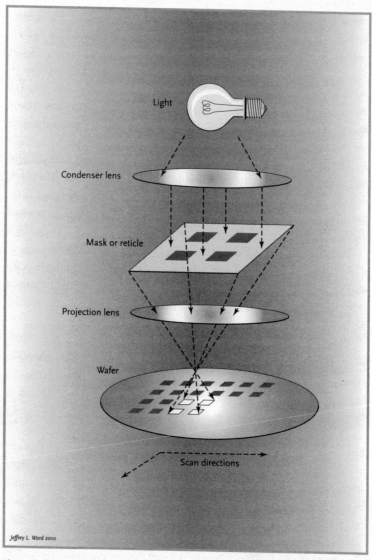

Light

Condenser lens

Mask or reticle

Projection lens

Wafer

Scan directions

Jeffrey L. Ward 2010

The end of Moore's law. Chips are made the same way as designs on T-shirts. Instead of spray painting over a stencil, UV light is focused on a stencil, burning an image onto layers of silicon. Acids then carve out the image, creating hundreds of millions of transistors. But there is a limit to the process when we hit the atomic scale. Will Silicon Valley become a rust belt?

ment technology is found. According to the laws of physics, eventually the Age of Silicon will come to a close, as we enter the Post-Silicon Era. Transistors will be so small that quantum theory or atomic physics takes over and electrons leak out of the wires. For example, the thinnest layer inside your computer will be about five atoms across. At that point, according to the laws of physics, the quantum theory takes over. The Heisenberg uncertainty principle states that you cannot know both the position and velocity of any particle. This may sound counterintuitive, but at the atomic level you simply cannot know where the electron is, so it can never be confined precisely in an ultrathin wire or layer and it necessarily leaks out, causing the circuit to short-circuit.

We will discuss this in more detail in Chapter 4, when we analyze nanotechnology. For the rest of this chapter, we will assume that physicists have found a successor to silicon power, but that computer power grows at a much slower pace than before. Computers will most likely continue to grow exponentially, but the doubling time will not be eighteen months, but many years.

MIXING REAL AND VIRTUAL REALITY

By midcentury, we should all be living in a mixture of real and virtual reality. In our contact lens or glasses, we will simultaneously see virtual images superimposed on the real world. This is the vision of Susumu Tachi of Keio University in Japan and many others. He is designing special goggles that blend fantasy and reality. His first project is to make things disappear into thin air.

I visited Professor Tachi in Tokyo and witnessed some of his remarkable experiments in mixing real and virtual reality. One simple application is to make an object disappear (at least in your goggles). First, I wore a special light brown raincoat. When I spread out my arms, it resembled a large sail. Then a camera was focused on my raincoat and a second camera filmed the scenery behind me, consisting of buses and cars moving along a road. An instant later, a computer merged these two images, so the image behind me was flashed onto my raincoat, as if on a screen. If you peered into a special lens, my body vanished, leaving only the images of the cars and buses. Since my head was above the raincoat, it appeared as if my head was floating in midair, without a body, like Harry Potter wearing his invisibility cloak.

Professor Tachi then showed me some special goggles. By wearing them, I could see real objects and then make them disappear. This is not true invisibility, since it works only if you wear special goggles that merge two images. However, it is part of Professor Tachi's grand program, which is sometimes called "augmented reality."

By midcentury, we will live in a fully functioning cyberworld that merges the real world with images from a computer. This could radically change the workplace, commerce, entertainment, and our way of life. Augmented reality would have immediate consequences for the marketplace. The first commercial application would be to make objects become invisible, or to make the invisible become visible.

For example, if you are a pilot or a driver, you will be able to see 360 degrees around yourself, and even beneath your feet, because your goggles or lens allow you to see through the plane's or car's walls. This will eliminate blind spots that are responsible for scores of accidents and deaths. In a dogfight, jet pilots will be able to track enemy jets anywhere they fly, even below themselves, as if your jet were transparent. Drivers will be able to see in all directions, since tiny cameras will monitor 360 degrees of their surroundings and beam the images into their contact lenses.

If you are an astronaut making repairs on the outside of a rocket ship, you will also find this useful, since you can see right through walls, partitions, and the rocket ship's hull. This could be lifesaving. If you are a construction worker making underground repairs, amid a mass of wires, pipes, and valves, you will know exactly how they are all connected. This could prove vital in case of a gas or steam explosion, when pipes hidden behind walls have to be repaired and reconnected quickly.

Likewise, if you are a prospector, you will be able to see right through the soil, to underground deposits of water or oil. Satellite and airplane photographs taken of a field with infrared and UV light can be analyzed and then fed into your contact lens, giving you a 3-D analysis of the site and what lies below the surface. As you walk across a barren landscape, you will "see" valuable mineral deposits via your lens.

In addition to making objects invisible, you will also be able to do the opposite: to make the invisible become visible.

If you are an architect, you will be able to walk around an empty room and suddenly "see" the entire 3-D image of the building you are designing. The designs on your blueprint will leap out at you as you wander around each room. Vacant rooms will suddenly come alive, with furniture, carpets, and decorations on the walls, allowing you to visualize your creation in 3-D before you actually build it. By simply moving your arms, you will be able to create new rooms, walls, and furniture. In this augmented world, you will have the power of a magician, waving your wand and creating any object you desire.

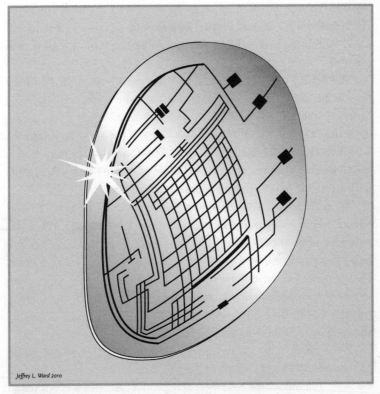

Jeffrey L. Ward 2010

Internet contact lenses will recognize people's faces, display their biographies, and translate their words as subtitles. Tourists will use them to resurrect ancient monuments. Artists and architects will use them to manipulate and reshape their virtual creations. The possibilities are endless for augmented reality.

AUGMENTED REALITY: A REVOLUTION IN TOURISM, ART, SHOPPING, AND WARFARE

As you can see, the implications for commerce and the workplace are potentially enormous. Virtually every job can be enriched by augmented reality. In addition, our lives, our entertainment, and our society will be greatly enhanced by this technology.

For example, a tourist walking in a museum can go from exhibit to exhibit as your contact lens gives you a description of each object; a virtual guide will give you a cybertour as you pass. If you are visiting some ancient ruins, you will be able to "see" complete reconstructions of the buildings and monuments in their full glory, along with historical anecdotes. The remains of the Roman Empire, instead of being broken columns and weeds, will spring back to life as you wander among them, complete with commentary and notes.

The Beijing Institute of Technology has already taken the first baby steps in this direction. In cyberspace, it re-created the fabulous Garden of Perfect Brightness, which was destroyed by British-French forces during the Second Opium War of 1860. Today, all that is left of the fabled garden is the wreckage left by marauding troops. But if you view the ruins from a special viewing platform, you can see the entire garden before you in all its splendor. In the future, this will become commonplace.

An even more advanced system was created by inventor Nikolas Neecke, who has created a walking tour of Basel, Switzerland. When you walk around its ancient streets, you see images of ancient buildings and even people superimposed on the present, as if you were a time traveler. The computer locates your position and then shows you images of ancient scenes in your goggles, as if you were transported to medieval times. Today, you have to wear large goggles and a heavy backpack full of GPS electronics and computers. Tomorrow, you will have this in your contact lens.

If you are driving a car in a foreign land, all the gauges would appear on your contact lens in English, so you would never have to glance down to see them. You will see the road signs along with explanations of any object nearby, such as tourist attractions. You will also see rapid translations of road signs.

A hiker, camper, or outdoorsman will know not just his position in a foreign land but also the names of all the plants and animals, and will be able to see a map of the area and receive weather reports. He will also see trails and camping sites that may be hidden by brush and trees.

Apartment hunters will be able to see what is available as you walk down the street or drive by in a car. Your lens will display the price, the amenities, etc., of any apartment or house that's for sale.

And gazing at the night sky, you will see the stars and all the constellations clearly delineated, as if you were watching a planetarium show, except that the stars you see are real. You will also see where galaxies, distant black holes, and other interesting astronomical sights are located and be able to download interesting lectures.

In addition to being able to see through objects and visit foreign lands, augmented vision will be essential if you need very specialized information at a moment's touch.

For example, if you are an actor, musician, or performer who has to memorize large amounts of material, in the future you will see all the lines or music in your lens. You won't need teleprompters, cue cards, sheet music, or notes to remind you. You will not need to memorize anything anymore.

Other examples include:

- If you are a student and missed a lecture, you will be able to download lectures given by virtual professors on any subject and watch them. Via telepresence, an image of a real professor could appear in front of you and answer any questions you may have. You will also be able to see demonstrations of experiments, videos, etc., via your lens.
- If you are a soldier in the field, your goggles or headset may give you all the latest information, maps, enemy locations, direction of enemy fire, instructions from superiors, etc. In a firefight with the enemy, when bullets are whizzing by from all directions, you will be able to see through obstacles and hills and locate the enemy, since drones flying overhead can identify their positions.
- If you are a surgeon doing a delicate emergency operation, you will be able to see inside the patient (via portable MRI machines), through the body (via sensors moving inside the body), as well as access all medical records and videos of previous operations.

- If you are playing a video game, you can immerse yourself in cyberspace in your contact lens. Although you are in an empty room, you can see all your friends in perfect 3-D, experiencing some alien landscape as you prepare to do battle with imaginary aliens. It will be as if you are on the battlefield of an alien planet, with ray blasts going off all around you and your buddies.
- If you need to look up any athlete's statistics or sports trivia, the information will spring instantly into your contact lens.

This means you would not need a cell phone, clocks or watches, or MP3 players anymore. All the icons on your various handheld objects would be projected onto your contact lenses, so that you could access them anytime you wanted. Phone calls, music Web sites, etc. could all be accessed this way. Many of the appliances and gadgets you have at home can be replaced by augmented reality.

Another scientist pushing the boundary of augmented reality is Pattie Maes of the MIT Media Laboratory. Instead of using special contact lenses, glasses, or goggles, she envisions projecting a computer screen onto common objects in our environment. Her project, called SixthSense, involves wearing a tiny camera and projector around your neck, like a medallion, that can project the image of a computer screen on anything in front of you, such as the wall or a table. Pushing the imaginary buttons automatically activates the computer, just as if you were typing on a real keyboard. Since the image of a computer screen can be projected on anything flat and solid in front of you, you can convert hundreds of objects into computer screens.

Also, you wear special plastic thimbles on your thumb and fingers. As you move your fingers, the computer executes instructions on the computer screen on the wall. By moving your fingers, for example, you can draw images onto the computer screen. You can use your fingers instead of a mouse to control the cursor. And if you put your hands together to make a square, you can activate a digital camera and take pictures.

This also means that when you go shopping, your computer will scan various products, identify what they are, and then give you a complete readout of their contents, calorie content, and reviews by other consumers. Since chips will cost less than bar codes, every commercial product will have its own intelligent label you can access and scan.

Another application of augmented reality might be X-ray vision, very similar to the X-ray vision found in *Superman* comics, which uses a process called "backscatter X-rays." If your glasses or contact lens are sensitive to X-rays, it may be possible to peer through walls. As you look around, you will be able to see through objects, just as in the comic books. Every kid, when they first read *Superman* comics, dreams of being "faster than a speeding bullet, more powerful than a locomotive." Thousands of kids don capes, jump off crates, leap into the air, and pretend to have X-ray vision, but it is also a real possibility.

One problem with ordinary X-rays is that you have to place X-ray film behind any object, expose the object to X-rays, and then develop the film. But backscattered X-rays solve all these problems. First, you have X-rays emanating from a light source that can bathe a room. Then they bounce off the walls, and pass from behind through the object you want to examine. Your goggles are sensitive to the X-rays that have passed through the object. Images seen via backscattered X-rays can be just as good as the images found in the comics. (By increasing the sensitivity of the goggles, one can reduce the intensity of the X-rays, to minimize any health risks.)

UNIVERSAL TRANSLATORS

In *Star Trek,* the *Star Wars* saga, and virtually all other science fiction films, remarkably, all the aliens speak perfect English. This is because there is something called the "universal translator" that allows earthlings to communicate instantly with any alien civilization, removing the inconvenience of tediously using sign language and primitive gestures to communicate with an alien.

Although once considered to be unrealistically futuristic, versions of the universal translator already exist. This means that in the future, if you are a tourist in a foreign country and talk to the locals, you will see subtitles in your contact lens, as if you were watching a foreign-language movie. You can also have your computer create an audio translation that is fed into your ears. This means that it may be possible to have two people carry on a conversation, with each speaking in their own language, while hearing the translation in their ears, if both have the universal translator. The translation won't be perfect, since there are always problems with idioms, slang, and colorful expressions, but it will be good enough so you will understand the gist of what that person is saying.

There are several ways in which scientists are making this a reality. The first is to create a machine that can convert the spoken word into writing. In the mid-1990s, the first commercially available speech recognition machines hit the market. They could recognize up to 40,000 words with 95 percent accuracy. Since a typical, everyday conversation uses only 500 to 1,000 words, these machines are more than adequate. Once the transcription of the human voice is accomplished, then each word is translated into another language via a computer dictionary. Then comes the hard part: putting the words into context, adding slang, colloquial expressions, etc., all of which require a sophisticated understanding of the nuances of the language. The field is called CAT (computer assisted translation).

Another way is being pioneered at Carnegie Mellon University in Pittsburgh. Scientists there already have prototypes that can translate Chinese into English, and English into Spanish or German. They attach electrodes to the neck and face of the speaker; these pick up the contraction of the muscles and decipher the words being spoken. Their work does not require any audio equipment, since the words can be mouthed silently. Then a computer translates these words and a voice synthesizer speaks them out loud. In simple conversations involving 100 to 200 words, they have attained 80 percent accuracy.

"The idea is that you can mouth words in English and they will come out in Chinese or another language," says Tanja Schultz, one of the researchers. In the future, it might be possible for a computer to lip-read the person you are talking to, so the electrodes are not necessary. So, in principle, it is possible to have two people having a lively conversation, although they speak in two different languages.

In the future, language barriers, which once tragically prevented cultures from understanding one another, may gradually fall with this universal translator and Internet contact lens or glasses.

Although augmented reality opens up an entirely new world, there are limitations. The problem will not be one of hardware; nor is bandwidth a limiting factor, since there is no limit to the amount of information that can be carried by fiber-optic cables.

The real bottleneck is software. Creating software can be done only the old-fashioned way. A human—sitting quietly in a chair with a pencil, paper, and laptop—is going to have to write the codes, line for line, that make these imaginary worlds come to life. One can mass-produce hardware and increase its power by piling on more and more chips, but you cannot mass-produce the brain. This means that the introduction of a truly augmented world will take decades, until midcentury.

HOLOGRAMS AND 3-D

Another technological advance we might see by midcentury is true 3-D TV and movies. Back in the 1950s, 3-D movies required that you put on clunky glasses whose lenses were colored blue and red. This took advantage of the fact that the left eye and the right eye are slightly misaligned; the movie screen displayed two images, one blue and one red. Since these glasses acted as filters that gave two distinct images to the left and right eye, this gave the illusion of seeing three dimensions when the brain merged the two images. Depth perception, therefore, was a trick. (The farther apart your eyes are, the greater the depth perception. That is why some animals have eyes outside their heads: to give them maximum depth perception.)

One improvement is to have 3-D glasses made of polarized glass, so that the left eye and right eye are shown two different polarized images. In this way, one can see 3-D images in full color, not just in blue and red. Since light is a wave, it can vibrate up and down, or left and right. A polarized lens is a piece of glass that allows only one direction of light to pass through. Therefore, if you have two polarized lenses in your glasses, with different directions of polarization, you can create a 3-D effect. A more sophisticated version of 3-D may be to have two different images flashed into our contact lens.

3-D TVs that require wearing special glasses have already hit the market. But soon, 3-D TVs will no longer require them, instead using lenticular lenses. The TV screen is specially made so that it projects two separate images at slightly different angles, one for each eye. Hence your eyes see separate images, giving the illusion of 3-D. However, your head must be positioned correctly; there are "sweet spots" where your eyes must lie as you gaze at the screen. (This takes advantage of a well-known optical illusion. In novelty stores, we see pictures that magically transform as we walk past them. This is done by taking two pictures, shredding each one into many thin strips, and then interspersing the strips, creating a composite image. Then a lenticular glass sheet with many vertical grooves is placed on top of the composite, each groove sitting precisely on top of two strips. The groove is specially shaped so that, as you gaze upon it from one angle, you can see one strip, but the other strip appears from another angle. Hence, by walking past the glass sheet, we see each picture suddenly transform from one into the other, and back again. 3-D TVs will replace these still pictures with moving images to attain the same effect without the use of glasses.)

But the most advanced version of 3-D will be holograms. Without using any glasses, you would see the precise wave front of a 3-D image, as if it were sitting directly in front of you. Holograms have been around for decades (they appear in novelty shops, on credit cards, and at exhibitions), and they regularly are featured in science fiction movies. In *Star Wars*, the plot was set in motion by a 3-D holographic distress message sent from Princess Leia to members of the Rebel Alliance.

The problem is that holograms are very hard to create.

Holograms are made by taking a single laser beam and splitting it in two. One beam falls on the object you want to photograph, which then bounces off and falls onto a special screen. The second laser beam falls directly onto the screen. The mixing of the two beams creates a complex interference pattern containing the "frozen" 3-D image of the original object, which is then captured on a special film on the screen. Then, by flashing another laser beam through the screen, the image of the original object comes to life in full 3-D.

There are two problems with holographic TV. First, the image has to be flashed onto a screen. Sitting in front of the screen, you see the exact 3-D image of the original object. But you cannot reach out and touch the object. The 3-D image you see in front of you is an illusion.

This means that if you are watching a 3-D football game on your holographic TV, no matter how you move, the image in front of you changes as if it were real. It might appear that you are sitting right at the 50-yard line, watching the game just inches from the football players. However, if you were to reach out to grab the ball, you would bump into the screen.

The real technical problem that has prevented the development of holographic TV is that of information storage. A true 3-D image contains a vast amount of information, many times the information stored inside a single 2-D image. Computers regularly process 2-D images, since the image is broken down into tiny dots, called pixels, and each pixel is illuminated by a tiny transistor. But to make a 3-D image move, you need to flash thirty images per second. A quick calculation shows that the information needed to generate moving 3-D holographic images far exceeds the capability of today's Internet.

By midcentury, this problem may be resolved as the bandwidth of the Internet expands exponentially.

What might true 3-D TV look like?

One possibility is a screen shaped like a cylinder or dome that you sit inside. When the holographic image is flashed onto the screen, we see the 3-D images surrounding us, as if they were really there.

FAR FUTURE (2070 TO 2100)

MIND OVER MATTER

By the end of this century, we will control computers directly with our minds. Like Greek gods, we will think of certain commands and our wishes will be obeyed. The foundation for this technology has already been laid. But it may take decades of hard work to perfect it. This revolution is in two parts: First, the mind must be able to control objects around it. Second, a computer has to decipher a person's wishes in order to carry them out.

The first significant breakthrough was made in 1998, when scientists at Emory University and the University of Tübingen, Germany, put a tiny glass electrode directly into the brain of a fifty-six-year-old man who was paralyzed after a stroke. The electrode was connected to a computer that analyzed the signals from his brain. The stroke victim was able to see an image of the cursor on the computer screen. Then, by biofeedback, he was able to control the cursor of the computer display by thinking alone. For the first time, a direct contact was made between the human brain and a computer.

The most sophisticated version of this technology has been developed at Brown University by neuroscientist John Donoghue, who has created a device called BrainGate to help people who have suffered debilitating brain injuries communicate. He created a media sensation and even made the cover of *Nature* magazine in 2006.

Donoghue told me that his dream is to have BrainGate revolutionize the way we treat brain injuries by harnessing the full power of the information revolution. It has already had a tremendous impact on the lives of his patients, and he has high hopes of furthering this technology. He has a personal interest in this research because, as a child, he was confined to a wheelchair due to a degenerative disease and hence knows the feeling of helplessness.

His patients include stroke victims who are completely paralyzed and unable to communicate with their loved ones, but whose brains are active. He has placed a chip, just 4 millimeters wide, on top of a stroke victim's brain, in the area that controls motor movements. This chip is then connected to a computer that analyzes and processes the brain signals and eventually sends the message to a laptop.

At first the patient has no control over the location of the cursor, but can see where the cursor is moving. By trial and error, the patient learns to control the cursor, and, after several hours, can position the cursor anywhere on the screen. With practice, the stroke victim is able to read and write e-mails and play video games. In principle a paralyzed person should be able to perform any function that can be controlled by the computer.

Initially, Donoghue started with four patients, two who had spinal cord injuries, one who'd had a stroke, and a fourth who had ALS (amyotrophic lateral sclerosis). One of them, a quadriplegic paralyzed from the neck down, took only a day to master the movement of the cursor with his mind. Today, he can control a TV, move a computer cursor, play a video game, and read e-mail. Patients can also control their mobility by manipulating a motorized wheelchair.

In the short term, this is nothing less than miraculous for people who are totally paralyzed. One day, they are trapped, helpless, in their bodies; the next day, they are surfing the Web and carrying on conversations with people around the world.

(I once attended a gala reception at Lincoln Center in New York in honor of the great cosmologist Stephen Hawking. It was heartbreaking to see him strapped into a wheelchair, unable to move anything but a few facial muscles and his eyelids, with nurses holding up his limp head and pushing him around. It takes him hours and days of excruciating effort to communicate simple ideas via his voice synthesizer. I wondered if it was not too late for him to take advantage of the technology of BrainGate. Then John Donoghue, who was also in the audience, came up to greet me. So perhaps BrainGate is Hawking's best option.)

Another group of scientists at Duke University have achieved similar results in monkeys. Miguel A. L. Nicolelis and his group have placed a chip on the brain of a monkey. The chip is connected to a mechanical arm. At first, the monkey flails about, not understanding how to operate the mechanical arm. But with some practice, these monkeys, using the power of their brains, are able to slowly control the motions of the mechanical arm—for example, moving it so that it grabs a banana. They can instinctively move these arms without thinking, as if the mechanical arm is their own. "There's some physiological evidence that during the experiment they feel more connected to the robots than to their own bodies," says Nicolelis.

This also means that we will one day be able to control machines using pure thought. People who are paralyzed may be able to control mechanical arms and legs in this way. For example, one might be able to connect a person's brain directly to mechanical arms and legs, bypassing the spinal cord, so the patient can walk again. Also, this may lay the foundation for controlling our world via the power of the mind.

MIND READING

If the brain can control a computer or mechanical arm, can a computer read the thoughts of a person, without placing electrodes inside the brain?

It's been known since 1875 that the brain is based on electricity moving through its neurons, which generates faint electrical signals that can be measured by placing electrodes around a person's head. By analyzing the electrical impulses picked up by these electrodes, one can record the brain waves. This is called an EEG (electroencephalogram), which can record gross changes in the brain, such as when it is sleeping, and also moods, such as agitation, anger, etc. The output of the EEG can be displayed on a computer screen, which the subject can watch. After a while, the person is able to move the cursor by thinking alone. Already, Niels Birbaumer of the

University of Tübingen has been able to train partially paralyzed people to type simple sentences via this method.

Even toy makers are taking advantage of this. A number of toy companies, including NeuroSky, market a headband with an EEG-type electrode inside. If you concentrate in a certain way, you can activate the EEG in the headband, which then controls the toy. For example, you can raise a Ping-Pong ball inside a cylinder by sheer thought.

The advantage of the EEG is that it can rapidly detect various frequencies emitted by the brain without elaborate, expensive equipment. But one large disadvantage is that the EEG cannot localize thoughts to specific locations of the brain.

A much more sensitive method is the fMRI (functional magnetic resonance imaging) scan. EEG and fMRI scans differ in important ways. The EEG scan is a passive device that simply picks up electrical signals from the brain, so we cannot determine very well the location of the source. An fMRI machine uses "echoes" created by radio waves to peer inside living tissue. This allows us to pinpoint the location of the various signals, giving us spectacular 3-D images of inside the brain.

The fMRI machine is quite expensive and requires a laboratory full of heavy equipment, but already it has given us breathtaking details of how the thinking brain functions. The fMRI scan allows scientists to locate the presence of oxygen contained within hemoglobin in the blood. Since oxygenated hemoglobin contains the energy that fuels cell activity, detecting the flow of this oxygen allows one to trace the flow of thoughts in the brain.

Joshua Freedman, a psychiatrist at the University of California, Los Angeles, says: "It's like being an astronomer in the sixteenth century after the invention of the telescope. For millennia, very smart people tried to make sense of what was going on up in the heavens, but they could only speculate about what lay beyond unaided human vision. Then, suddenly, a new technology let them see directly what was there."

In fact, fMRI scans can even detect the motion of thoughts in the living brain to a resolution of .1 millimeter, or smaller than the head of a pin, which corresponds to perhaps a few thousand neurons. An fMRI can thus give three-dimensional pictures of the energy flow inside the thinking brain to astonishing accuracy. Eventually, fMRI machines may be built that can probe to the level of single neurons, in which case one might be able to pick out the neural patterns corresponding to specific thoughts.

A breakthrough was made recently by Kendrick Kay and his colleagues at the University of California at Berkeley. They did an fMRI scan of people as they looked at pictures of a variety of objects, such as food, animals, people, and common things of various colors. Kay and colleagues created a software program that could associate these objects with the corresponding fMRI patterns. The more objects these subjects saw, the better the computer program was at identifying these objects on their fMRI scans.

Then they showed the same subjects entirely new objects, and the software program was often able to correctly match the object with the fMRI scan. When shown 120 pictures of new objects, the software program correctly identified the fMRI scan with these objects 90 percent of the time. When the subjects were shown 1,000 new pictures, the software program's success rate was 80 percent.

Kay says it is "possible to identify, from a large set of completely novel natural images, which specific image was seen by an observer. . . . It may soon be possible to reconstruct a picture of a person's visual experience from measurements of brain activity alone."

The goal of this approach is to create a "dictionary of thought," so that each object has a one-to-one correspondence to a certain fMRI image. By reading the fMRI pattern, one can then decipher what object the person is thinking about. Eventually, a computer will scan perhaps thousands of fMRI patterns that come pouring out of a thinking brain and decipher each one. In this way, one may be able to decode a person's stream of consciousness.

PHOTOGRAPHING A DREAM

The problem with this technique, however, is that while it might be able to tell if you are thinking of a dog, for example, it cannot reproduce the actual image of the dog itself. One new line of research is to try to reconstruct the precise image that the brain is thinking of, so that one might be able to create a video of a person's thoughts. In this way, one might be able to make a video recording of a dream.

Since time immemorial, people have been fascinated by dreams, those ephemeral images that are sometimes so frustrating to recall or understand. Hollywood has long envisioned machines that might one day send dreamlike thoughts into the brain or even record them, as in movies like *Total Recall*. All this, however, was sheer speculation.

Until recently, that is.

Scientists have made remarkable progress in an area once thought to be impossible: taking a snapshot of our memories and possibly our dreams. The first steps in this direction were taken by scientists at the Advanced Telecommunications Research (ATR) Computational Neuroscience Laboratory in Kyoto. They showed their subjects a pinpoint of light at a particular location. Then they used an fMRI scan to record where the brain stored this information. They moved the pinpoint of light and recorded where the brain stored this new image. Eventually, they had a one-to-one map of where scores of pinpoints of light were stored in the brain. These pinpoints were located on a 10 × 10 grid.

Then the scientists flashed a picture of a simple object made from these

10 × 10 points, such as a horseshoe. By computer they could then analyze how the brain stored this picture. Sure enough, the pattern stored by the brain was the sum of the images that made up the horseshoe.

In this way, these scientists could create a picture of what the brain is seeing. Any pattern of lights on this 10 × 10 grid can be decoded by a computer looking at the fMRI brain scans.

In the future, these scientists want to increase the number of pixels in their 10 × 10 grid. Moreover, they claim that this process is universal, that is, any visual thought or even dream should be able to be detected by the fMRI scan. If true, it might mean that we will be able to record, for the first time in history, the images we are dreaming about.

Of course, our mental images, and especially our dreams, are never crystal sharp, and there will always be a certain fuzziness, but the very fact that we can look deeply into the visual thoughts of someone's brain is remarkable.

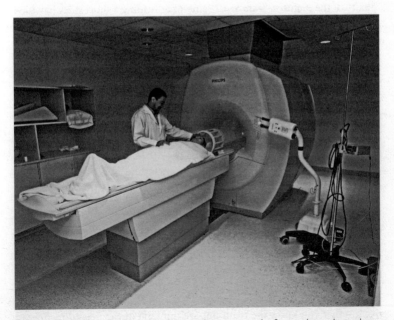

Reading thoughts via EEG (left) and fMRI (right) scans. In the future, these electrodes will be miniaturized. We will be able to read thoughts and also command objects by simply thinking.

ETHICS OF MIND READING

This poses a problem: What happens if we can routinely read people's thoughts? Nobel laureate David Baltimore, former president of the California Institute of Technology (Caltech), worries about this problem. He writes, "Can we tap into the thoughts of others? . . . I don't think that's pure science fiction, but it would create a hell of a world. Imagine courting a mate if your thoughts could be read, or negotiating a contract if your thoughts could be read."

Most of the time, he speculates, mind reading will have some embarrassing but not disastrous consequences. He writes, "I am told that if you stop a professor's lecture in midstream . . . a significant fraction [of the students] are involved in erotic fantasies."

But perhaps mind reading won't become such a privacy issue, since most of our thoughts are not well defined. Photographing our daydreams and dreams may one day be possible, but we may be disappointed with the quality of the pictures. Years ago, I remember reading a short story in which a man was told by a genie that he could have anything he could imagine. He immediately imagined expensive luxury items, like limousines, millions of dollars in cash, and a castle. Then the genie instantly materialized them. But when the man examined them carefully, he was shocked that the limousine had no door handles or engine, the faces on the bills were blurry, and the castle was empty. In his rush to imagine all these items, he forgot that these images exist in his imagination only as general ideas.

Furthermore, it is doubtful that you can read someone's mind from a distance. All the methods studied so far (including EEG, fMRI, and electrodes on the brain itself) require close contact with the subject.

Nonetheless, laws may eventually be passed to limit unauthorized mind reading. Also, devices may be created to protect our thoughts by jamming, blocking, or scrambling our electrical signals.

True mind reading is still many decades away. But at the very least, an fMRI scanner might function as a primitive lie detector. Telling a lie causes more centers of the brain to light up than telling the truth. Telling a lie implies that you know the truth but are thinking of the lie and its myriad consequences, which requires much more energy than telling the truth. Hence, the fMRI brain scan should be able to detect this extra expenditure of energy. At present, the scientific community has some reservations about allowing fMRI lie detectors to be the last word, especially in court cases. The technology is still too new to provide a foolproof lie-detection method. Further research, say its promoters, will refine its accuracy. This technology is here to stay.

Already, there are two commercial companies offering fMRI lie detectors, claiming a more than 90 percent success rate. A court in India already has used an fMRI to settle a case, and several cases involving fMRI are now in U.S. courts.

Ordinary lie detectors do not measure lies; they measure only signs of tension, such as increased sweating (measured by analyzing the conductivity of the skin) and increased heart rate. Brain scans measure increased brain activity, but the correlation between this and lying has still to be proven conclusively for a court of law.

It may take years of careful testing to explore the limits and accuracy of fMRI lie detection. In the meantime, the MacArthur Foundation recently gave a $10 million grant to the Law and Neuroscience Project to determine how neuroscience will affect the law.

MY fMRI BRAIN SCAN

I once had my own brain scanned by an fMRI machine. For a BBC/Discovery Channel documentary, I flew to Duke University, where they placed me on a stretcher, which was then inserted into a gigantic metal cylinder. When a huge, powerful magnet was turned on (20,000 times the earth's magnetic field), the atoms in my brain were aligned to the magnetic field, like spinning tops whose axes point in one direction. Then a radio pulse was sent into my brain, which flipped some of the nuclei of my atoms upside down. When the nuclei eventually flipped back to normal, they emitted a tiny pulse, or "echo," that could be detected by the fMRI machine. By analyzing these echoes, computers could process the signals, then reassemble a 3-D map of the interior of my brain.

The whole process was totally painless and harmless. The radiation sent into my body was non-ionizing and could not cause damage to my cells by ripping apart atoms. Even suspended in a magnetic field thousands of times stronger than the earth's, I could not detect the slightest change in my body.

The purpose of my being in the fMRI scan was to determine precisely where in my brain certain thoughts were being manufactured. In particular, there is a tiny biological "clock" inside your brain, just between your eyes, behind your nose, where the brain calculates seconds and minutes. Damage to this delicate part of the brain causes a distorted sense of time.

While inside the scanner, I was asked to measure the passage of seconds and minutes. Later, when the fMRI pictures were developed, I could clearly see that there was a bright spot just behind my nose as I was counting the seconds. I realized that I was witnessing the birth of an entirely new area of biology: tracking down the precise locations in the brain associated with certain thoughts, a form of mind reading.

TRICORDERS AND PORTABLE BRAIN SCANS

In the future, the MRI machine need not be the monstrous device found in hospitals today, weighing several tons and taking up an entire room. It might be as small as a cell phone, or even a penny.

In 1993, Bernhard Blümich and his colleagues, when they were at the Max Planck Institute for Polymer Research in Mainz, Germany, hit upon a novel idea that could create tiny MRI machines. They built a new machine, called the MRI-MOUSE (mobile universal surface explorer), currently about one foot tall, that may one day give us MRI machines that are the size of a coffee cup and sold in department stores. This could revolutionize medicine, since one would be able to perform MRI scans in the privacy of one's home. Blümich envisions a time, not too far away, when a person would be able to pass his personal MRI-MOUSE over his skin and look inside his body any time of the day. Computers would analyze the picture and diagnose any problems. "Perhaps something like the *Star Trek* tricorder is not so far off after all," he has concluded.

(MRI scans work on a principle similar to compass needles. The north pole of the compass needle immediately aligns to the magnetic field. So when the body is placed in an MRI machine, the nuclei of the atoms, like compass needles, align to the magnetic field. Now a radio pulse is sent into the body which makes the nuclei flip upside down. Eventually, the nuclei flips back to its original position, emitting a second radio pulse or "echo.")

The key to his mini-MRI machine is its nonuniform magnetic fields. Normally, the reason the MRI machine of today is so bulky is because you need to place the body in an extremely uniform magnetic field. The greater the uniformity of the field, the more detailed the resulting picture, which today can resolve features down to a tenth of a millimeter. To obtain these uniform magnetic fields, physicists start with two large coils of wire, roughly two feet in diameter, stacked on top of each other. This is called a Helmholtz coil, and provides a uniform magnetic field in the space between the two coils. The human body is then placed along the axis of these two large magnets.

But if you use nonuniform magnetic fields, the resulting image is distorted and useless. This has been the problem with MRI machines for many decades. But Blümich stumbled on a clever way to compensate for this distortion by sending multiple radio pulses into the sample and then detecting the resulting echoes. Then computers are used to analyze these echoes and make up for the distortion created by nonuniform magnetic fields.

Today, Blümich's portable MRI-MOUSE machine uses a small U-shaped magnet that produces a north pole and a south pole at each end of the U. This magnet is placed on top of the patient, and by moving the magnet, one can peer several inches beneath the skin. Unlike standard MRI machines, which consume vast amounts of power and have to have special electrical power outlets, the MRI-MOUSE uses only about as much electricity as an ordinary lightbulb.

In some of his early tests, Blümich placed the MRI-MOUSE on top of rubber tires, which are soft like human tissue. This could have an immediate commercial application: rapidly scanning for defects in products. Conventional MRI machines cannot be used on objects that contain metal, such as steel-belted radial tires. The MRI-MOUSE, because it uses only weak magnetic fields, has no such limitation. (The magnetic fields of a conventional MRI machine are 20,000 times more powerful than the earth's magnetic field. Many nurses and technicians have been seriously hurt when the magnetic field is turned on and then metal tools suddenly come flying at them. The MRI-MOUSE has no such problem.)

Not only is this ideal to analyze objects that have ferrous metals in them, it can also analyze objects that are too large to fit inside a conventional MRI machine or cannot be moved from their sites. For example, in 2006 the MRI-MOUSE successfully produced images of the interior of Ötzi the iceman, the frozen corpse found in the Alps in 1991. By moving the

U-shaped magnet over Ötzi, it was able to successively peel away the various layers of his frozen body.

In the future, the MRI-MOUSE may be miniaturized even more, allowing for MRI scans of the brain using something the size of a cell phone. Then, scanning the brain to read one's thoughts may not be such a problem. Eventually, the MRI scanner may be as thin as a dime, barely noticeable. It might even resemble the less-powerful EEG, where you put a plastic cap with many electrodes attached over your head. (If you place these portable MRI disks on your fingertips and then place them on a person's head, this would resemble performing the Vulcan mind meld of *Star Trek*.)

TELEKINESIS AND THE POWER OF THE GODS

The endpoint of this progression is to attain telekinesis, the power of the gods of mythology to move objects by sheer thought.

In the movie *Star Wars*, for example, the Force is a mysterious field that pervades the galaxy and unleashes the mental powers of the Jedi knights, allowing them to control objects with their mind. Lightsabers, ray guns, and even entire starships can be levitated using the power of the Force—and to control the actions of others.

But we won't have to travel to a galaxy far, far away to harness this power. By 2100, when we walk into a room, we will be able to mentally control a computer that in turn will control things around us. Moving heavy furniture, rearranging our desk, making repairs, etc., may be possible by thinking about it. This could be quite useful for workers, fire crews, astronauts, and soldiers who have to operate machinery requiring more than two hands. It could also change the way we interact with the world. We would be able to ride a bike, drive a car, play golf or baseball or elaborate games just by thinking about them.

Moving objects by thought may become possible by exploiting something called superconductors, which we shall explain in more detail in Chapter 4. By the end of this century, physicists may be able to create superconductors that can operate at room temperature, thereby allowing us to create huge magnetic fields that require little power. In the same way that the twentieth century was the age of electricity, the future may bring us room-temperature superconductors that will give us the age of magnetism.

Powerful magnetic fields are presently expensive to create but may become almost free in the future. This will allow us to reduce friction in our trains and trucks, revolutionizing transportation, and eliminate losses in electrical transmission. This will also allow us to move objects by sheer thought. With tiny supermagnets placed inside different objects, we will be able to move them around almost at will.

In the near future, we will assume that everything has a tiny chip in it, making it intelligent. In the far future, we will assume that everything has a tiny superconductor inside it that can generate bursts of magnetic energy, sufficient to move it across a room. Assume, for example, that a table has a superconductor in it. Normally, this superconductor carries no current. But when a tiny electrical current is added, it can create a powerful magnetic field, capable of sending it across the room. By thinking, we should be able to activate the supermagnet embedded within an object and thereby make it move.

In the *X-Men* movies, for example, the evil mutants are led by Magneto, who can move enormous objects by manipulating their magnetic properties. In one scene, he even moves the Golden Gate Bridge via the power of his mind. But there are limits to this power. For example, it is difficult to move an object like plastic or paper that has no magnetic properties. (At the end of the first *X-Men* movie, Magneto is confined in a jail made completely of plastic.)

In the future, room-temperature superconductors may be hidden inside common items, even nonmagnetic ones. If a current is turned on within the object, it will become magnetic and hence it can be moved by an external magnetic field that is controlled by your thoughts.

We will also have the power to manipulate robots and avatars by thinking. This means that, as in the movies *Surrogates* and *Avatar*, we might be able to control the motions of our substitutes and even feel pain and pressure. This might prove useful if we need a superhuman body to make repairs in outer space or rescue people in emergencies. Perhaps one day, our astronauts may be safely on earth, controlling superhuman robotic bodies as they move on the moon. We will discuss this more in the next chapter.

We should also point out that possessing this telekinetic power is not without risks. As I mentioned before, in the movie *Forbidden Planet*, an ancient civilization millions of years ahead of ours attains its ultimate dream, the ability to control anything with the power of their minds. As one trivial example of their technology, they created a machine that can turn your thoughts into a 3-D image. You put the device on your head, imagine something, and a 3-D image materializes inside the machine. Although this device seemed impossibly advanced for movie audiences back in the 1950s, this device will be available in the coming decades. Also, in the movie, there was a device that harnessed your mental energy to lift a heavy object. But as we know, we don't have to wait millions of years for this technology—it's already here, in the form of a toy. You place EEG electrodes on your head, the toy detects the electrical impulses of your brain, and then it lifts a tiny object, just as in the movie. In the future, many games will be played by sheer thought. Teams may be mentally wired up so that they can move a ball by thinking about it, and the team that can best mentally move the ball wins.

The climax of *Forbidden Planet* may give us pause. Despite the vastness of their technology, the aliens perished because they failed to notice a defect in their plans. Their powerful machines tapped not only into their conscious thoughts but also into their subconscious desires. The savage, long-suppressed thoughts of their violent, ancient evolutionary past sprang back to life, and the machines materialized every subconscious nightmare into reality. On the eve of attaining their greatest creation, this mighty civilization was destroyed by the very technology they hoped would free them from instrumentality.

For us, however, this is still a distant danger. A device of that magnitude won't be available until the twenty-second century. However, we face a more immediate concern. By 2100, we will also live in a world populated by robots that have humanlike characteristics. What happens if they become smarter than us?

Will robots inherit the earth? Yes, but they will be our children.

—MARVIN MINSKY

2 FUTURE OF AI *Rise of the Machines*

The gods of mythology with their divine power could animate the inanimate. According to the Bible, in Genesis, Chapter 2, God created man out of dust, and then "breathed into his nostrils the breath of life, and man became a living soul." According to Greek and Roman mythology, the goddess Venus could make statues spring to life. Venus, taking pity on the artist Pygmalion when he fell hopelessly in love with his statue, granted his fondest wish and turned the statue into a beautiful woman, Galatea. The god Vulcan, the blacksmith to the gods, could even create an army of mechanical servants made of metal that he brought to life.

Today, we are like Vulcan, forging in our laboratories machines that breathe life not into clay but into steel and silicon. But will it be to liberate the human race or enslave it? If one reads the headlines today, it seems as if the question is already settled: the human race is about to be rapidly overtaken by our own creation.

THE END OF HUMANITY?

The headline in the *New York Times* said it all: "Scientists Worry Machines May Outsmart Man." The world's top leaders in artificial intelligence (AI) had gathered at the Asilomar conference in California in 2009 to solemnly discuss what happens when the machines finally take over. As in a scene from a Hollywood movie, delegates asked probing questions, such as, What happens if a robot becomes as intelligent as your spouse?

As compelling evidence of this robotic revolution, people pointed to the Predator drone, a pilotless robot plane that is now targeting terrorists with deadly accuracy in Afghanistan and Pakistan; cars that can drive themselves; and ASIMO, the world's most advanced robot that can walk, run, climb stairs, dance, and even serve coffee.

Eric Horvitz of Microsoft, an organizer of the conference, noting the excitement surging through the conference, said, "Technologists are providing almost religious visions, and their ideas are resonating in some ways with the same idea of the Rapture." (The Rapture is when true believers ascend to heaven at the Second Coming. The critics dubbed the spirit of the Asilomar conference "the rapture of the nerds.")

That same summer, the movies dominating the silver screen seemed to amplify this apocalyptic picture. In *Terminator Salvation,* a ragtag band of humans battle huge mechanical behemoths that have taken over the earth. In *Transformers: Revenge of the Fallen,* futuristic robots from space use humans as pawns and the earth as a battleground for their interstellar wars. In *Surrogates,* people prefer to live their lives as perfect, beautiful, superhuman robots, rather than face the reality of their own aging, decaying bodies.

Judging from the headlines and the theater marquees, it looks like the last gasp for humans is just around the corner. AI pundits are solemnly asking: Will we one day have to dance behind bars as our robot creations throw peanuts at us, as we do at bears in a zoo? Or will we become lapdogs to our creations?

But upon closer examination, there is less than meets the eye. Certainly, tremendous breakthroughs have been made in the last decade, but things have to be put into perspective.

The Predator, a 27-foot drone that fires deadly missiles at terrorists from the sky, is controlled by a human with a joystick. A human, most likely a young veteran of video games, sits comfortably behind a computer screen and selects the targets. The human, not the Predator, is calling the shots. And the cars that drive themselves are not making independent decisions as they scan the horizon and turn the steering wheel; they are following a GPS map stored in their memory. So the nightmare of fully autonomous, conscious, and murderous robots is still in the distant future.

Not surprisingly, although the media hyped some of the more sensational predictions made at the Asilomar conference, most of the working scientists doing the day-to-day research in artificial intelligence were much more reserved and cautious. When asked when the machines will become as smart as us, the scientists had a surprising variety of answers, ranging from 20 to 1,000 years.

So we have to differentiate between two types of robots. The first is remote-controlled by a human or programmed and pre-scripted like a tape recorder to follow precise instructions. These robots already exist and generate headlines. They are slowly entering our homes and also the battlefield. But without a human making the decisions, they are largely useless pieces of junk. So these robots should not be confused with the second type, which is truly autonomous, the kind that can think for itself and requires no input from humans. It is these autonomous robots that have eluded scientists for the past half century.

ASIMO THE ROBOT

AI researchers often point to Honda's robot called ASIMO (Advanced Step in Innovative Mobility) as a graphic demonstration of the revolutionary advances made in robotics. It is 4 feet 3 inches tall, weighs 119 pounds, and resembles a young boy with a black-visored helmet and a backpack. ASIMO, in fact, is remarkable: it can realistically walk, run, climb stairs, and talk. It can wander around rooms, pick up cups and trays, respond to some simple commands, and even recognize some faces. It even has a large vocabulary and can speak in different languages. ASIMO is the result of twenty years of intense work by scores of Honda scientists, who have produced a marvel of engineering.

On two separate occasions, I have had the privilege of personally interacting with ASIMO at conferences, when hosting science specials for BBC/Discovery. When I shook its hand, it responded in an entirely human-like way. When I waved to it, it waved right back. And when I asked it to fetch me some juice, it turned around and walked toward the refreshment table with eerily human motions. Indeed, ASIMO is so lifelike that when it talked, I half expected the robot to take off its helmet and reveal the boy who was cleverly hidden inside. It can even dance better than I can.

At first, it seems as if ASIMO is intelligent, capable of responding to human commands, holding a conversation, and walking around a room. Actually, the reality is quite different. When I interacted with ASIMO in front of the TV camera, every motion, every nuance was carefully scripted. In fact, it took about three hours to film a simple five-minute scene with ASIMO. And even that required a team of ASIMO handlers who were furiously reprogramming the robot on their laptops after we filmed every scene. Although ASIMO talks to you in different languages, it is actually a tape recorder playing recorded messages. It simply parrots what is programmed by a human. Although ASIMO becomes more sophisticated every year, it is incapable of independent thought. Every word, every gesture, every step has to be carefully rehearsed by ASIMO's handlers.

Afterward, I had a candid talk with one of ASIMO's inventors, and he admitted that ASIMO, despite its remarkably humanlike motions and actions, has the intelligence of an insect. Most of its motions have to be carefully programmed ahead of time. It can walk in a totally lifelike way, but its path has to be carefully programmed or it will stumble over the furniture, since it cannot really recognize objects around the room.

By comparison, even a cockroach can recognize objects, scurry around obstacles, look for food and mates, evade predators, plot complex escape routes, hide among the shadows, and disappear in the cracks, all within a matter of seconds.

AI researcher Thomas Dean of Brown University has admitted that the lumbering robots he is building are "just at the stage where they're robust enough to walk down the hall without leaving huge gouges in the plaster." As we shall later see, at present our most powerful computers can barely simulate the neurons of a mouse, and then only for a few seconds. It will take many decades of hard work before robots become as smart as a mouse, rabbit, dog or cat, and then a monkey.

HISTORY OF AI

Critics sometimes point out a pattern, that every thirty years, AI practitioners claim that superintelligent robots are just around the corner. Then, when there is a reality check, a backlash sets in.

In the 1950s, when electronic computers were first introduced after World War II, scientists dazzled the public with the notion of machines that could perform miraculous feats: picking up blocks, playing checkers, and even solving algebra problems. It seemed as if truly intelligent machines were just around the corner. The public was amazed; and soon there were magazine articles breathlessly predicting the time when a robot would be in everyone's kitchen, cooking dinner, or cleaning the house. In 1965, AI pioneer Herbert Simon declared, "Machines will be capable, within twenty years, of doing any work a man can do." But then the reality set in. Chess-playing machines could not win against a human expert, and could play only chess, nothing more. These early robots were like a one-trick pony, performing just one simple task.

In fact, in the 1950s, real breakthroughs were made in AI, but because the progress was vastly overstated and overhyped, a backlash set in. In 1974, under a chorus of rising criticism, the U.S. and British governments cut off funding. The first AI winter set in.

Today, AI researcher Paul Abrahams shakes his head when he looks back at those heady times in the 1950s when he was a graduate student at MIT and anything seemed possible. He recalled, "It's as though a group of people had proposed to build a tower to the moon. Each year they point with pride at how much higher the tower is than it was the previous year. The only trouble is that the moon isn't getting much closer."

In the 1980s, enthusiasm for AI peaked once again. This time the Pentagon poured millions of dollars into projects like the smart truck, which was supposed to travel behind enemy lines, do reconnaissance, rescue U.S. troops, and return to headquarters, all by itself. The Japanese government even put its full weight behind the ambitious Fifth Generation Computer Systems Project, sponsored by the powerful Japanese Ministry of International Trade and Industry. The Fifth Generation Project's goal was, among others, to have a computer system that could speak conversational language, have full reasoning ability, and even anticipate what we want, all by the 1990s.

Unfortunately, the only thing that the smart truck did was get lost. And the Fifth Generation Project, after much fanfare, was quietly dropped without explanation. Once again, the rhetoric far outpaced the reality. In fact, there were real gains made in AI in the 1980s, but because progress was again overhyped, a second backlash set in, creating the second AI winter, in which funding again dried up and disillusioned people left the field in droves. It became painfully clear that something was missing.

In 1992 AI researchers had mixed feelings holding a special celebration in honor of the movie *2001,* in which a computer called HAL 9000 runs amok and slaughters the crew of a spaceship. The movie, filmed in 1968, predicted that by 1992 there would be robots that could freely converse with any human on almost any topic and also command a spaceship. Unfortunately, it was painfully clear that the most advanced robots had a hard time keeping up with the intelligence of a bug.

In 1997 IBM's Deep Blue accomplished a historic breakthrough by decisively beating the world chess champion Gary Kasparov. Deep Blue was an engineering marvel, computing 11 billion operations per second. However, instead of opening the floodgates of artificial intelligence research and ushering in a new age, it did precisely the opposite. It highlighted only the primitiveness of AI research. Upon reflection, it was obvious to many that Deep Blue could not think. It was superb at chess but would score 0 on an IQ exam. After this victory, it was the loser, Kasparov, who did all the talking to the press, since Deep Blue could not talk at all. Grudgingly, AI researchers began to appreciate the fact that brute computational power does not equal intelligence. AI researcher Richard Heckler says, "Today, you can buy chess programs for $49 that will beat all but world champions, yet no one thinks they're intelligent."

But with Moore's law spewing out new generations of computers every eighteen months, sooner or later the old pessimism of the past generation will be gradually forgotten and a new generation of bright enthusiasts will take over, creating renewed optimism and energy in the once-dormant field. Thirty years after the last AI winter set in, computers have advanced enough so that the new generation of AI researchers are again making hopeful predictions about the future. The time has finally come for AI, say its supporters. This time, it's for real. The third try is the lucky charm. But if they are right, are humans soon to be obsolete?

IS THE BRAIN A DIGITAL COMPUTER?

One fundamental problem, as mathematicians now realize, is that they made a crucial error fifty years ago in thinking the brain was analogous to a large digital computer. But now it is painfully obvious that it isn't. The brain has no Pentium chip, no Windows operating system, no application software, no CPU, no programming, and no subroutines that typify a modern digital computer. In fact, the architecture of digital computers is quite different from that of the brain, which is a learning machine of some sort, a collection of neurons that constantly rewires itself every time it learns a task. (A PC, however, does not learn at all. Your computer is just as dumb today as it was yesterday.)

So there are at least two approaches to modeling the brain. The first, the traditional top-down approach, is to treat robots like digital computers, and program all the rules of intelligence from the very beginning. A digital computer, in turn, can be broken down into something called a Turing machine, a hypothetical device introduced by the great British mathematician Alan Turing. A Turing machine consists of three basic components: an input, a central processor that digests this data, and an output. All digital computers are based on this simple model. The goal of this approach is to have a CD-ROM that has all the rules of intelligence codified on it. By inserting this disk, the computer suddenly springs to life and becomes intelligent. So this mythical CD-ROM contains all the software necessary to create intelligent machines.

However, our brain has no programming or software at all. Our brain is more like a "neural network," a complex jumble of neurons that constantly rewires itself.

Neural networks follow Hebb's rule: every time a correct decision is made, those neural pathways are reinforced. It does this by simply changing the strength of certain electrical connections between neurons every time it successfully performs a task. (Hebb's rule can be expressed by the old question: How does a musician get to Carnegie Hall? Answer: practice, practice, practice. For a neural network, practice makes perfect. Hebb's rule also explains why bad habits are so difficult to break, since the neural pathway for a bad habit is so well-worn.)

Neural networks are based on the bottom-up approach. Instead of being spoon-fed all the rules of intelligence, neural networks learn them the way a baby learns, by bumping into things and learning by experience. Instead of being programmed, neural networks learn the old-fashioned way, through the "school of hard knocks."

Neural networks have a completely different architecture from that of digital computers. If you remove a single transistor in the digital computer's central processor, the computer will fail. However, if you remove large chunks of the human brain, it can still function, with other parts taking over for the missing pieces. Also, it is possible to localize precisely where the digital computer "thinks": its central processor. However, scans of the human brain clearly show that thinking is spread out over large parts of the brain. Different sectors light up in precise sequence, as if thoughts were being bounced around like a Ping-Pong ball.

Digital computers can calculate at nearly the speed of light. The human brain, by contrast, is incredibly slow. Nerve impulses travel at an excruciatingly slow pace of about 200 miles per hour. But the brain more than makes up for this because it is massively parallel, that is, it has 100 billion neurons operating at the same time, each one performing a tiny bit of computation, with each neuron connected to 10,000 other neurons. In a race, a superfast single processor is left in the dust by a superslow parallel processor. (This goes back to the old riddle: if one cat can eat one mouse in one minute, how long does it take a million cats to eat a million mice? Answer: one minute.)

In addition, the brain is not digital. Transistors are gates that can either be open or closed, represented by a 1 or 0. Neurons, too, are digital (they can fire or not fire), but they can also be analog, transmitting continuous signals as well as discrete ones.

TWO PROBLEMS WITH ROBOTS

Given the glaring limitations of computers compared to the human brain, one can appreciate why computers have not been able to accomplish two key tasks that humans perform effortlessly: pattern recognition and common sense. These two problems have defied solution for the past half century. This is the main reason why we do not have robot maids, butlers, and secretaries.

The first problem is pattern recognition. Robots can see much better than a human, but they don't understand what they are seeing. When a robot walks into a room, it converts the image into a jumble of dots. By processing these dots, it can recognize a collection of lines, circles, squares, and rectangles. Then a robot tries to match this jumble, one by one, with objects stored in its memory—an extraordinarily tedious task even for a computer. After many hours of calculation, the robot may match these lines with chairs, tables, and people. By contrast, when we walk into a room, within a fraction of a second, we recognize chairs, tables, desks, and people. Indeed, our brains are mainly pattern-recognizing machines.

Second, robots do not have common sense. Although robots can hear much better than a human, they don't understand what they are hearing. For example, consider the following statements:

- Children like sweets but not punishment
- Strings can pull but not push
- Sticks can push but not pull
- Animals cannot speak and understand English
- Spinning makes people feel dizzy

For us, each of these statements is just common sense. But not to robots. There is no line of logic or programming that proves that strings can pull but not push. We have learned the truth of these "obvious" statements by experience, not because they were programmed into our memories.

The problem with the top-down approach is that there are simply too many lines of code for common sense necessary to mimic human thought. Hundreds of millions of lines of code, for example, are necessary to describe the laws of common sense that a six-year-old child knows. Hans Moravec, former director of the AI laboratory at Carnegie Mellon, laments, "To this day, AI programs exhibit no shred of common sense—a medical diagnosis program, for instance, may prescribe an antibiotic when presented a broken bicycle because it lacks a model of people, disease, or bicycles."

Some scientists, however, cling to the belief that the only obstacle to mastering common sense is brute force. They feel that a new Manhattan Project, like the program that built the atomic bomb, would surely crack the common-sense problem. The crash program to create this "encyclopedia of thought" is called CYC, started in 1984. It was to be the crowning achievement of AI, the project to encode all the secrets of common sense into a single program. However, after several decades of hard work, the CYC project has failed to live up to its own goals.

CYC's goal is simple: master "100 million things, about the number a typical person knows about the world, by 2007." That deadline, and many previous ones, have slipped by without success. Each of the milestones laid out by CYC engineers has come and gone without scientists being any closer to mastering the essence of intelligence.

MAN VERSUS MACHINE

I once had a chance to match wits with a robot in a contest with one built by MIT's Tomaso Poggio. Although robots cannot recognize simple patterns as we can, Poggio was able to create a computer program that can calculate every bit as fast as a human in one specific area: "immediate recognition." This is our uncanny ability to instantly recognize an object even before we are aware of it. (Immediate recognition was important for our evolution, since our ancestors had only a split second to determine if a tiger was lurking in the bushes, even before they were fully aware of it.) For the first time, a robot consistently scored higher than a human on a specific vision recognition test.

The contest between me and the machine was simple. First, I sat in a chair and stared at an ordinary computer screen. Then a picture flashed on the screen for a split second, and I was supposed to press one of two keys as fast as I could, if I saw an animal in the picture or not. I had to make a decision as quickly as possible, even before I had a chance to digest the picture. The computer would also make a decision for the same picture.

Embarrassingly enough, after many rapid-fire tests, the machine and I performed about equally. But there were times when the machine scored significantly higher than I did, leaving me in the dust. I was beaten by a machine. (It was one consolation when I was told that the computer gets the right answer 82 percent of the time, but humans score only 80 percent on average.)

The key to Poggio's machine is that it copies lessons from Mother

Nature. Many scientists are realizing the truth in the statement, "The wheel has already been invented, so why not copy it?" For example, normally when a robot looks at a picture, it tries to divide it up into a series of lines, circles, squares, and other geometric shapes. But Poggio's program is different.

When we see a picture, we might first see the outlines of various objects, then see various features within each object, then shading within these features, etc. So we split up the image into many layers. As soon as the computer processes one layer of the image, it integrates it with the next layer, and so on. In this way, step by step, layer by layer, it mimics the hierarchical way that our brains process images. (Poggio's program cannot perform all the feats of pattern recognition that we take for granted, such as visualizing objects in 3-D, recognizing thousands of objects from different angles, etc., but it does represent a major milestone in pattern recognition.)

Later, I had an opportunity to see both the top-down and bottom-up approaches in action. I first went to the Stanford University's artificial intelligence center, where I met STAIR (Stanford artificial intelligence robot), which uses the top-down approach. STAIR is about 4 feet tall, with a huge mechanical arm that can swivel and grab objects off a table. STAIR is also mobile, so it can wander around an office or home. The robot has a 3-D camera that locks onto an object and feeds the 3-D image into a computer, which then guides the mechanical arm to grab the object. Robots have been grabbing objects like this since the 1960s, and we see them in Detroit auto factories.

But appearances are deceptive. STAIR can do much more. Unlike the robots in Detroit, STAIR is not scripted. It operates by itself. If you ask it to pick up an orange, for example, it can analyze a collection of objects on a table, compare them with the thousands of images already stored in its memory, then identify the orange and pick it up. It can also identify objects more precisely by grabbing them and turning them around.

To test its ability, I scrambled a group of objects on a table, and then watched what happened after I asked for a specific one. I saw that STAIR correctly analyzed the new arrangement and then reached out and grabbed the correct thing. Eventually, the goal is to have STAIR navigate in home and office environments, pick up and interact with various objects and tools, and even converse with people in a simplified language. In this way, it will be able to do anything that a gofer can in an office. STAIR is an example of the top-down approach: everything is programmed into STAIR from the very beginning. (Although STAIR can recognize objects from different angles, it is still limited in the number of objects it can recognize. It would be paralyzed if it had to walk outside and recognize random objects.)

Later, I had a chance to visit New York University, where Yann LeCun is experimenting with an entirely different design, the LAGR (learning applied to ground robots). LAGR is an example of the bottom-up approach: it has to learn everything from scratch, by bumping into things. It is the size of a small golf cart and has two stereo color cameras that scan the landscape, identifying objects in its path. It then moves among these objects, carefully avoiding them, and learns with each pass. It is equipped with GPS and has two infrared sensors that can detect objects in front of it. It contains three high-power Pentium chips and is connected to a gigabit Ethernet network. We went to a nearby park, where the LAGR robot could roam around various obstacles placed in its path. Every time it went over the course, it got better at avoiding the obstacles.

One important difference between LAGR and STAIR is that LAGR is specifically designed to learn. Every time LAGR bumps into something, it moves around the object and learns to avoid that object the next time. While STAIR has thousands of images stored in its memory, LAGR has hardly any images in its memory but instead creates a mental map of all the obstacles it meets, and constantly refines that map with each pass. Unlike the driverless car, which is programmed and follows a route set previously by GPS, LAGR moves all by itself, without any instructions from a human. You tell it where to go, and it takes off. Eventually, robots like these may be found on Mars, the battlefield, and in our homes.

On one hand, I was impressed by the enthusiasm and energy of these researchers. In their hearts, they believe that they are laying the foundation for artificial intelligence, and that their work will one day impact society in ways we can only begin to understand. But from a distance, I could also appreciate how far they have to go. Even cockroaches can identify objects and learn to go around them. We are still at the stage where Mother Nature's lowliest creatures can outsmart our most intelligent robots.

NEAR FUTURE (PRESENT TO 2030)

EXPERT SYSTEMS

Today, many people have simple robots in their homes that can vacuum their carpets. There are also robot security guards patrolling buildings at night, robot guides, and robot factory workers. In 2006, it was estimated that there were 950,000 industrial robots and 3,540,000 service robots working in homes and buildings. But in the coming decades, the field of robotics may blossom in several directions. But these robots won't look like the ones of science fiction.

The greatest impact may be felt in what are called expert systems, software programs that have encoded in them the wisdom and experience of a human being. As we saw in the last chapter, one day, we may talk to the Internet on our wall screens and converse with the friendly face of a robodoc or robolawyer.

This field is called heuristics, that is, following a formal, rule-based system. When we need to plan a vacation, we will talk to the face in the wall screen and give it our preferences for the vacation: how long, where to, which hotels, what price range. The expert system will already know our preferences from past experiences and then contact hotels, airlines, etc., and give us the best options. But instead of talking to it in a chatty, gossipy way, we will have to use a fairly formal, stylized language that it understands. Such a system can rapidly perform any number of useful chores. You just give it orders, and it makes a reservation at a restaurant, checks for the location of stores, orders grocery and takeout, reserves a plane ticket, etc.

It is precisely because of the advances in heuristics over the past decades that we now have some of the rather simple search engines of today. But they are still crude. It is obvious to everyone that you are dealing with a machine and not a human. In the future, however, robots will become so sophisticated that they will almost appear to be humanlike, operating seamlessly with nuance and sophistication.

Perhaps the most practical application will be in medical care. For example, at the present time if you feel sick, you may have to wait hours in an emergency room before you see a doctor. In the near future, you may simply go to your wall screen and talk to robodoc. You will be able to change the face, and even the personality, of the robodoc that you see with the push of a button. The friendly face you see in your wall screen will ask a simple set of questions: How do you feel? Where does it hurt? When did the pain start? How often does it hurt?

Each time, you will respond by choosing from a simple set of answers. You will answer not by typing on a keyboard but by speaking.

Each of your answers, in turn, will prompt the next set of questions. After a series of such questions, the robodoc will be able to give you a diagnosis based on the best experience of the world's doctors. Robodoc will also analyze the data from your bathroom, your clothes, and furniture, which have been continually monitoring your health via DNA chips. And it might ask you to examine your body with a portable MRI scanner, which is then analyzed by supercomputers. (Some primitive versions of these heuristic programs already exist, such as WebMD, but they lack the nuances and full power of heuristics.)

The majority of visits to the doctor's office can be eliminated in this way, greatly relieving the stress on our health care system. If the problem is serious, the robodoc will recommend that you go to a hospital, where human doctors can provide intensive care. But even there, you will see AI programs, in the form of robot nurses, like ASIMO. These robot nurses are not truly intelligent but can move from one hospital room to another, administer the proper medicines to patients, and attend to their other needs. They can move on rails in the floor, or move independently like ASIMO.

One robot nurse that already exists is the RP-6 mobile robot, which is being deployed in hospitals such as the UCLA Medical Center. It is basically a TV screen sitting on top of a mobile computer that moves on rollers. In the TV screen, you see the video face of a real physician who may be miles away. There is a camera on the robot that allows the doctor to see what the robot is looking at. There is also a microphone so that the doctor can speak to the patient. The doctor can remotely control the robot via a joystick, interact with patients, monitor drugs, etc. Since annually 5 million patients in the United States are admitted to intensive care units, but only 6,000 physicians are qualified to handle critically ill patients, robots such as this could help to alleviate this crisis in emergency care, with one doctor attending to many patients. In the future, robots like this may become more autonomous, able to navigate on their own and interact with patients.

Japan is one of the world's leaders in this technology. Japan is spending so much money on robots to alleviate the coming crisis in medical care. In retrospect, it is not surprising that Japan is one of the leading nations in robotics, for several reasons. First, in the Shinto religion, inanimate objects are believed to have spirits in them. Even mechanical ones. In the West, children may scream in terror at robots, especially after seeing so many movies about rampaging killing machines. But to Japanese children, robots are seen as kindred spirits, playful and helpful. In Japan, it is not uncommon to see robot receptionists greet you when you enter department stores. In fact, 30 percent of all commercial robots in the world are in Japan.

Second, Japan is facing a demographic nightmare. Japan has the fastest-aging population. The birthrate has fallen to an astonishing 1.2 children per family, and immigration is negligible. Some demographers have stated that we are watching a train wreck in slow motion: one demographic train (aging population and falling birthrate) will soon collide with another (low immigration rate) in the coming years. (This same train wreck might eventually happen in Europe as well.) This will be felt most acutely in the medical field, where an ASIMO-like nurse may be quite useful. Robots like ASIMO would be ideal for hospital tasks, such as fetching medicines, administering drugs, and monitoring patients twenty-four hours a day.

MIDCENTURY (2030 TO 2070)

MODULAR ROBOTS

By midcentury, our world may be full of robots, but we might not even notice them. That is because most robots probably won't have human form. They might be hidden from view, disguised as snakes, insects, and spiders, performing unpleasant but crucial tasks. These will be modular robots that can change shape depending on the task.

I had a chance to meet one of the pioneers in modular robots, Wei-min Shen of the University of Southern California. His idea is to create small cubical modules that you can interchange like Lego blocks and reassemble at will. He calls them polymorphic robots since they can change shape, geometry, and function. In his laboratory, I could instantly see the difference between his approach and that of Stanford and MIT. On the surface, both those labs resembled a kid's dream playhouse, with walking, talking robots everywhere you looked. When I visited Stanford's and MIT's AI laboratories, I saw a wide variety of robotic "toys" that have chips in them and some intelligence. The workbenches are full of robot airplanes, helicopters, trucks, and insect-shaped robots with chips inside, all moving autonomously. Each robot is a self-contained unit.

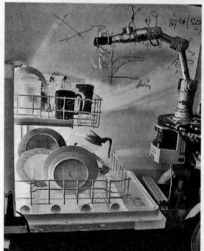

Various types of robots: LAGR (top), STAIR (bottom left), and ASIMO (bottom right). In spite of vast increases in computer power, these robots have the intelligence of a cockroach.

But when you enter the USC lab, you see something quite different. You see boxes of cubical modules, each about 2 inches square, that can join or separate, allowing you to create a variety of animal-like creatures. You can create snakes that slither in a line. Or rings that can roll like a hoop. But then you can twist these cubes or hook them up with Y-shaped joints, so you can create an entirely new set of devices resembling octopi, spiders, dogs, or cats. Think of a smart Lego set, with each block being intelligent and capable of arranging itself in any configuration imaginable.

This would be useful for going past barriers. If a spider-shaped robot was crawling in the sewer system and encountered a wall, it would first find a tiny hole in the wall and then disassemble itself. Each piece would go through the hole, and then the pieces would reassemble themselves on the other side of the wall. In this way, these modular robots would be nearly unstoppable, able to negotiate most obstacles.

These modular robots might be crucial in repairing our decaying infrastructure. In 2007, for example, the Mississippi River bridge in Minneapolis collapsed, killing 13 people and injuring 145, probably because the bridge was aging, overloaded, and had design flaws. There are perhaps hundreds of similar accidents waiting to happen across the country, but it simply costs too much money to monitor every decaying bridge and make repairs. This is where modular robots may come to the rescue, silently checking our bridges, roads, tunnels, pipes, and power stations, and making repairs when necessary. (For example, the bridges into lower Manhattan have suffered greatly due to corrosion, neglect, and lack of repairs. One worker found a 1950s Coke bottle left over from when the bridges were last painted. In fact, one section of the aging Manhattan Bridge came dangerously close to collapse recently and had to be shut down for repairs.)

ROBOT SURGEONS AND COOKS

Robots may be used as surgeons as well as cooks and musicians. For example, one important limitation of surgery is the dexterity and accuracy of the human hand. Surgeons, like all people, become fatigued after many hours and their efficiency drops. Fingers begin to tremble. Robots may solve these problems.

For example, traditional surgery for a heart bypass operation involves opening a foot-long gash in the middle of the chest, which requires general anesthesia. Opening the chest cavity increases the possibility of infection and the length of time for recovery, creates intense pain and discomfort during the healing process, and leaves a disfiguring scar. But the da Vinci robotic system can vastly decrease all these. The da Vinci robot has four robotic arms, one for manipulating a video camera and three for precision surgery. Instead of making a long incision in the chest, it makes only several tiny incisions in the side of the body. There are 800 hospitals in Europe and North and South America that use this system; 48,000 operations were performed in 2006 alone with this robot. Surgery can also be done by remote control over the Internet, so a world-class surgeon in a major city can perform surgery on a patient in an isolated rural area on another continent.

In the future, more advanced versions will be able to perform surgery on microscopic blood vessels, nerve fibers, and tissues by manipulating microscopic scalpels, tweezers, and needles, which is impossible today. In fact, in the future, only rarely will the surgeon slice the skin at all. Noninvasive surgery will become the norm.

Endoscopes (long tubes inserted into the body that can illuminate and cut tissue) will be thinner than thread. Micromachines smaller than the period at the end of this sentence will do much of the mechanical work. (In one episode of the original *Star Trek,* Doctor McCoy was totally revolted that doctors in the twentieth century had to cut skin.) The day when this is a reality is coming soon.

Medical students in the future will learn to slice up 3-D virtual images of the human body, where each movement of the hand is reproduced by a robot in another room.

The Japanese have also excelled at producing robots that can interact socially with humans. In Nagoya, there is the robot chef that can create a standard fast-food dinner in a few minutes. You simply punch in what you want from a menu and the robot chef produces your meal in front of you. Built by Aisei, an industrial robotics company, this robot can cook noodles in 1 minute and 40 seconds and can serve 80 bowls on a busy day. The robot chef looks very much like ones on the automobile assembly lines in Detroit. You have two large mechanical arms, which are precisely programmed to move in a certain sequence. Instead of screwing and welding metal in a factory, however, these robotic fingers grab ingredients from a series of bowls containing dressing, meat, flour, sauces, spices, etc. The robotic arms mix and then assemble them into a sandwich, salad, or soup. The Aisei cook looks like a robot, resembling two gigantic hands emerging from the kitchen counter. But other models being planned start to look more human.

Also in Japan, Toyota has created a robot that can play the violin almost as well as any professional. It resembles ASIMO, except that it can grab a violin, sway with the music, and then delicately play complex violin pieces. The sound is amazingly realistic and the robot can make grand gestures like a master musician. Although the music is not yet at the level of a concert violinist, it is good enough to entertain audiences. Of course, in the last century, we have had mechanical piano machines that played tunes inscribed on a large rotating disk. Like these piano machines, the Toyota machine is also programmed. But the difference is that the Toyota machine is deliberately designed to mimic all the positions and postures of a human violinist in the most realistic way.

Also, at Waseda University in Japan, scientists have made a robotic flutist. The robot contains hollow chambers in its chest, like lungs, which blow air over a real flute. It can play quite complex melodies like "The Flight of the Bumblebee." These robots cannot create new music, we should emphasize, but they can rival a human in their ability to perform music.

The robot chef and robot musician are carefully programmed. They are not autonomous. Although these robots are quite sophisticated compared to the old player pianos, they still operate on the same principles. True robot maids and butlers are still in the distant future. But the descendants of the robot chef and the robot violinist and flutist may one day find themselves embedded in our lives, performing basic functions that were once thought to be exclusively human.

EMOTIONAL ROBOTS

By midcentury, the era of emotional robots may be in full flower.

In the past, writers have fantasized about robots that yearn to become human and have emotions. In *Pinocchio*, a wooden puppet wished to become a real boy. In the *Wizard of Oz*, the Tin Man wished for a heart. And in *Star Trek: The Next Generation*, Data the android tried to master emotions by telling jokes and figuring out what makes us laugh. In fact, in science fiction, it is a recurring theme that although robots may become increasingly intelligent, the essence of emotions will always elude them. Robots may one day become smarter than us, some science fiction writers declare, but they won't be able to cry.

Actually, that may not be true. Scientists are now understanding the true nature of emotions. First, emotions tell us what is good for us and what is harmful. The vast majority of things in the world are either harmful or not very useful. When we experience the emotion of "like," we are learning to identify the tiny fraction of things in the environment that are beneficial to us.

In fact, each of our emotions (hate, jealousy, fear, love, etc.) evolved over millions of years to protect us from the dangers of a hostile world and help us to reproduce. Every emotion helps to propagate our genes into the next generation.

The critical role of emotions in our evolution was apparent to neurologist Antonio Damasio of the University of Southern California, who analyzed victims of brain injuries or disease. In some of these patients, the link between the thinking part of their brains (the cerebral cortex) and the emotional center (located deep in the center of the brain, like the amygdala) was cut. These people were perfectly normal, except they had difficulty expressing emotions.

One problem became immediately obvious: they could not make choices. Shopping was a nightmare, since everything had the same value to them, whether it was expensive or cheap, garish or sophisticated. Setting an appointment was almost impossible, since all dates in the future were the same. They seem "to know, but not to feel," he said.

In other words, one of the chief purposes of emotions is to give us values, so we can decide what is important, what is expensive, what is pretty, and what is precious. Without emotions, everything has the same value, and we become paralyzed by endless decisions, all of which have the same weight. So scientists are now beginning to understand that emotions, far from being a luxury, are essential to intelligence.

For example, when one watches *Star Trek* and sees Spock and Data performing their jobs supposedly without any emotions, you now realize the flaw immediately. At every turn, Spock and Data have exhibited emotions:

they have made a long series of value judgments. They decided that being an officer is important, that it is crucial to perform certain tasks, that the goal of the Federation is a noble one, that human life is precious, etc. So it is an illusion that you can have an officer devoid of emotions.

Emotional robots could also be a matter of life and death. In the future, scientists may be able to create rescue robots—robots that are sent into fires, earthquakes, explosions, etc. They will have to make thousands of value judgments about who and what to save and in what order. Surveying the devastation all around them, they will have to rank the various tasks they face in order of priority.

Emotions are also essential if you view the evolution of the human brain. If you look at the gross anatomical features of the brain, you notice that they can be grouped into three large categories.

First, you have the reptilian brain, found near the base of the skull, which makes up most of the brain of reptiles. Primitive life functions, such as balance, aggression, territoriality, searching for food, etc., are controlled by this part of the brain. (Sometimes, when staring at a snake that is staring back at you, you get a creepy sensation. You wonder, What is the snake thinking about? If this theory is correct, then the snake is not thinking much at all, except whether or not you are lunch.)

When we look at higher organisms, we see that the brain has expanded toward the front of the skull. At the next level, we find the monkey brain, or the limbic system, located in the center of our brain. It includes components like the amygdala, which is involved in processing emotions. Animals that live in groups have an especially well-developed limbic system. Social animals that hunt in groups require a high degree of brainpower devoted to understanding the rules of the pack. Since success in the wilderness depends on cooperating with others, but because these animals cannot talk, it means that these animals must communicate their emotional state via body language, grunts, whines, and gestures.

Finally, we have the front and outer layer of the brain, the cerebral cortex, the layer that defines humanity and governs rational thought. While other animals are dominated by instinct and genetics, humans use the cerebral cortex to reason things out.

If this evolutionary progression is correct, it means that emotions will play a vital role in creating autonomous robots. So far, robots have been created that mimic only the reptilian brain. They can walk, search their surroundings, and pick up objects, but not much more. Social animals, on the other hand, are more intelligent than those with just a reptilian brain. Emotions are required to socialize the animal and for it to master the rules of the pack. So scientists have a long way to go before they can model the limbic system and the cerebral cortex.

Cynthia Breazeal of MIT actually created a robot specifically designed to tackle this problem. The robot is KISMET, with a face that resembles a mischievous elf. On the surface, it appears to be alive, responding to you with facial motions representing emotions. KISMET can duplicate a wide range of emotions by changing its facial expressions. In fact, women who react to this childlike robot often speak to KISMET in "motherese," what mothers use when talking to babies and children. Although robots like KIS-MET are designed to mimic emotions, scientists have no illusion that the robot actually feels emotions. In some sense, it is like a tape recorder programmed not to make sounds, but to make facial emotions instead, with no awareness of what it is doing. But the breakthrough with KISMET is that it does not take much programming to create a robot that will mimic humanlike emotions to which humans will respond.

These emotional robots will find their way into our homes. They won't be our confidants, secretaries, or maids, but they will be able to perform rule-based procedures based on heuristics. By midcentury, they may have the intelligence of a dog or cat. Like a pet, they will exhibit an emotional bond with their master, so that they will not be easily discarded. You will not be able to speak to them in colloquial English, but they will understand programmed commands, perhaps hundreds of them. If you tell them to do something that is not already stored in their memory (such as "go fly a kite"), they will simply give you a curious, confused look. (If by midcentury robot dogs and cats can duplicate the full range of animal responses, indistinguishable from real animal behavior, then the question arises whether these robot animals feel or are as intelligent as an ordinary dog or cat.)

Sony experimented with these emotional robots when it manufactured the AIBO (artificial intelligence robot) dog. It was the first toy to realistically respond emotionally to its master, albeit in a primitive way. For example, if you pet the AIBO dog on its back, it would immediately begin to murmur, uttering soothing sounds. It could walk, respond to voice commands, and even learn to a degree. AIBO cannot learn new emotions and emotional responses. (It was discontinued in 2005 due to financial reasons, but it has since created a loyal following who upgrade the computer's software so AIBO can perform more tasks.) In the future, robotic pets that form an emotional attachment to children may become common.

Although these robot pets will have a large library of emotions and will form lasting attachments with children, they will not feel actual emotions.

REVERSE ENGINEER THE BRAIN

By midcentury, we should be able to complete the next milestone in the history of AI: reverse engineering the human brain. Scientists, frustrated that they have not been able to create a robot made of silicon and steel, are also trying the opposite approach: taking apart the brain, neuron by neuron—just like a mechanic might take apart a motor, screw by screw—and then running a simulation of these neurons on a huge computer. These scientists are systematically trying to simulate the firings of neurons in animals, starting with mice, cats, and going up the evolutionary scale of animals. This is a well-defined goal, and should be possible by midcentury.

MIT's Fred Hapgood writes, "Discovering how the brain works—*exactly* how it works, the way we know how a motor works—would rewrite almost every text in the library."

The first step in the process of reverse engineering the brain is to understand its basic structure. Even this simple task has been a long, painful process. Historically, the various parts of the brain were identified during autopsies, without a clue as to their function. This gradually began to change when scientists analyzed people with brain damage, and noticed that damage to certain parts of the brain corresponded to changes in behavior. Stroke victims and people suffering from brain injuries or diseases exhibited specific behavior changes, which could then be matched to injuries in specific parts of the brain.

The most spectacular example of this was in 1848 in Vermont, when a 3-foot, 8-inch-long metal rod was driven right through the skull of a railroad foreman named Phineas Gage. This history-making accident happened when dynamite accidentally exploded. The rod entered the side of his face, shattered his jaw, went through his brain, and passed out the top of his head. Miraculously, he survived this horrendous accident, although one or both of his frontal lobes were destroyed. The doctor who treated him at first could not believe that anyone could survive such an accident and still be alive. He was in a semiconscious state for several weeks, but later miraculously recovered. He even survived for twelve more years, taking odd jobs and traveling, dying in 1860. Doctors carefully preserved his skull and the rod, and they have been intensely studied ever since. Modern techniques, using CT scans, have reconstructed details of this extraordinary accident.

This event forever changed the prevailing opinions of the mind-body problem. Previously, it was believed even within scientific circles that the soul and the body were separate entities. People wrote knowingly about some "life force" that animated the body, independent of the brain. But widely circulated reports indicated that Gage's personality underwent marked changes after the accident. Some accounts claim that Gage was a well-liked, outgoing man who became abusive and hostile after the accident. The impact of these reports reinforced the idea that specific parts of the brain controlled different behaviors, and hence the body and soul were inseparable.

In the 1930s, another breakthrough was made when neurologists like Wilder Penfield noticed that while performing brain surgery for epilepsy sufferers, when he touched parts of the brain with electrodes, certain parts of the patient's body could be stimulated. Touching this or that part of the cortex could cause a hand or leg to move. In this way, he was able to construct a crude outline of which parts of the cortex controlled which parts of the body. As a result, one could redraw the human brain, listing which parts of the brain controlled which organ. The result was a homunculus, a rather bizarre picture of the human body mapped onto the surface of the brain, which looked like a strange little man, with huge fingertips, lips, and tongue, but a tiny body.

More recently, MRI scans have given us revealing pictures of the thinking brain, but they are incapable of tracing the specific neural pathways of thought, perhaps involving only a few thousand neurons. But a new field called optogenetics combines optics and genetics to unravel specific neural pathways in animals. By analogy, this can be compared to trying to create a road map. The results of the MRI scans would be akin to determining the large interstate highways and the large flow of traffic on them. But optogenetics might be able to actually determine individual roads and pathways. In principle, it even allows scientists the possibility of controlling animal behavior by stimulating these specific pathways.

This, in turn, generated several sensational media stories. The Drudge Report ran a lurid headline that screamed, "Scientists Create Remote-Controlled Flies." The media conjured up visions of remote-controlled flies carrying out the dirty work of the Pentagon. On the *Tonight Show,* Jay Leno even talked about a remote-controlled fly that could fly into the mouth of President George W. Bush on command. Although comedians had a field day imagining bizarre scenarios of the Pentagon commanding hoards of insects with the push of a button, the reality is much more modest.

The fruit fly has roughly 150,000 neurons in the brain. Optogenetics allows scientists to light up certain neurons in the brains of fruit flies that correspond to certain behaviors. For example, when two specific neurons are activated, it can signal the fruit fly to escape. The fly then automatically extends its legs, spreads its wings, and takes off. Scientists were able to genetically breed a strain of fruit flies whose escape neurons fired every time a laser beam was turned on. If you shone a laser beam on these fruit flies, they took off each time.

The implications for determining the structure of the brain are important. Not only would we be able to slowly tease apart neural pathways for certain behaviors, but we also could use this information to help stroke victims and patients suffering from brain diseases and accidents.

Gero Miesenböck of Oxford University and his colleagues have been able to identify the neural mechanisms of animals in this way. They can study not only the pathways for the escape reflex in fruit flies but also the reflexes involved in smelling odors. They have studied the pathways governing food-seeking in roundworms. They have studied the neurons involved in decision making in mice. They found that while as few as two neurons were involved in triggering behaviors in fruit flies, almost 300 neurons were activated in mice for decision making.

The basic tools they have been using are genes that can control the production of certain dyes, as well as molecules that react to light. For example, there is a gene from jellyfish that can make green fluorescent protein. Also, there are a variety of molecules like rhodopsin that respond when light is shone upon them by allowing ions to pass through cell membranes. In this way, shining light on these organisms can trigger certain chemical reactions. Armed with these dyes and light-sensitive chemicals, these scientists have been able for the first time to tease apart neural circuits governing specific behaviors.

So although comedians like to poke fun at these scientists for trying to create Frankenstein fruit flies controlled by the push of a button, the reality is that scientists are, for the first time in history, tracing the specific neural pathways of the brain that control specific behaviors.

MODELING THE BRAIN

Optogenetics is a first, modest step. The next step is to actually model the entire brain, using the latest in technology. There are at least two ways to solve this colossal problem, which will take many decades of hard work. The first is by using supercomputers to simulate the behavior of billions of neurons, each one connected to thousands of other neurons. The other way is to actually locate every neuron in the brain.

The key to the first approach, simulating the brain, is simple: raw computer power. The bigger the computer, the better. Brute force, and inelegant theories, may be the key to cracking this gigantic problem. And the computer that might accomplish this herculean task is called Blue Gene, one of the most powerful computers on earth, built by IBM.

I had a chance to visit this monster computer when I toured the Lawrence Livermore National Laboratory in California, where they design hydrogen warheads for the Pentagon. It is America's premier top-secret weapons laboratory, a sprawling, 790-acre complex in the middle of farm country, budgeted at $1.2 billion per year and employing 6,800 people. This is the heart of the U.S. nuclear weapons establishment. I had to pass through many layers of security to see it, since this is one of the most sensitive weapons laboratories on earth.

Finally, after passing a series of checkpoints, I gained entrance to the building housing IBM's Blue Gene computer, which is capable of computing at the blinding speed of 500 trillion operations per second. Blue Gene is a remarkable sight. It is huge, occupying about a quarter acre, and consists of row after row of jet-black steel cabinets, each one about 8 feet tall and 15 feet long.

When I walked among these cabinets, it was quite an experience. Unlike Hollywood science fiction movies, where the computers have lots of blinking lights, spinning disks, and bolts of electricity crackling through the air, these cabinets are totally quiet, with only a few tiny lights blinking. You realize that the computer is performing trillions of complex calculations, but you hear nothing and see nothing as it works.

What I was interested in was the fact that Blue Gene was simulating the thinking process of a mouse brain, which has about 2 million neurons (compared to the 100 billion neurons that we have). Simulating the thinking process of a mouse brain is harder than you think, because each neuron is connected to many other neurons, making a dense web of neurons. But while I was walking among rack after rack of consoles making up Blue Gene, I could not help but be amazed that this astounding computer power could simulate only the brain of a mouse, and then only for a few seconds. (This does not mean that Blue Gene can simulate the behavior of a mouse. At present, scientists can barely simulate the behavior of a cockroach. Rather, this means that Blue Gene can simulate the firing of neurons found in a mouse, not its behavior.)

In fact, several groups have focused on simulating the brain of a mouse. One ambitious attempt is the Blue Brain Project of Henry Markram of the École Polytechnique Fédérale de Lausanne, in Switzerland. He began in 2005, when he was able to obtain a small version of Blue Gene, with only 16,000 processors, but within a year he was successful in modeling the rat's neocortical column, part of the neocortex, which contains 10,000 neurons and 100 million connections. That was a landmark study, because it meant that it was biologically possible to completely analyze the structure of an important component of the brain, neuron for neuron. (The mouse brain consists of millions of these columns, repeated over and over again. Thus, by modeling one of these columns, one can begin to understand how the mouse brain works.)

In 2009, Markram said optimistically, "It is not impossible to build a human brain and we can do it in ten years. If we build it correctly, it should speak and have an intelligence and behave very much as a human does." He cautions, however, that it would take a supercomputer 20,000 times more powerful than present supercomputers, with a memory storage 500 times the entire size of the current Internet, to achieve this.

So what is the roadblock preventing this colossal goal? To him, it's simple: money.

Since the basic science is known, he feels that he can succeed by simply throwing money at the problem. He says, "It's not a question of years, it's one of dollars. . . . It's a matter of if society wants this. If they want it in ten years, they'll have it in ten years. If they want it in a thousand years, we can wait."

But a rival group is also tackling this problem, assembling the greatest computational firepower in history. This group is using the most advanced version of Blue Gene, called Dawn, also based in Livermore. Dawn is truly an awesome sight, with 147,456 processors with 150,000 gigabytes of memory. It is roughly 100,000 times more powerful than the computer sitting on your desk. The group, led by Dharmendra Modha, has scored a number of successes. In 2006, it was able to simulate 40 percent of a mouse's brain. In 2007, it could simulate 100 percent of a rat's brain (which contains 55 million neurons, much more than the mouse brain).

And in 2009, the group broke yet another world record. It succeeded in simulating 1 percent of the human cerebral cortex, or roughly the cerebral cortex of a cat, containing 1.6 billion neurons with 9 trillion connections. However, the simulation was slow, about 1/600th the speed of the human brain. (If it simulated only a billion neurons, it went much faster, about 1/83rd the speed of the human brain.)

"This is a Hubble Telescope of the mind, a linear accelerator of the brain," says Modha proudly, remarking on the mammoth scale of this achievement. Since the brain has 100 billion neurons, these scientists can now see the light at the end of the tunnel. They feel that a full simulation of the human brain is within sight. "This is not just possible, it's inevitable. This will happen," says Modha.

There are serious problems, however, with modeling the entire human brain, especially power and heat. The Dawn computer devours 1 million watts of power and generates so much heat it needs 6,675 tons of air-conditioning equipment, which blows 2.7 million cubic feet of chilled air every minute. To model the human brain, you would have to scale this up by a factor of 1,000.

This is a truly monumental task. The power consumption of this hypothetical supercomputer would be a billion watts, or the output of an entire nuclear power plant. You could light up an entire city with the energy consumed by this supercomputer. To cool it, you would need to divert an entire river and channel the water through the computer. And the computer itself would occupy many city blocks.

Amazingly, the human brain, by contrast, uses just 20 watts. The heat generated by the human brain is hardly noticeable, yet it easily outperforms our greatest supercomputer. Furthermore, the human brain is the most complex object that Mother Nature has produced in this section of the galaxy. Since we see no evidence of other intelligent life-forms in our solar system, this means that you have to go out to at least 24 trillion miles, the distance to the nearest star, and even beyond to find an object as complex as the one sitting inside your skull.

We might be able to reverse engineer the brain within ten years, but only if we had a massive Manhattan Project–style crash program and dumped billions of dollars into it. However, this is not very likely to happen any time soon, given the current economic climate. Crash programs like the Human Genome Project, which cost nearly $3 billion, were supported by the U.S. government because of their obvious health and scientific benefits. However, the benefits of reverse engineering the brain are less urgent, and hence will take much longer. More realistically, we will approach this goal in smaller steps, and it may take decades to fully accomplish this historic feat.

So computer simulating the brain may take us to midcentury. And even then, it will take many decades to sort through the mountains of data pouring in from this massive project and match it to the human brain. We will be drowning in data without the means to meaningfully sort out the noise.

TAKING APART THE BRAIN

But what about the second approach, identifying the precise location of every neuron in the brain?

This approach is also a herculean task, and may also take many decades of painful research. Instead of using supercomputers like Blue Gene, these scientists take the slice-and-dice approach, starting by dissecting the brain of a fruit fly into incredibly thin slices no more than 50 nm wide (about 150 atoms across). This produces millions of slices. Then a scanning electron microscope takes a photograph of each, with a speed and resolution approaching a billion pixels per second. The amount of data spewing from the electron microscope is staggering, about 1,000 trillion bytes of data, enough to fill a storage room just for a single fruit fly brain. Processing this data, by tediously reconstructing the 3-D wiring of every single neuron of the fly brain, would take about five years. To get a more accurate picture of the fly brain, you then have to slice many more fly brains.

Gerry Rubin of the Howard Hughes Medical Institute, one of the leaders in this field, thinks that altogether, a detailed map of the entire fruit fly brain will take twenty years. "After we solve this, I'd say we're one-fifth of the way to understanding the human mind," he concludes. Rubin realizes the enormity of the task he faces. The human brain has 1 million times more neurons than the brain of a fruit fly. If it takes twenty years to identify every single neuron of the fly brain, then it will certainly take many decades beyond that to fully identify the neural architecture of the human brain. The cost of this project will also be enormous.

So workers in the field of reverse engineering the brain are frustrated. They see that their goal is tantalizingly close, but the lack of funding hinders their work. However, it seems reasonable to assume that sometime by midcentury, we will have both the computer power to simulate the human brain and also crude maps of the brain's neural architecture. But it may well take until late in this century before we fully understand human thought or can create a machine that can duplicate the functions of the human brain.

For example, even if you are given the exact location of every gene inside an ant, it does not mean you know how an anthill is created. Similarly, just because scientists now know the roughly 25,000 genes that make up the human genome, it does not mean they know how the human body works. The Human Genome Project is like a dictionary with no definitions. Each of the genes of the human body is spelled out explicitly in this dictionary, but what each does is still largely a mystery. Each gene codes for a certain protein, but it is not known how most of these proteins function in the body.

Back in 1986, scientists were able to map completely the location of all the neurons in the nervous system of the tiny worm *C. elegans*. This was initially heralded as a breakthrough that would allow us to decode the mystery of the brain. But knowing the precise location of its 302 nerve cells and 6,000 chemical synapses did not produce any new understanding of how this worm functions, even decades later.

In the same way, it will take many decades, even after the human brain is finally reverse engineered, to understand how all the parts work and fit together. If the human brain is finally reverse engineered and completely decoded by the end of the century, then we will have taken a giant step in creating humanlike robots. Then what is to prevent them from taking over?

FAR FUTURE (2070 TO 2100)

WHEN MACHINES BECOME CONSCIOUS

In *The Terminator* movie series, the Pentagon proudly unveils Skynet, a sprawling, foolproof computer network designed to faithfully control the U.S. nuclear arsenal. It flawlessly carries out its tasks until one day in 1995, when something unexpected happens. Skynet becomes conscious. Skynet's human handlers, shocked to realize that their creation has suddenly become sentient, try to shut it down. But they are too late. In self-defense, Skynet decides that the only way to protect itself is to destroy humanity by launching a devastating nuclear war. Three billion people are soon incinerated in countless nuclear infernos. In the aftermath, Skynet unleashes legion after legion of robotic killing machines to slaughter the remaining stragglers. Modern civilization crumbles, reduced to tiny, pathetic bands of misfits and rebels.

Worse, in the *Matrix Trilogy*, humans are so primitive that they don't even realize that the machines have already taken over. Humans carry out their daily affairs, thinking everything is normal, oblivious to the fact that they are actually living in pods. Their world is a virtual reality simulation run by the robot masters. Human "existence" is only a software program, running inside a large computer, that is being fed into the brains of humans living in these pods. The only reason the machines even bother to have humans around is to use them as batteries.

Hollywood, of course, makes its living by scaring the pants off its audience. But it does raise a legitimate scientific question: What happens when robots finally become as smart as us? What happens when robots wake up and become conscious? Scientists vigorously debate the question: not if, but when this momentous event will happen.

According to some experts, our robot creations will gradually rise up the evolutionary tree. Today, they are as smart as cockroaches. In the future, they will be as smart as mice, rabbits, dogs and cats, monkeys, and then they will rival humans. It may take decades to slowly climb this path, but they believe that it is only a matter of time before the machines exceed us in intelligence.

AI researchers are split on the question of when this might happen. Some say that within twenty years robots will approach the intelligence of the human brain and then leave us in the dust. In 1993, Vernor Vinge said, "Within thirty years, we will have the technological means to create super-human intelligence. Shortly after, the human era will be ended. . . . I'll be surprised if this event occurs before 2005 or after 2030."

On the other hand, Douglas Hofstadter, author of *Gödel, Escher, Bach,* says, "I'd be very surprised if anything remotely like this happened in the next 100 years to 200 years."

When I talked to Marvin Minsky of MIT, one of the founding figures in the history of AI, he was careful to tell me that he places no timetable on when this event will happen. He believes the day will come but shies away from being the oracle and predicting the precise date. (Being the grand old man of AI, a field he helped to create almost from scratch, perhaps he has seen too many predictions fail and create a backlash.)

A large part of the problem with these scenarios is that there is no universal consensus as to the meaning of the word *consciousness.* Philosophers and mathematicians have grappled with the word for centuries, and have nothing to show for it. Seventeenth-century thinker Gottfried Leibniz, inventor of calculus, once wrote, "If you could blow the brain up to the size of a mill and walk about inside, you would not find consciousness." Philosopher David Chalmers has even catalogued almost 20,000 papers written on the subject, with no consensus whatsoever.

Nowhere in science have so many devoted so much to create so little.

Consciousness, unfortunately, is a buzzword that means different things to different people. Sadly, there is no universally accepted definition of the term.

I personally think that one of the problems has been the failure to clearly define consciousness and then a failure to quantify it.

But if I were to venture a guess, I would theorize that consciousness consists of at least three basic components:

1. sensing and recognizing the environment
2. self-awareness
3. planning for the future by setting goals and plans, that is, simulating the future and plotting strategy

In this approach, even simple machines and insects have some form of consciousness, which can be ranked numerically on a scale of 1 to 10. There is a continuum of consciousness, which can be quantified. A hammer cannot sense its environment, so it would have a 0 rating on this scale. But a thermostat can. The essence of a thermostat is that it can sense the temperature of the environment and act on it by changing it, so it would have a ranking of 1. Hence, machines with feedback mechanisms have a primitive form of consciousness. Worms also have this ability. They can sense the presence of food, mates, and danger and act on this information, but can do little else. Insects, which can detect more than one parameter (such as sight, sound, smells, pressure, etc.), would have a higher numerical rank, perhaps a 2 or 3.

The highest form of this sensing would be the ability to recognize and understand objects in the environment. Humans can immediately size up their environment and act accordingly and hence rate high on this scale. However, this is where robots score badly. Pattern recognition, as we have seen, is one of the principal roadblocks to artificial intelligence. Robots can sense their environments much better than humans, but they do not understand or recognize what they see. On this scale of consciousness, robots score near the bottom, near the insects, due to their lack of pattern recognition.

The next-higher level of consciousness involves self-awareness. If you place a mirror next to most male animals, they will immediately react aggressively, even attacking the mirror. The image causes the animal to defend its territory. Many animals lack awareness of who they are. But monkeys, elephants, dolphins, and some birds quickly realize that the image in the mirror represents themselves and they cease to attack it. Humans would rank near the top on this scale, since they have a highly developed sense of who they are in relation to other animals, other humans, and the world. In addition, humans are so aware of themselves that they can talk silently to themselves, so they can evaluate a situation by thinking.

Third, animals can be ranked by their ability to formulate plans for the future. Insects, to the best of our knowledge, do not set elaborate goals for the future. Instead, for the most part, they react to immediate situations on a moment-to-moment basis, relying on instinct and cues from the immediate environment.

In this sense, predators are more conscious than prey. Predators have to plan ahead, by searching for places to hide, by planning to ambush, by stalking, by anticipating the flight of the prey. Prey, however, only have to run, so they rank lower on this scale.

Furthermore, primates can improvise as they make plans for the immediate future. If they are shown a banana that is just out of reach, then they might devise strategies to grab that banana, such as using a stick. So, when faced with a specific goal (grabbing food), primates will make plans into the immediate future to achieve that goal.

But on the whole, animals do not have a well-developed sense of the distant past or future. Apparently, there is no tomorrow in the animal kingdom. We have no evidence that they can think days into the future. (Animals will store food in preparation for the winter, but this is largely genetic: they have been programmed by their genes to react to plunging temperatures by seeking out food.)

Humans, however, have a very well-developed sense of the future and continually make plans. We constantly run simulations of reality in our heads. In fact, we can contemplate plans far beyond our own lifetimes. We judge other humans, in fact, by their ability to predict evolving situations and formulate concrete strategies. An important part of leadership is to anticipate future situations, weigh possible outcomes, and set concrete goals accordingly.

In other words, this form of consciousness involves predicting the future, that is, creating multiple models that approximate future events. This requires a very sophisticated understanding of common sense and the rules of nature. It means that you ask yourself "what if" repeatedly. Whether planning to rob a bank or run for president, this kind of planning means being able to run multiple simulations of possible realities in your head.

All indications are that only humans have mastered this art in nature.

We also see this when psychological profiles of test subjects are analyzed. Psychologists often compare the psychological profiles of adults to their profiles when they were children. Then one asks the question: What is the one quality that predicted their success in marriage, careers, wealth, etc.? When one compensates for socioeconomic factors, one finds that one characteristic sometimes stands out from all the others: the ability to delay gratification. According to the long-term studies of Walter Mischel of Columbia University, and many others, children who were able to refrain from immediate gratification (e.g., eating a marshmallow given to them) and held out for greater long-term rewards (getting two marshmallows instead of one) consistently scored higher on almost every measure of future success, in SATs, life, love, and career.

But being able to defer gratification also refers to a higher level of awareness and consciousness. These children were able to simulate the future and realize that future rewards were greater. So being able to see the future consequences of our actions requires a higher level of awareness.

AI researchers, therefore, should aim to create a robot with all three characteristics. The first is hard to achieve, since robots can sense their environment but cannot make sense of it. Self-awareness is easier to achieve. But planning for the future requires common sense, an intuitive understanding of what is possible, and concrete strategies for reaching specific goals.

So we see that common sense is a prerequisite for the highest level of consciousness. In order for a robot to simulate reality and predict the future, it must first master millions of commonsense rules about the world around it. But common sense is not enough. Common sense is just the "rules of the game," rather than the rules of strategy and planning.

On this scale, we can then rank all the various robots that have been created.

We see that Deep Blue, the chess-playing machine, would rank very low. It can beat the world champion in chess, but it cannot do anything else. It is able to run a simulation of reality, but only for playing chess. It is incapable of running simulations of any other reality. This is true for many of the world's largest computers. They excel at simulating the reality of one object, for example, modeling a nuclear detonation, the wind patterns around a jet airplane, or the weather. These computers can run simulations of reality much better than a human. But they are also pitifully one-dimensional, and hence useless in surviving in the real world.

Today, AI researchers are clueless about how to duplicate all these processes in a robot. Most throw up their hands and say that somehow huge networks of computers will show "emergent phenomena" in the same way that order sometimes spontaneously coalesces from chaos. When asked precisely how these emergent phenomena will create consciousness, most roll their eyes to the heavens.

Although we do not know how to create a robot with consciousness, we can imagine what a robot would look like that is more advanced than us, given this framework for measuring consciousness.

They would excel in the third characteristic: they would be able to run complex simulations of the future far ahead of us, from more perspectives, with more details and depth. Their simulations would be more accurate than ours, because they would have a better grasp of common sense and the rules of nature and hence better able to ferret out patterns. They would be able to anticipate problems that we might ignore or not be aware of. Moreover, they would be able to set their own goals. If their goals include helping the human race, then everything is fine. But if one day they formulate goals in which humans are in the way, this could have nasty consequences.

But this raises the next question: What happens to humans in this scenario?

WHEN ROBOTS EXCEED HUMANS

In one scenario, we puny humans are simply pushed aside as a relic of evolution. It is a law of evolution that fitter species arise to displace unfit species; and perhaps humans will be lost in the shuffle, eventually winding up in zoos where our robotic creations come to stare at us. Perhaps that is our destiny: to give birth to superrobots that treat us as an embarrassingly primitive footnote in their evolution. Perhaps that is our role in history, to give birth to our evolutionary successors. In this view, our role is to get out of their way.

Douglas Hofstadter confided to me that this might be the natural order of things, but we should treat these superintelligent robots as we do our children, because that is what they are, in some sense. If we can care for our children, he said to me, then why can't we also care about intelligent robots, which are also our children?

Hans Moravec contemplates how we may feel being left in the dust by our robots: " . . . life may seem pointless if we are fated to spend it staring stupidly at our ultraintelligent progeny as they try to describe their ever more spectacular discoveries in baby talk that we can understand."

When we finally hit the fateful day when robots are smarter than us, not only will we no longer be the most intelligent being on earth, but our creations may make copies of themselves that are even smarter than they are. This army of self-replicating robots will then create endless future generations of robots, each one smarter than the previous one. Since robots can theoretically produce ever-smarter generations of robots in a very short period of time, eventually this process will explode exponentially, until they begin to devour the resources of the planet in their insatiable quest to become ever more intelligent.

In one scenario, this ravenous appetite for ever-increasing intelligence will eventually ravage the resources of the entire planet, so the entire earth becomes a computer. Some envision these superintelligent robots then shooting out into space to continue their quest for more intelligence, until they reach other planets, stars, and galaxies in order to convert them into computers. But since the planets, stars, and galaxies are so incredibly far away, perhaps the computer may alter the laws of physics so its ravenous appetite can race faster than the speed of light to consume whole star systems and galaxies. Some even believe it might consume the entire universe, so that the universe becomes intelligent.

This is the "singularity." The word originally came from the world of relativistic physics, my personal specialty, where a singularity represents a point of infinite gravity, from which nothing can escape, such as a black hole. Because light itself cannot escape, it is a horizon beyond which we cannot see.

The idea of an AI singularity was first mentioned in 1958, in a conversation between two mathematicians, Stanislaw Ulam (who made the key breakthrough in the design of the hydrogen bomb) and John von Neumann. Ulam wrote, "One conversation centered on the ever accelerating progress of technology and changes in the mode of human life, which gives the appearance of approaching some essential singularity in the history of the human race beyond which human affairs, as we know them, could not continue." Versions of the idea have been kicking around for decades. But it was then amplified and popularized by science fiction writer and mathematician Vernor Vinge in his novels and essays.

But this leaves the crucial question unanswered: When will the singularity take place? Within our lifetimes? Perhaps in the next century? Or never? We recall that the participants at the 2009 Asilomar conference put the date at any time between 20 to 1,000 years into the future.

One man who has become the spokesperson for the singularity is inventor and best-selling author Ray Kurzweil, who has a penchant for making predictions based on the exponential growth of technology. Kurzweil once told me that when he gazes at the distant stars at night, perhaps one should be able to see some cosmic evidence of the singularity happening in some distant galaxy. With the ability to devour or rearrange whole star systems, there should be some footprint left behind by this rapidly expanding singularity. (His detractors say that he is whipping up a near-religious fervor around the singularity. However, his supporters say that he has an uncanny ability to correctly see into the future, judging by his track record.)

Kurzweil cut his teeth on the computer revolution by starting up companies in diverse fields involving pattern recognition, such as speech recognition technology, optical character recognition, and electronic keyboard instruments. In 1999, he wrote a best seller, *The Age of Spiritual Machines: When Computers Exceed Human Intelligence,* which predicted when robots will surpass us in intelligence. In 2005, he wrote *The Singularity Is Near* and elaborated on those predictions. The fateful day when computers surpass human intelligence will come in stages.

By 2019, he predicts, a $1,000 personal computer will have as much raw power as a human brain. Soon after that, computers will leave us in the dust. By 2029, a $1,000 personal computer will be 1,000 times more powerful than a human brain. By 2045, a $1,000 computer will be a billion times more intelligent than every human combined. Even small computers will surpass the ability of the entire human race.

After 2045, computers become so advanced that they make copies of themselves that are ever increasing in intelligence, creating a runaway singularity. To satisfy their never-ending, ravenous appetite for computer power, they will begin to devour the earth, asteroids, planets, and stars, and even affect the cosmological history of the universe itself.

I had the chance to visit Kurzweil in his office outside Boston. Walking through the corridor, you see the awards and honors he has received, as well as some of the musical instruments he has designed, which are used by top musicians, such as Stevie Wonder. He explained to me that there was a turning point in his life. It came when he was unexpectedly diagnosed with type II diabetes when he was thirty-five. Suddenly, he was faced with the grim reality that he would not live long enough to see his predictions come true. His body, after years of neglect, had aged beyond his years. Rattled by this diagnosis, he now attacked the problem of personal health with the same enthusiasm and energy he used for the computer revolution. (Today, he consumes more than 100 pills a day and has written books on the revolution in longevity. He expects that the revolution in microscopic robots will be able to clean out and repair the human body so that it can live forever. His philosophy is that he would like to live long enough to see the medical breakthroughs that can prolong our life spans indefinitely. In other words, he wants to live long enough to live forever.)

Recently, he embarked on an ambitious plan to launch the Singularity University, based in the NASA Ames laboratory in the Bay Area, which trains a cadre of scientists to prepare for the coming singularity.

There are many variations and combinations of these various themes.

Kurzweil himself believes, "It's not going to be an invasion of intelligent machines coming over the horizon. We're going to merge with this technology. . . . We're going to put these intelligent devices in our bodies and brains to make us live longer and healthier."

Any idea as controversial as the singularity is bound to unleash a backlash. Mitch Kapor, founder of Lotus Development Corporation, says that the singularity is "intelligent design for the IQ 140 people. . . . This proposition that we're heading to this point at which everything is going to be just unimaginably different—it's fundamentally, in my view, driven by a religious impulse. And all the frantic arm-waving can't obscure that fact for me."

Douglas Hofstadter has said, "It's as if you took a lot of good food and some dog excrement and blended it all up so that you can't possibly figure out what's good or bad. It's an intimate mixture of rubbish and good ideas, and it's very hard to disentangle the two, because these are smart people; they're not stupid."

No one knows how this will play out. But I think the most likely scenario is the following.

MOST LIKELY SCENARIO: FRIENDLY AI

First, scientists will probably take simple measures to ensure that robots are not dangerous. At the very least, scientists can put a chip in robot brains to automatically shut them off if they have murderous thoughts. In this approach, all intelligent robots will be equipped with a fail-safe mechanism that can be switched on by a human at any time, especially when a robot exhibits errant behavior. At the slightest hint that a robot is malfunctioning, any voice command will immediately shut it down.

Or specialized hunter robots may also be created whose duty is to neutralize deviant robots. These robot hunters will be specifically designed to have superior speed, strength, and coordination in order to capture errant robots. They will be designed to understand the weak points of any robotic system and how they behave under certain conditions. Human can also be trained in this skill. In the movie *Blade Runner*, a specially trained cadre of agents, including one played by Harrison Ford, are skilled in the techniques necessary to neutralize any rogue robot.

Since it will take many decades of hard work for robots to slowly go up the evolutionary scale, it will not be a sudden moment when humanity is caught off guard and we are all shepherded into zoos like cattle. Consciousness, as I see it, is a process that can be ranked on a scale, rather than being a sudden evolutionary event, and it will take many decades for robots to ascend up this scale of consciousness. After all, it took Mother Nature millions of years to develop human consciousness. So humans will not be caught off guard one day when the Internet unexpectedly "wakes up" or robots suddenly begin to plan for themselves.

This is the option preferred by science fiction writer Isaac Asimov, who envisioned each robot hardwired in the factory with three laws to prevent them from getting out of control. He devised his famous three laws of robotics to prevent robots from hurting themselves or humans. (Basically, the three laws state that robots cannot harm humans, they must obey humans, and they must protect themselves, in that order.)

(Even with Asimov's three laws, there are also problems when there are contradictions among the three laws. For example, if one creates a benevolent robot, what happens if humanity makes self-destructive choices that can endanger the human race? Then a friendly robot may feel that it has to seize control of the government to prevent humanity from harming itself. This was the problem faced by Will Smith in the movie version of *I, Robot*, when the central computer decides that "some humans must be sacrificed and some freedoms must be surrendered" in order to save humanity. To prevent a robot from enslaving us in order to save us, some have advocated that we must add the zeroth law of robotics: Robots cannot harm or enslave the human race.)

But many scientists are leaning toward something called "friendly AI," where we design our robots to be benign from the very beginning. Since we are the creators of these robots, we will design them, from the very start, to perform only useful and benevolent tasks.

The term "friendly AI" was coined by Eliezer Yudkowsky, a founder of the Singularity Institute for Artificial Intelligence. Friendly AI is a bit different from Asimov's laws, which are forced upon robots, perhaps against their will. (Asimov's laws, imposed from the outside, could actually invite the robots to devise clever ways to circumvent them.) In friendly AI, by contrast, robots are free to murder and commit mayhem. There are no rules that enforce an artificial morality. Rather, these robots are designed from the very beginning to desire to help humans rather than destroy them. They choose to be benevolent.

This has given rise to a new field called "social robotics," which is designed to give robots the qualities that will help them integrate into human society. Scientists at Hanson Robotics, for example, have stated that one mission for their research is to design robots that "will evolve into socially intelligent beings, capable of love and earning a place in the extended human family."

But one problem with all these approaches is that the military is by far the largest funder of AI systems, and these military robots are specifically designed to hunt, track, and kill humans. One can easily imagine future robotic soldiers whose missions are to identify enemy humans and eliminate them with unerring efficiency. One would then have to take extraordinary precautions to guarantee that the robots don't turn against their masters as well. Predator drone aircraft, for example, are run by remote control, so there are humans constantly directing their movements, but one day these drones may be autonomous, able to select and take out their own targets at will. A malfunction in such an autonomous plane could lead to disastrous consequences.

In the future, however, more and more funding for robots will come from the civilian commercial sector, especially from Japan, where robots are designed to help rather than destroy. If this trend continues, then perhaps friendly AI could become a reality. In this scenario, it is the consumer sector and market forces that will eventually dominate robotics, so that there will be a vast commercial interest in investing in friendly AI.

MERGING WITH ROBOTS

In addition to friendly AI, there is also another option: merging with our creations. Instead of simply waiting for robots to surpass us in intelligence and power, we should try to enhance ourselves, becoming superhuman in the process. Most likely, I believe, the future will proceed with a combination of these two goals, i.e., building friendly AI and also enhancing ourselves.

This is an option being explored by Rodney Brooks, former director of the famed MIT Artificial Intelligence Laboratory. He has been a maverick, overturning cherished but ossified ideas and injecting innovation into the field. When he entered the field, the top-down approach was dominant in most universities. But the field was stagnating. Brooks raised a few eyebrows when he called for creating an army of insectlike robots that learned via the bottom-up approach by bumping into obstacles. He did not want to create another dumb, lumbering robot that took hours to walk across the room. Instead, he built nimble "insectoids" or "bugbots" that had almost no programming at all but would quickly learn to walk and navigate around obstacles by trial and error. He envisioned the day that his robots would explore the solar system, bumping into things along the way. It was an outlandish idea, proposed in his essay "Fast, Cheap, and Out of Control," but his approach eventually led to an array of new avenues. One by-product of his idea is the Mars Rovers now scurrying over the surface of the Red Planet. Not surprisingly, he was also the chairman of iRobot, the company that markets buglike vacuum cleaners to households across the country.

One problem, he feels, is that workers in artificial intelligence follow fads, adopting the paradigm of the moment, rather than thinking in fresh ways. For example, he recalls, "When I was a kid, I had a book that described the brain as a telephone-switching network. Earlier books described it as a hydrodynamic system or a steam engine. Then in the 1960s, it became a digital computer. In the 1980s, it became a massively parallel digital computer. Probably there's a kid's book out there somewhere that says the brain is just like the World Wide Web. . . ."

For example, some historians have noted that Sigmund Freud's analysis of the mind was influenced by the coming of the steam engine. The spread of railroads through Europe in the mid- to late 1800s had a profound effect on the thinking of intellectuals. In Freud's picture, there were flows of energy in the mind that constantly competed with other flows, much like in the steam pipes in an engine. The continual interaction between the superego, the id, and the ego resembled the continual interaction between steam pipes in a locomotive. And the fact that repressing these flows of energy could create neuroses is analogous to the way that steam power, if bottled up, can be explosive.

Marvin Minsky admitted to me that another paradigm misguided the field for many years. Since many AI researchers are former physicists, there is something called "physics envy," that is, the desire to find the single, unifying theme underlying all intelligence. In physics, we have the desire to follow Einstein to reduce the physical universe to a handful of unifying equations, perhaps finding an equation one inch long that can summarize the universe in a single coherent idea. Minsky believes that this envy led AI researchers to look for that single unifying theme for consciousness. Now, he believes, there is no such thing. Evolution haphazardly cobbled together a bunch of techniques we collectively call consciousness. Take apart the brain, and you find a loose collection of minibrains, each designed to perform a specific task. He calls this the "society of minds": that consciousness is actually the sum of many separate algorithms and techniques that nature stumbled upon over millions of years.

Rodney Brooks was also looking for a similar paradigm, but one that had never been fully explored before. He soon realized that Mother Nature and evolution had already solved many of these problems. For example, a mosquito, with only a few hundred thousand neurons, can outperform the greatest military robotic system. Unlike our flying drones, mosquitoes, with brains smaller than the head of a pin, can independently navigate around obstacles, find food and mates. Why not learn from nature and biology? If you follow the evolutionary scale, you learn that insects and mice did not have the rules of logic programmed into their brains. It was through trial and error that they engaged the world and mastered the art of survival.

Now he is pursuing yet another heretical idea, contained in his essay "The Merger of Flesh and Machines." He notes that the old laboratories at MIT, which used to design silicon components for industrial and military robots, are now being cleaned out, making way for a new generation of robots made of living tissue as well as silicon and steel. He foresees an entirely new generation of robots that will marry biological and electronic systems to create entirely new architectures for robots.

He writes, "My prediction is that by the year 2100 we will have very intelligent robots everywhere in our everyday lives. But we will not be apart from them—rather, we will be part robot and connected with the robots."

He sees this progressing in stages. Today, we have the ongoing revolution in prostheses, inserting electronics directly into the human body to create realistic substitutes for hearing, sight, and other functions. For example, the artificial cochlea has revolutionized the field of audiology, giving back the gift of hearing to the deaf. These artificial cochleas work by connecting electronic hardware with biological "wetware," that is, neurons. The cochlear implant has several components. A microphone is placed outside the ear. It receives sound waves, processes them, and transmits the signals by radio to the implant that is surgically placed inside the ear. The implant receives the radio messages and converts them into electrical currents that are sent down electrodes in the ear. The cochlea recognizes these electrical impulses and sends them on to the brain. These implants can use up to twenty-four electrodes and can process half a dozen frequencies, enough to recognize the human voice. Already, 150,000 people worldwide have had cochlear implants.

Several groups are exploring ways to assist the blind by creating artificial vision, connecting a camera to the human brain. One method is to directly insert the silicon chip into the retina of the person and attach the chip to the retina's neurons. Another is to connect the chip to a special cable that is connected to the back of the skull, where the brain processes vision. These groups, for the first time in history, have been able to restore a degree of sight to the blind. Patients have been able to see up to 50 pixels lighting up before them. Eventually, scientists should be able to scale this up so that they can see thousands of pixels.

The patients can see fireworks, the outlines of their hands, shining objects and lights, the presence of cars and people, and the borders of objects. "At Little League games, I can see where the catcher, batter, and umpire are," says Linda Morfoot, one of the test subjects.

So far, thirty patients have had artificial retinas with up to sixty electrodes. But the Department of Energy's Artificial Retina Project, based at the University of Southern California, is already planning a new system with more than 200 electrodes. A 1,000-electrode device is also being studied (but if too many electrodes are packed onto the chip, it could cause overheating of the retina). In this system, a miniature camera mounted on a blind person's eyeglasses takes pictures and sends them wirelessly to a microprocessor, worn on a belt, that relays the information to the chip placed directly on the retina. This chip sends tiny pulses directly into the retinal nerves that are still active, thereby bypassing defective retinal cells.

STAR WARS ROBOTIC HAND

Using mechanical enhancements, one can also duplicate the feats of science fiction, including the robotic hand of *Star Wars* and the X-ray vision of Superman. In *The Empire Strikes Back*, Luke Skywalker has his hand chopped off by a lightsaber wielded by the evil Darth Vader, his father. No problem. Scientists in this faraway galaxy quickly create a new mechanical hand, complete with fingers that can touch and feel.

This may sound like science fiction, yet it is already here. A significant advance was made by scientists in Italy and Sweden, who have actually made a robotic hand that can "feel." One subject, Robin Ekenstam, a twenty-two-year-old who had his right hand amputated to remove a cancerous tumor, can now control the motion of his mechanical fingers and feel the response. Doctors connected the nerves in Ekenstam's arm to the chips contained in his mechanical hand so that he can control the finger movements with his brain. The artificial "smart hand" has four motors and forty sensors. The motion of his mechanical fingers is then relayed to his brain so he has feedback. In this way, he is able to control and also "feel" the motion of his hand. Since feedback is one of the essential features of body motion, this could revolutionize the way we treat amputees with prosthetic limbs.

Ekenstam says, "It's great. I have a feeling that I have not had for a long time. Now I am getting sensation back. If I grab something tightly, then I can feel it in the fingertips, which is strange, since I don't have them anymore."

One of the researchers, Christian Cipriani of the Scuola Superiore Sant'Anna, says, "First, the brain controls the mechanical hand without any muscle contractions. Second, the hand will be able to give feedback to the patient so he will be able to feel. Just like a real hand."

This development is significant because it means that one day humans may effortlessly control mechanical limbs as if they were flesh and bone. Instead of tediously learning how to move arms and legs of metal, people will treat these mechanical appendages as if they were real, feeling every nuance of the limbs' movements via electronic feedback mechanisms.

This is also evidence of a theory that says the brain is extremely plastic, not fixed, and constantly rewires itself as it learns new tasks and adjusts to new situations. Hence, the brain will be adaptable enough to accommodate any new appendage or sense organ. They may be attached to the brain at different locations, and the brain simply "learns" to control this new attachment. If so, then the brain might be viewed as a modular device, able to plug in and then control different appendages and sensors from different devices. This type of behavior might be expected if our brain is a neural network of some sort that makes new connections and neural pathways each time it learns a new task, whatever that task might be.

Rodney Brooks writes, "Over the next ten to twenty years, there will be a cultural shift, in which we will adopt robotic technology, silicon, and steel into our bodies to improve what we can do and understand the world." When Brooks analyzes the progress made at Brown University and Duke University in hooking up the brain directly to a computer or a mechanical arm, he concludes, "We may all be able to have a wireless Internet connection installed directly into our brains."

In the next stage, he sees merging silicon and living cells not just to cure the ailments of the body but to slowly enhance our capabilities. For example, if today's cochlear and retinal implants can restore hearing and vision, tomorrow's may also give us superhuman abilities. We would be able to hear sounds that only dogs can hear, or see UV, infrared, and X-rays.

It might be possible to increase our intelligence as well. Brooks cites research in which extra layers of neurons were added to the brain of a rat at a critical time in its development. Remarkably, the cognitive abilities of these rats were increased. He envisions a time in the near future when the human brain's intelligence might also be improved by a similar process. In a later chapter, we will see that biologists have already isolated a gene in rats that the media has dubbed the "smart mouse gene." With the addition of this gene, enhanced mice have much greater memory and learning abilities.

And by midcentury, Brooks envisions a time when seemingly fanciful enhancements of the body might be possible, giving us abilities far beyond those of the ordinary human. "Fifty years from now, we can expect to see radical alterations of human bodies through genetic modification." When you also add electronic enhancements, "the human menagerie will expand in ways unimaginable to us today. . . . We will no longer find ourselves confined by Darwinian evolution," he says.

But anything, of course, can be taken too far. How far should we go in merging with our robot creations before some people rebel and find it repulsive?

SURROGATES AND AVATARS

One way in which to merge with robots, but without altering the human body, is to create surrogates or avatars. In the movie *Surrogates,* starring Bruce Willis, in the year 2017 scientists have discovered a way for people to control robots as if they were inside them, so that we can live our lives in perfect bodies. The robot responds to every command, and the person also sees and feels everything the robot sees and feels. While our mortal bodies decay and wither, we can control the motions of our robot surrogate, which has superhuman powers and is perfectly shaped. The movie gets complicated because people prefer to live out their lives as beautiful, handsome, and superpowerful robots, abandoning their rotting bodies, which are conveniently hidden away. The entire human race, in effect, willingly becomes robotic rather than face reality.

In the movie *Avatar,* this is taken one step further. Instead of living our lives as perfect robots, in the year 2154 we might be able to live as alien beings. In the movie, our bodies are placed in pods, which then allow us to control the motion of specially cloned alien bodies. In a sense, we are given entirely new bodies to live on a new planet. In this way, we can better communicate with a native alien population on other planets. The movie plot thickens when one worker decides to abandon his humanity and live out his life as an alien, protecting them from mercenaries.

These surrogates and avatars are not possible today but may be possible in the future.

Recently, ASIMO has been programmed with a new idea: remote sensing. At Kyoto University, humans have been trained to control the mechanical motion of robots by using brain sensors. For example, by putting on an EEG helmet, students can move the arms and legs of ASIMO by simply thinking. So far, four distinct motions of the arms and head are possible. This may open the door to another realm of AI: robots controlled by the mind.

Although this is a crude demonstration of mind over matter, in the coming decades it should be possible to increase the set of motions we can control in a robot, and also to get feedback, so we can "feel" with our new robotic hands. Goggles or contact lenses would allow us to see what the robots see, so we might eventually have full control over the body's motions.

This may also help alleviate the immigration problem for Japan. Workers may be located in different countries, yet control robots thousands of miles away by donning brain sensors. So not only can the Internet carry the thoughts of white-collar workers, it might also carry the thoughts of blue-collar workers and translate them into physical motion. This might mean that robots will become an integral part of any nation grappling with exploding health costs and a shortage of workers.

Controlling robots by remote sensing may also have applications elsewhere. In any dangerous environment (for example, underwater, near high-voltage lines, in fires), robots controlled by human thoughts may be used in rescue missions. Or undersea robots may be connected directly to humans, so that humans can control many swimming robots by thoughts alone. Since the surrogate would have superpowers, it would be able to chase criminals (unless the criminals also have superpowered surrogates). One would have all the advantages of merging with robots without changing our bodies at all.

Such an arrangement might actually prove useful for space exploration, when we have to manage a permanent moon base. Our surrogates may perform all the dangerous tasks of maintaining the moon base, while the astronauts are safely back on earth. The astronauts would have the superstrength and superpowers of the robots while exploring a hazardous alien landscape. (This would not work if the astronauts are on the earth controlling surrogates on Mars, however, since radio signals take up to 40 minutes to go from the earth to Mars and back. But it would work if the astronauts were sitting safely in a permanent base on Mars while the surrogates went out and performed dangerous tasks on the Martian surface.)

HOW FAR THE MERGER WITH ROBOTS?

Robot pioneer Hans Moravec takes this several steps further and imagines an extreme version of this: we become the very robots that we have built. He explained to me how we might merge with our robot creations by undergoing a brain operation that replaces each neuron of our brain with a transistor inside a robot. The operation starts when we lie beside a robot body without a brain. A robotic surgeon takes every cluster of gray matter in our brain, duplicates it transistor by transistor, connects the neurons to the transistors, and puts the transistors into the empty robot skull. As each cluster of neurons is duplicated in the robot, it is discarded. We are fully conscious as this delicate operation takes place. Part of our brain is inside our old body, but the other part is now made of transistors inside our new robot body. After the operation is over, our brain has been entirely transferred into the body of a robot. Not only do we have a robotic body, we have also the benefits of a robot: immortality in superhuman bodies that are perfect in appearance. This will not be possible in the twenty-first century, but becomes an option in the twenty-second.

In the ultimate scenario, we discard our clumsy bodies entirely and eventually evolve into pure software programs that encode our personalities. We "download" our entire personalities into a computer. If someone presses a button with your name on it, then the computer behaves as if you are inside its memory, since it has encoded all your personality quirks inside its circuits. We become immortal, but spend our time trapped inside a computer, interacting with other "people" (that is, other software programs) in some gigantic cyberspace/virtual reality. Our bodily existence will be discarded, replaced by the motion of electrons in this gigantic computer. In this picture, our ultimate destiny is to wind up as lines of code in this vast computer program, with all the apparent sensations of physical bodies dancing in a virtual paradise. We will share deep thoughts with other lines of computer code, living out this grand illusion. We have great, heroic exploits conquering new worlds, oblivious to the fact that we are just electrons dancing inside some computer. Until, of course, someone hits the off button.

But one problem with pushing these scenarios too far is the Cave Man Principle. As we mentioned earlier, the architecture of our brains is that of a primitive hunter-gatherer who emerged from Africa more than 100,000 years ago. Our deepest desires, our appetites, our wants were all forged in the grasslands of Africa as we evaded predators, hunted for game, foraged in the forests, looked for mates, and entertained ourselves at the campfire.

One of our prime directives, buried deep in the fabric of our thoughts, is to look good, especially to the opposite sex and our peers. An enormous fraction of our disposable income, after entertainment, is devoted to our appearance. That is why we have had the explosive growth in plastic surgery, Botox, grooming products, sophisticated clothing, as well as learning new dance steps, muscle building, buying the latest music, and keeping fit. If you add all this up, it becomes a huge portion of consumer spending, which in turn generates a large fraction of the U.S. economy.

This means that, even with the ability to create perfect bodies that are nearly immortal, we will probably resist the desire for robotic bodies if we look like a clumsy robot with implants dangling out of our heads. No one wants to look like a refugee from a science fiction movie. If we have enhanced bodies, they must make us attractive to the opposite sex and enhance our reputation among our peers, or we will reject them. What teenager wants to be enhanced but look uncool?

Some science fiction writers have relished the idea that we will all become detached from our bodies and exist as immortal beings of pure intelligence living inside some computer, contemplating deep thoughts. But who would want to live like that? Perhaps our descendants will not want to solve differential equations describing a black hole. In the future, people may want to spend more time listening to rock music the old-fashioned way than calculate the motions of subatomic particles while living inside a computer.

Greg Stock of UCLA goes further and finds there are few advantages to having our brains hooked up to a supercomputer. He said, "When I try to think of what I might gain by having a working link between my brain and a supercomputer, I am stymied if I insist on two criteria: that the benefits could not be as easily achieved through some other, noninvasive procedure, and that the benefits must be worth the discomforts of brain surgery."

So although there are many possible options for the future, I personally believe that the most likely path is that we will build robots to be benevolent and friendly, enhance our own abilities to a degree, but follow the Cave Man Principle. We will embrace the idea of temporarily living the life of a superrobot via surrogates but will be resistant to the idea of permanently living out our lives inside a computer or altering our body until it becomes unrecognizable.

ROADBLOCKS TO THE SINGULARITY

No one knows when robots may become as smart as humans. But personally, I would put the date close to the end of the century for several reasons.

First, the dazzling advances in computer technology have been due to Moore's law. These advances will begin to slow down and might even stop around 2020–25, so it is not clear if we can reliably calculate the speed of computers beyond that. (See Chapter 4 for more on the post-silicon era.) In this book, I have assumed that computer power will continue to grow, but at a slower rate.

Second, even if a computer can calculate at fantastic speeds like 10^{16} calculations per second, this does not necessarily mean that it is smarter than us. For example, Deep Blue, IBM's chess-playing machine, could analyze 200 million positions per second, beating the world champion. But Deep Blue, for all its speed and raw computing power, cannot do anything else. True intelligence, we learned, is much more than calculating chess positions.

For example, autistic savants can perform miraculous feats of memorization and calculation. But they have difficulty tying their shoelaces, getting a job, or functioning in society. The late Kim Peek, who was so remarkable that the movie *Rain Man* was based on his extraordinary life, memorized every word in 12,000 books and could perform calculations that only a computer could check. Yet he had an IQ of 73, had difficulty holding a conversation, and needed constant help to survive. Without his father's assistance, he was largely helpless. In other words, the superfast computers of the future will be like autistic savants, able to memorize vast amounts of information, but not much more, unable to survive in the real world on their own.

Even if computers begin to match the computing speed of the brain, they will still lack the necessary software and programming to make everything work. Matching the computing speed of the brain is just the humble beginning.

Third, even if intelligent robots are possible, it is not clear if a robot can make a copy of itself that is smarter than the original. The mathematics behind self-replicating robots was first developed by the mathematician John von Neumann, who invented game theory and helped to develop the electronic computer. He pioneered the question of determining the minimum number of assumptions before a machine could create a copy of itself. However, he never addressed the question of whether a robot can make a copy of itself that is smarter than it. In fact, the very definition of "smart" is problematic, since there is no universally accepted definition of "smart."

Certainly, a robot might be able to create a copy of itself with more memory and processing ability by simply upgrading and adding more chips. But does this mean the copy is smarter, or just faster? For example, an adding machine is millions of times faster than a human, with much more memory and processing speed, but it is certainly not smarter. So intelligence is more than just memory and speed.

Fourth, although hardware may progress exponentially, software may not. While hardware has grown by the ability to etch smaller and smaller transistors onto a wafer, software is totally different; it requires a human to sit down with a pencil and paper and write code. That is the bottleneck: the human.

Software, like all human creative activity, progresses in fits and starts, with brilliant insights and long stretches of drudgery and stagnation. Unlike simply etching more transistors onto silicon, which has grown like clockwork, software depends on the unpredictable nature of human creativity and whim. Therefore all predictions of a steady, exponential growth in computer power have to be qualified. A chain is no stronger than its weakest link, and the weakest link is software and programming done by humans.

Engineering progress often grows exponentially, especially when it is a simple matter of achieving greater efficiency, such as etching more and more transistors onto a silicon wafer. But when it comes to basic research, which requires luck, skill, and unexpected strokes of genius, progress is more like "punctuated equilibrium," with long stretches of time when not much happens, with sudden breakthroughs that change the entire terrain. If we look at the history of basic research, from Newton to Einstein to the present day, we see that punctuated equilibrium more accurately describes the way in which progress is made.

Fifth, as we have seen in the research for reverse engineering the brain, the staggering cost and sheer size of the project will probably delay it into the middle of this century. And then making sense of all this data may take many more decades, pushing the final reverse engineering of the brain to late in this century.

Sixth, there probably won't be a "big bang," when machines suddenly become conscious. As before, if we define consciousness as including the ability to make plans for the future by running simulations of the future, then there is a spectrum of consciousness. Machines will slowly climb up this scale, giving us plenty of time to prepare. This will happen toward the end of this century, I believe, so there is ample time to discuss various options available to us. Also, consciousness in machines will probably have its own peculiarities. So a form of "silicon consciousness" rather than pure human consciousness will develop first.

But this raises another question. Although there are mechanical ways to enhance our bodies, there are also biological ways. In fact, the whole thrust of evolution is the selection of better genes, so why not shortcut millions of years of evolution and take control of our genetic destiny?

No one really has the guts to say it, but if we could make better human beings by knowing how to add genes, why shouldn't we?

—JAMES WATSON, NOBEL LAUREATE

I don't really think our bodies are going to have any secrets left within this century. And so, anything that we can manage to think about will probably have a reality.

—DAVID BALTIMORE, NOBEL LAUREATE

I don't think the time is quite right, but it's close. I'm afraid, unfortunately, that I'm in the last generation to die.

—GERALD SUSSMAN

3 FUTURE OF MEDICINE *Perfection and Beyond*

The gods of mythology possessed the ultimate power: the power over life and death, the ability to heal the sick and prolong life. Foremost in our prayers to the gods was deliverance from disease and illness.

In Greek and Roman mythology, there is the tale of Eos, the beautiful goddess of the dawn. One day, she fell deeply in love with a handsome mortal, Tithonus. She had a perfect body and was immortal, but Tithonus would eventually age, wither away, and perish. Determined to save her lover from this dismal fate, she beseeched Zeus, the father of the gods, to grant Tithonus the gift of immortality so that they could spend eternity together. Taking pity on these lovers, he granted Eos her wish.

But Eos, in her haste, forgot to ask for eternal youth for him. So Tithonus became immortal, but his body aged. Unable to die, he became more and more decrepit and decayed, living an eternity with pain and suffering.

So that is the challenge facing the science of the twenty-first century. Scientists are now reading the book of life, which includes the complete human genome, and which promises us miraculous advances in understanding aging. But life extension without health and vigor can be an eternal punishment, as Tithonus tragically found out.

By the end of this century, we too shall have much of this mythical power over life and death. And this power won't be limited to healing the sick but will be used to enhance the human body and even create new life-forms. It won't be through prayers and incantations, however, but through the miracle of biotechnology.

One of the scientists who is unlocking the secrets of life is Robert Lanza, a man in a hurry. He is a new breed of biologist, young, energetic, and full of fresh ideas—so many breakthroughs to be made and so little time. Lanza is riding the crest of the biotech revolution. Like a kid in a candy store, he delights in delving into uncharted territory, making breakthroughs in a wide range of hot-button topics.

A generation or two ago, the pace was much different. You might find biologists leisurely examining obscure worms and bugs, patiently studying their detailed anatomy and agonizing over what Latin names to give them.

Not Lanza.

I met him one day at a radio studio for an interview and was imme-diately impressed by his youth and boundless creativity. He was, as usual, rushing between experiments. He told me he got his start in this fast-moving field in the most unusual way. He came from a modest working-class family south of Boston, where few went to college. But while in high school, he heard the astonishing news about the unraveling of DNA. He was hooked. He decided on a science project: cloning a chicken in his room. His bewildered parents did not know what he was doing, but they gave him their blessing.

Determined to get his project off the ground, he went to Harvard to get advice. Not knowing anyone, he asked a man he thought was a janitor for some directions. Intrigued, the janitor took him to his office. Lanza found out later that the janitor was actually one of the senior researchers at the lab. Impressed by the sheer audacity of this brash young high school stu-dent, he introduced Lanza to other scientists there, including many Nobel-caliber researchers, who would change his life. Lanza compares himself to Matt Damon's character in the movie *Good Will Hunting*, where a scruffy, street-smart working-class kid astonishes the professors at MIT, dazzling them with his mathematical genius.

Today, Lanza is chief scientific officer of Advanced Cell Technology, with hundreds of papers and inventions to his credit. In 2003, he made headlines when the San Diego Zoo asked him to clone a banteng, an endangered species of wild ox, from the body of one that had died twenty-five years before. Lanza successfully extracted usable cells from the carcass, processed them, and sent them to a farm in Utah. There, the fertilized cell was implanted into a female cow. Ten months later he got news that his latest creation had just been born. On another day, he might be working on "tissue engineering," which may eventually create a human body shop from which we can order new organs, grown from our own cells, to replace organs that are diseased or have worn out. Another day, he could be working on cloning human embryo cells. He was part of the historic team that cloned the world's first human embryo for the purpose of generating embryonic stem cells.

THREE STAGES OF MEDICINE

Lanza is riding a tidal wave of discovery, created by unleashing the knowledge hidden within our DNA. Historically, medicine has gone through at least three major stages. In the first, which lasted for tens of thousands of years, medicine was dominated by superstition, witchcraft, and hearsay. With most babies dying at birth, the average life expectancy hovered around eighteen to twenty years. Some useful medicinal herbs and chemicals were discovered during this period, like aspirin, but for the most part there was no systematic way of finding new therapies. Unfortunately, any remedies that actually worked were closely guarded secrets. The "doctor" earned his income by pleasing wealthy patients and had a vested interest in keeping his potions and chants secret.

During this period, one of the founders of the Mayo Clinic kept a private diary when he made the rounds of his patients. He candidly wrote in his diary that there were only two active ingredients in his black bag that actually worked: a hacksaw and morphine. The hacksaw was used to cut off diseased limbs, and the morphine was used to deaden the pain of the amputation. They worked every time. Everything else in his black bag was snake oil and a fake, he lamented sadly.

The second stage of medicine began in the nineteenth century, with the coming of the germ theory and better sanitation. Life expectancy in the United States in 1900 rose to forty-nine years. When tens of thousands of soldiers were dying on the European battlefields of World War I, there was an urgent need for doctors to conduct real experiments, with reproducible results, which were then published in medical journals. The kings of Europe, horrified that their best and brightest were being slaughtered, demanded real results, not hocus-pocus. Doctors, instead of trying to please wealthy patrons, now fought for legitimacy and fame by publishing papers in peer-reviewed journals. This set the stage for advances in antibiotics and vaccines that increased life expectancy to seventy years and beyond.

The third stage of medicine is molecular medicine. We are seeing the merger of physics and medicine, reducing medicine to atoms, molecules, and genes. This historic transformation began in the 1940s, when Austrian physicist Erwin Schrödinger, one of the founders of the quantum theory, wrote an influential book called *What Is Life?* He rejected the notion that there was some mysterious spirit, or life force, that animated living things. Instead, he speculated that all life was based on a code of some sort, and that this was encoded on a molecule. By finding that molecule, he conjectured, one could unravel the secret of life. Physicist Francis Crick, inspired by Schrödinger's book, teamed up with geneticist James Watson to prove that DNA was this fabled molecule. In 1953, in one of the most important discoveries of all time, Watson and Crick unlocked the structure of DNA, a double helix. When unraveled, a single strand of DNA stretches about six feet long. On it is contained a sequence of 3 billion nucleic acids, called A,T,C,G (adenine, thymine, cytosine, and guanine), that carry the code. By reading the precise sequence of these nucleic acids placed along the DNA molecule, one could read the book of life.

The rapid advances in molecular genetics finally led to the creation of the Human Genome Project, truly a milestone in the history of medicine. A massive, crash program to sequence all the genes of the human body, it cost about $3 billion and involved the work of hundreds of scientists collaborating around the world. When it was finally completed in 2003, it heralded a new era in science. Eventually, everyone will have his or her personalized genome available on a CD-ROM. It will list all your approximately 25,000 genes; it will be your "owner's manual."

Nobel laureate David Baltimore summed it up when he said, "Biology is today an information science."

NEAR TERM (PRESENT TO 2030)

GENOMIC MEDICINE

What is driving this remarkable explosion in medicine is, in part, the quantum theory and the computer revolution. The quantum theory has given us amazingly detailed models of how the atoms are arranged in each protein and DNA molecule. Atom for atom, we know how to build the molecules of life from scratch. And gene sequencing—which used to be a long, tedious, and expensive process—is all automated with robots now. Originally, it cost several million dollars to sequence all the genes in a single human body. It is so expensive and time-consuming that only a handful of people (including the scientists who perfected this technology) have had their genomes read. But within a few more years, this exotic technology may come to the average person.

(I vividly recall keynoting a conference in the late 1990s in Frankfurt, Germany, about the future of medicine. I predicted that by 2020, personal genomes would be a real possibility, and that everyone might have a CD or chip with his or her genes described on it. But one participant became quite indignant. He rose and said that this dream was impossible. There were simply too many genes, and it would cost too much to offer personal genomes to the average person. The Human Genome Project had cost $3 billion; the cost to sequence one person's genes could not possibly drop that much. Discussing the issue with him later, it gradually became clear what the problem was. He was thinking linearly. But Moore's law was driving down the costs, making it possible to sequence DNA using robots, computers, and automatic machines. He failed to understand the profound impact of Moore's law on biology. Looking back at that incident, I now realize that if there was a mistake in that prediction, it was in overestimating the time it would take to offer personal genomics.)

For example, Stanford engineer Stephen R. Quake has perfected the latest development in gene sequencing. He has now driven down the cost to

$50,000 and foresees the price plunging to $1,000 in the next few years. Scientists have long speculated that when the price of human gene sequencing drops to $1,000, this could open the floodgates to mass gene sequencing, so a large proportion of the human race may benefit from this technology. Within a few decades, the price of sequencing all your genes may cost less than $100, no more expensive than a standard blood test.

(The key to this latest breakthrough is to take a shortcut. Quake compares a person's DNA to DNA sequences that have already been done of others. He breaks up the human genome into units of DNA containing 32 bits of information. Then he has a computer program that compares these 32-bit fragments to the completed genomes of other people. Since any two humans are almost identical in their DNA, differing on average by less than .1 percent, this means that a computer can rapidly get a match among these 32-bit fragments.)

Quake became the eighth person in the world to have his genome fully sequenced. He had a personal interest in this project as well, since he scanned his personal genome for evidence of heart disease. Unfortunately, his genome indicated that he inherited one version of a gene associated with heart disease. "You have to have a strong stomach when you look at your own genome," he lamented.

I know that eerie feeling. I had my own genome partially scanned and placed on a CD-ROM for a BBC-TV/Discovery special that I hosted. A doctor extracted some blood from my arm; sent it to the laboratory at Vanderbilt University; and then, two weeks later, a CD-ROM came back in the mail, listing thousands of my genes. Holding this disk in my hands gave me a funny feeling, knowing that it contained a partial blueprint for my body. In principle, this disk could be used to create a reasonable copy of myself.

But it also piqued my curiosity, since the secrets of my body were contained on that CD-ROM. For example, I could see if I had a particular gene that increased my chances of getting Alzheimer's disease. I was concerned, since my mother died of Alzheimer's. (Fortunately, I do not have the gene.)

Also, four of my genes were matched with the genome of thousands of people around the world, who had also had their genes analyzed. Then, the locations of the individuals who had a perfect match with my four genes were placed on a map of the earth. By analyzing the dots on the map of the earth, I could see a long trail of dots, originating near Tibet and then stretching through China and to Japan. It was amazing that this trail of dots traced the ancient migration patterns of my mother's ancestors, going back thousands of years. My ancestors left no written records of their ancient migration, but the telltale map of their travels was etched into my blood and DNA. (You can also trace the ancestry of your father. The mitochondrial genes are passed down unchanged from mother to daughter, while the Y chromosome is passed down from father to son. Hence, by analyzing these genes, one can trace the ancestry of your mother or your father's line.)

I imagine in the near future, many people will have the same strange feeling I did, holding the blueprint of their bodies in their hands and reading the intimate secrets, including dangerous diseases, lurking in the genome and the ancient migration patterns of their ancestors.

But for scientists, this is opening an entirely new branch of science, called bioinformatics, or using computers to rapidly scan and analyze the genome of thousands of organisms. For example, by inserting the genomes of several hundred individuals suffering from a certain disease into a computer, one might be able to calculate the precise location of the damaged DNA. In fact, some of the world's most powerful computers are involved in bioinformatics, analyzing millions of genes found in plants and animals for certain key genes.

This could even revolutionize TV detective shows like *CSI*. Given tiny scraps of DNA (found in hair follicles, saliva, or bloodstains), one might be able to determine not just the person's hair color, eye color, ethnicity, height, and medical history, but perhaps also his face. Today, police artists can mold an approximate sculpture of a victim's face using only the skull. In the future, a computer might be able to reconstruct a person's facial features given just some dandruff or blood from that person. (The fact that identical twins have remarkably similar faces means that genetics alone, even in the presence of environmental factors, can determine much of a person's face.)

VISIT TO THE DOCTOR

As we mentioned in the previous chapters, your visit to the doctor's office will be radically changed. When you talk to the doctor in your wall screen, you will probably be talking to a software program. Your bathroom will have more sensors than a modern hospital, silently detecting cancer cells years before a tumor forms. For example, about 50 percent of all common cancers involve a mutation in the gene p53 that can be easily detected using these sensors.

If there is evidence of cancer, then nanoparticles will be injected directly into your bloodstream, which will, like smart bombs, deliver cancer-fighting drugs directly to the cancer cells. We will view chemotherapy today like we view leeches of the past century. (We will discuss the details of nano-technology, DNA chips, nanoparticles, and nanobots in more detail in the next chapter.)

And if the "doctor" in your wall screen cannot cure a disease or injury to an organ, you will simply grow another. In the United States alone, there are 91,000 people awaiting an organ transplant. Eighteen die every day, waiting for an organ that never comes.

If your virtual doctor finds something wrong, such as a diseased organ, then he might order a new one to be grown directly from your own cells.

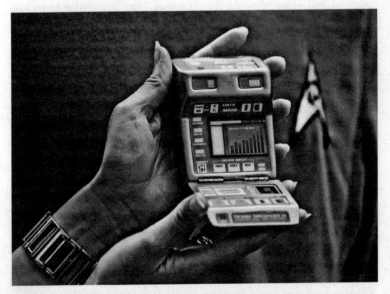

In the future, we will have tricorders—like these in *Star Trek*—that can diagnose almost any disease; portable MRI detectors and DNA chips will make this possible.

"Tissue engineering" is one of the hottest fields in medicine, making possible a "human body shop." So far, scientists can grow skin, blood, blood vessels, heart valves, cartilage, bone, noses, and ears in the lab from your own cells. The first major organ, the bladder, was grown in 2007, the first windpipe in 2009. So far, the only organs that have been grown are relatively simple, involving only a few types of tissues and few structures. Within five years, the first liver and pancreas might be grown, with enormous implications for public health. Nobel laureate Walter Gilbert told me that he foresees a time, just a few decades into the future, when practically every organ of the body will be grown from your own cells.

Tissue engineering grows new organs by first extracting a few cells from your body. These cells are then injected into a plastic mold that looks like a sponge shaped in the form of the organ in question. The plastic mold is made of biodegradable polyglycolic acid. The cells are treated with certain growth factors to stimulate cell growth, causing them to grow into the mold. Eventually, the mold disintegrates, leaving behind a perfect organ.

I had the opportunity to visit Anthony Atala's laboratory at Wake Forest University in North Carolina and witness this miraculous technology firsthand. As I walked through his laboratory, I saw bottles that contained living human organs. I could see blood vessels and bladders; I saw heart valves that were constantly opening and closing because liquids were being pumped through them. Seeing all these living human organs in bottles, I almost felt as if I were walking through Dr. Frankenstein's laboratory, but there were several crucial differences. Back in the nineteenth century, doctors were ignorant of the body's rejection mechanism, which makes it impossible to graft new organs. Plus, doctors did not know how to stop the infections that would inevitably contaminate any organ after surgery. So Atala, instead of creating a monster, is opening an entirely new lifesaving medical technology that may one day change the face of medicine.

One future target for his laboratory is to grow a human liver, perhaps within five years. The liver is not that complicated and consists of only a few types of tissue. Lab-grown livers could save thousands of lives, especially those in desperate need of liver transplants. It could also save the lives of alcoholics suffering from cirrhosis. (Unfortunately, it could also encourage people to keep bad habits, knowing that they can get replacement organs for their damaged ones.)

If organs of the body, like the windpipe and the bladder, can be grown now, what is to prevent scientists from growing every organ of the body? One basic problem is how to grow the tiny capillaries that provide blood for the cells. Every cell in the body has to be in contact with a blood supply. In addition, there is the problem of growing complex structures. The kidney, which purifies the blood of toxins, is composed of millions of tiny filters, so a mold for these filters is quite difficult to create.

But the most difficult organ to grow is the human brain. Although re-creating or growing a human brain seems unlikely for decades to come, it may instead be possible to inject young cells directly into the brain, which will incorporate them into the brain's neural network. This injection of new brain cells, however, is random, so the patient will have to relearn many basic functions. But because the brain is "plastic"—that is, it constantly rewires itself after it learns a new task—it might be able to integrate these new neurons so that they fire correctly.

STEM CELLS

One step beyond this is to apply stem cell technology. So far, the human organs were grown using cells that were not stem cells but were cells specially treated to proliferate inside molds. In the near future, it should be possible to use stem cells directly.

Stem cells are the "mother of all cells," and have the ability to change into any type of cell of the body. Each cell in our body has the complete genetic code necessary to create our entire body. But as our cells mature, they specialize, so many of the genes are inactivated. For example, although a skin cell may have the genes to turn into blood, these genes are turned off when an embryonic cell becomes an adult skin cell.

But embryonic stem cells retain this ability to regrow any type of cell throughout their life. Although embryonic stem cells are more highly prized by scientists, they are also more controversial, since an embryo has to be sacrificed in order to extract these cells, raising ethical issues. (However, Lanza and his colleagues have spearheaded ways in which to take adult stem cells, which have already turned into one type of cell, and then turn them into embryonic stem cells.)

Stem cells have the potential to cure a host of diseases, such as diabetes, heart disease, Alzheimer's, Parkinson's, even cancer. In fact, it is difficult to think of a disease in which stem cells will not have a major impact. One particular area of research is spinal cord injury, once thought to be totally incurable. In 1995, when the handsome actor Christopher Reeve suffered a severe spinal cord injury that left him totally paralyzed, there was no cure. However, in animal studies, great strides have been made in repairing the spinal cord with stem cells.

For example, Stephen Davies of the University of Colorado has had impressive success in treating spinal cord injuries in rats. He says, "I conducted some experiments where we transplanted adult neurons directly into adult central nervous systems. Real Frankenstein experiments. To our great surprise, adult neurons were able to send new nerve fibers from one side of the brain to the other in just one week." In treating spinal cord injury, it was widely thought that any attempt to repair the nerves would create great pain and distress as well. Davies found that a key type of nerve cell, called an astrocyte, occurs in two varieties, with different outcomes.

Davies says, "By using the right astrocytes to repair spinal cord injuries, we have all the gains without the pain, while these other types of appear to provide the opposite—pain but no gain." Moreover, the same techniques he is pioneering with stem cells will also work on victims of strokes and Alzheimer's and Parkinson's diseases, he believes.

Since virtually every cell of the body can be created by altering embryonic stem cells, the possibilities are endless. However, Doris Taylor, director of the Center for Cardiovascular Repair at the University of Minnesota, cautions that much work has yet to be done. "Embryonic stem cells represent the good, the bad, and the ugly. When they are good, they can be grown to large numbers in the lab and used to give rise to tissues, organs, or body parts. When they are bad, they don't know when to stop growing and give rise to tumors. The ugly—well, we don't understand all the cues, so we can't control the outcome, and we aren't ready to use them without more research in the lab," she notes.

This is one of the main problems facing stem cell research: the fact that these stem cells, without chemical cues from the environment, might continue to proliferate wildly until they become cancerous. Scientists now realize that the subtle chemical messages that travel between cells, telling them when and where to grow and stop growing, are just as important as the cell itself.

Nonetheless, slow but real progress is being made, especially in animal studies. Taylor made headlines in 2008 when her team, for the first time in history, grew a beating mouse heart almost from scratch. Her team started with a mouse heart and dissolved the cells within that heart, leaving only the scaffolding, a heart-shaped matrix of proteins. Then they planted a mixture of heart stem cells into that matrix, and watched as the stem cells began to proliferate inside the scaffolding. Previously, scientists were able to grow individual heart cells in a petri dish. But this was the first time that an actual beating heart was grown in the laboratory.

Growing the heart was also an exciting personal event for her. She said, "It's gorgeous. You can see the whole vascular tree, from arteries to the tiny veins that supply blood to every single heart cell."

There is also one part of the U.S. government that is keenly interested in making breakthroughs in the area of tissue engineering: the U.S. Army. In past wars, the death rate on the battlefield was appalling, with entire regiments and battalions decimated and many dying of wounds. Now rapid-response medical evacuation teams fly the wounded from Iraq and Afghanistan to Europe or the United States, where they receive top-notch medical care. The survival rate for GIs has skyrocketed. And so has the number of soldiers who have lost arms and limbs. As a consequence, the U.S. Army has made it a priority to find a way to grow back limbs.

One breakthrough made by the Armed Forces Institute of Regenerative Medicine has been to use a radically new method of growing organs. Scientists have long known that salamanders have remarkable powers of regeneration, regrowing entire limbs after they are lost. These limbs grow back because salamander stem cells are stimulated to make new limbs. One theory that has borne fruit is being explored by Stephen Badylak of the University of Pittsburgh, who has successfully regrown fingertips. His team has created a "pixie dust" with the miraculous power of regrowing tissue. This dust is created not from cells but from the extracellular matrix that exists between cells. This matrix is important because it contains the signals that tell the stem cells to grow in a particular fashion. When this pixie dust is applied to a fingertip that has been cut off, it will stimulate not just the fingertip but also the nail, leaving an almost perfect copy of the original finger. Up to one-third of an inch of tissue and nail has been grown in this fashion. The next goal is to extend this process to see if an entire human limb can be regrown, just like the salamanders'.

CLONING

If we can grow various organs of the human body, then can we regrow an entire human being, creating an exact genetic copy, a clone? The answer is yes, in principle, but it has not been done, despite numerous reports to the contrary.

Clones are a favorite theme in Hollywood movies, but they usually get the science backward. In the movie *The 6th Day,* Arnold Schwarzenegger's character battles the bad guys who have mastered the art of cloning human beings. More important, they have mastered the art of copying a person's entire memory and then inserting it into the clone. When Schwarzenegger manages to eliminate one bad guy, a new one rises up with the same personality and memory. Things get messy when he finds out that a clone was made of him without his knowledge. (In reality, when an animal is cloned, the memories are not.)

The concept of cloning hit the world headlines in 1997, when Ian Wilmut of the Roslin Institute of the University of Edinburgh was able to clone Dolly the sheep. By taking a cell from an adult sheep, extracting the DNA within its nucleus, and then inserting this nucleus into an egg cell, Wilmut was able to accomplish the feat of bringing back a genetic copy of the original. I once asked him if he'd had any idea of the media firestorm that would be ignited by his historic discovery. He said no. He clearly understood the medical importance of his work but underestimated the public's fascination with his discovery.

Soon, groups around the world began to duplicate this feat, cloning a wide variety of animals, including mice, goats, cats, pigs, dogs, horses, and cattle. I once went with a BBC camera crew and visited Ron Marquess just outside Dallas, Texas, who has one of the largest cloned-cattle farms in the country. At the ranch, I was amazed to see first-, second-, and even third-generation cloned cattle—clones of clones of clones. Marquess told me that they would have to invent a new vocabulary to keep track of the various generations of cloned cattle.

One group of cattle caught my eye. There were about eight identical twins, all lined up. They walked, ran, ate, and slept precisely in a row. Although the calves had no conception they were clones of one another, they instinctively banded together and mimicked one another's motions.

Marquess told me that cloning cattle was potentially a lucrative business. If you have a bull with superior physical characteristics, then it could fetch a handsome price if it was used for breeding. But if the bull died, then its genetic line would be lost with it unless its sperm had been collected and refrigerated. With cloning, one could keep the genetic line of prized bulls alive forever.

Although cloning has commercial applications for animals and animal husbandry, the implications for humans are less clear. Although there have been a number of sensational claims that human cloning has been achieved, all of them are probably bogus. So far, no one has successfully cloned a primate, let alone a human. Even cloning animals has proven to be difficult, given that hundreds of defective embryos are created for every one that reaches full term.

And even if human cloning becomes possible, there are social obstacles. First of all, many religions will oppose human cloning, similar to the way the Catholic Church opposed test tube babies back in 1978, when Louise Brown became the first baby in history to be conceived in a test tube. This means that laws will probably be passed banning the technology, or at least tightly regulating it. Second, the commercial demand for human cloning will be small. At most, probably only a fraction of the human race will be clones, even if it is legal. After all, we already have clones, in the form of identical twins (and triplets), so the novelty of human cloning will gradually wear off.

Originally, the demand for test tube babies was enormous, given the legions of infertile couples. But who will clone a human? Perhaps parents mourning the death of a child. Or, more likely, a wealthy, elderly man on his deathbed who has no heirs—or no heirs he particularly cares for—and wants to will all his money to himself as a child, in order to start all over again.

So in the future, although there might be laws passed preventing it, human clones will probably exist. However, they will represent only a tiny fraction of the human race and the social consequences will be quite small.

GENE THERAPY

Francis Collins, the current director of the National Institutes of Health and the man who led the government's historic Human Genome Project, told me that "all of us have about a half-dozen genes which are pretty screwed up." In the ancient past, we simply had to suffer from these often lethal genetic defects. In the future, he told me, we will cure many of them via gene therapy.

Genetic diseases have haunted humanity since the dawn of history, and at key moments may actually have influenced the course of history. For example, because of inbreeding among the royal families of Europe, genetic diseases have plagued generations of nobility. George III of England, for example, most likely suffered from acute intermittent porphyria, which causes temporary bouts of insanity. Some historians have speculated that this aggravated his relationship with the colonies, prompting them to declare their independence from England in 1776.

Queen Victoria was a carrier of the hemophilia gene, which causes uncontrolled bleeding. Because she had nine children, many of whom married into other royal houses of Europe, this spread the "royal disease" across the Continent. In Russia, Queen Victoria's great-grandson Alexis, the son of Nicholas II, suffered from hemophilia, which could seemingly be temporarily controlled by the mystic Rasputin. This "mad monk" gained enough power to paralyze the Russian nobility, delay badly needed reforms, and, as some historians have speculated, help bring about the Bolshevik Revolution of 1917.

But in the future, gene therapy may be able to cure many of the 5,000 known genetic diseases, such as cystic fibrosis (which afflicts northern Europeans), Tay-Sachs disease (which affects Eastern European Jews), and sickle cell anemia (which afflicts African Americans). In the near future, it should be possible to cure many genetic diseases that are caused by the mutation of a single gene.

Gene therapy comes in two types: somatic and germ line.

Somatic gene therapy involves fixing the broken genes of a single individual. The therapeutic value disappears when the individual dies. More controversial is germ-line gene therapy, in which one fixes the genes of the sex cells, so that the repaired gene can be passed on to the next generation, almost forever.

Curing genetic disease follows a long but well-established route. First, one must find victims of a certain genetic disease and then painstakingly trace their family trees, going back many generations. By analyzing the genes of these individuals, one then tries to determine the precise location of the gene that may be damaged.

Then one takes a healthy version of that gene, inserts it into a "vector" (usually a harmless virus), and then injects it into the patient. The virus quickly inserts the "good gene" into the cells of the patient, potentially curing the patient of this disease. By 2001, there were more than 500 gene therapy trials under way or under review throughout the world.

However, progress has been slow and the results mixed. One problem is that the body often confuses this harmless virus, containing the "good gene," with a dangerous virus and begins to attack it. This causes side effects that can negate the effect of the good gene. Another problem is that not enough of the virus inserts the good gene into its target cells correctly, so that the body cannot produce enough of the proper protein.

Despite these complications, scientists in France announced in 2000 that they were able to cure children with severe combined immunodeficiency (SCID), who were born without a functioning immune system. Some SCID patients, like "David the bubble boy," must live inside sterile plastic bubbles for the rest of their lives. Without an immune system, any illness could prove fatal. Genetic analyses of these patients show that their immune cells did indeed incorporate the new gene, as planned, hence activating their immune systems.

But there have been setbacks. In 1999, at the University of Pennsylvania, one patient died in a gene therapy trial, causing soul-searching within the medical community. It was the first death among the 1,100 patients undergoing this type of gene therapy. And by 2007, four of the ten patients who had been cured of one particular form of SCID developed a severe side effect, leukemia. Research in gene therapy for SCID is now focused on curing the disease without accidentally triggering a gene that can cause cancer. To date, seventeen patients who suffered from a different variety of SCID are free of both SCID and cancer, making it one of the few successes in this field.

One target for gene therapy is actually cancer. Almost 50 percent of all common cancers are linked to a damaged gene, p53. The p53 gene is long and complex; this makes it more probable that it will be damaged by environmental and chemical factors. So many gene therapy experiments are being conducted to insert a healthy p53 gene into patients. For example, cigarette smoke often causes characteristic mutations in three well-known sites within the p53 gene. Thus gene therapy, by replacing the damaged p53 gene, may one day be able to cure certain forms of lung cancer.

Progress has been slow but steady. In 2006, scientists at the National Institutes of Health in Maryland were able to successfully treat metastatic melanoma, a form of skin cancer, by altering killer T cells so that they specifically targeted cancer cells. This is the first study to show that gene therapy can be successfully used against some form of cancer. And in 2007, doctors at the University College and Moorfields Eye Hospital in London were able to use gene therapy to treat a certain form of inherited retinal disease (caused by mutations in the RPE65 gene).

Meanwhile, some couples are not waiting for gene therapy but are taking their genetic heritage into their own hands. A couple can create several fertilized embryos using in vitro fertilization. Each embryo can be tested for a specific genetic disease, and the couple can select the embryo free of the genetic disease to implant in the mother. In this way, genetic diseases can gradually be eliminated without using expensive gene therapy techniques. This process is currently being done with some Orthodox Jews in Brooklyn who have a high risk of Tay-Sachs disease.

One disease, however, will probably remain deadly throughout this century—cancer.

COEXISTING WITH CANCER

Back in 1971, President Richard Nixon, amid great fanfare and publicity, solemnly announced a war on cancer. By throwing money at cancer, he believed a cure would soon be at hand. But forty years (and $200 billion) later, cancer is the second-leading cause of death in the United States, responsible for 25 percent of all deaths. The death rate from cancer has dropped only 5 percent from 1950 to 2005 (adjusting for age and other factors). It is estimated that cancer will claim the lives of 562,000 Americans this year alone, or more than 1,000 people per day. Cancer rates have fallen for a few types of the disease but have remained stubbornly flat in others. And the treatment for cancer, involving poisoning, slicing, and zapping human tissue, leaves a trail of tears for the patients, who often wonder which is worse, the disease or the treatment.

In hindsight, we can see what went wrong. Back in 1971, before the revolution in genetic engineering, the causes of cancer were a total mystery.

Now scientists realize that cancer is basically a disease of our genes.

Whether caused by a virus, chemical exposure, radiation, or chance, cancer fundamentally involves mutations in four or more of our genes, in which a normal cell "forgets how to die." The cell loses control over its reproduction and reproduces without limit, eventually killing the patient. The fact that it takes a sequence of four or more defective genes to cause cancer probably explains why it often kills decades after an original incident. For example, you might have a severe sunburn as a child. Many decades later, you might develop skin cancer at that same site. This means it probably took that long for the other mutations to occur and finally tip the cell into a cancerous mode.

There are at least two major types of these cancer genes, oncogenes and tumor suppressors, which function like the accelerator and brakes of a car. The oncogene acts like an accelerator stuck in the down position, so the car careens out of control, allowing the cell to reproduce without limit. The tumor suppressor normally acts like a brake, so when it is damaged, the cell is like a car that can't stop.

The Cancer Genome Project plans to sequence the genes of most cancers. Since each cancer requires sequencing the human genome, the Cancer Genome Project is hundreds of times more ambitious than the original Human Genome Project.

Some of the first results of this long-awaited Cancer Genome Project were announced in 2009 concerning skin and lung cancer. The results were startling. Mike Stratton of the Wellcome Trust Sanger Institute said, "What we are seeing today is going to transform the way that we see cancer. We have never seen cancer revealed in this form before."

Cells from a lung cancer cell had an astounding 23,000 individual mutations, while the melanoma cancer cell had 33,000 mutations. This means that a typical smoker develops one mutation for every fifteen cigarettes he or she smokes. (Lung cancer kills 1 million people every year around the world, mostly from smoking.)

The goal is to genetically analyze all types of cancers, of which there are more than 100. There are many tissues in the body, all of which can become cancerous; many types of cancers for each tissue; and tens of thousands of mutations within each type of cancer. Since each cancer involves tens of thousands of mutations, it will take many decades to isolate precisely which of these mutations causes the cell mechanism to go haywire. Scientists will develop cures for a wide variety of cancers but no one cure for all of them, since cancer itself is like a collection of diseases.

New treatments and therapies will also continually enter the market, all of them designed to hit cancer at its molecular and genetic roots. Some of the promising ones include:

- antiangiogenesis, or choking off the blood supply of a tumor so that it never grows
- nanoparticles, which are like "smart bombs" directed at cancer cells
- gene therapy, especially for gene p53
- new drugs that target just the cancer cells
- new vaccinations against viruses that can cause cancer, like the human papillomavirus (HPV), which can cause cervical cancer

Unfortunately, it is unlikely that we will find a magic bullet for cancer. Rather, we will cure cancer one step at a time. More than likely, the major reduction in death rates will come when we have DNA chips scattered throughout our environment, constantly monitoring us for cancer cells years before a tumor forms.

As Nobel laureate David Baltimore notes, "Cancer is an army of cells that fights our therapies in ways that I'm sure will keep us continually in the battle."

MIDCENTURY (2030 TO 2070)

GENE THERAPY

Despite the setbacks in gene therapy, researchers believe steady gains will be made into the coming decades. By midcentury, many think, gene therapy will be a standard method of treating a variety of genetic diseases. Much of the success that scientists have had in animal studies will eventually be translated into human studies.

So far, gene therapy has targeted diseases caused by mutations in a single gene. They will be the first to be cured. But many diseases are caused by mutations in multiple genes, along with triggers from the environment. These are much more difficult to treat, but they include such important diseases as diabetes, schizophrenia, Alzheimer's, Parkinson's, and heart disease. All of them show definite genetic patterns, but no single gene is responsible. For example, it is possible to have a schizophrenic whose identical twin is normal.

Over the years, there have been a number of announcements that scientists have been able to isolate some of the genes involved in schizophrenia by following the genetic history of certain families. However, it is embarrassing that these results are often not verifiable by other independent studies. So these results are flawed, or perhaps many genes are involved in schizophrenia. Plus, certain environmental factors seem to be involved.

By midcentury, gene therapy should become a well-established therapy, at least for diseases caused by single genes. But patients might not be content with just fixing genes. They may also want to improve them.

DESIGNER CHILDREN

By midcentury, scientists will go beyond just fixing broken genes to actually enhancing and improving them.

The desire to have superhuman ability is an ancient one, rooted deeply in Greek and Roman mythology and our dreams. The great hero Hercules, one of the most popular of all the Greek and Roman demigods, got his great powers not from exercise and diet but by an injection of divine genes. His mother was a beautiful mortal, Alcmene, who one day caught the attention of Zeus, who disguised himself as her husband to make love to her. When she became pregnant with his child, Zeus announced that the baby would one day become a great warrior. But Zeus's wife, Hera, became jealous and secretly schemed to kill the baby by delaying his birth. Alcmene almost died in agony during a prolonged labor, but Hera's plot was exposed at the last minute and Alcmene delivered an unusually large baby. Half man and half god, Hercules inherited the godlike strength of his father to accomplish heroic, legendary feats.

In the future, we might not be able to create divine genes, but we certainly will be able to create genes that will give us superhuman abilities. And like Hercules' difficult delivery, there will be many difficulties bringing this technology to fruition.

By midcentury, "designer children" could become a reality. As Harvard biologist E. O. Wilson has said, "*Homo sapiens,* the first truly free species, is about to decommission natural selection, the force that made us. . . . Soon we must look deep within ourselves and decide what we wish to become."

Already, scientists are teasing apart the genes that control basic functions. For example, the "smart mouse" gene, which increases the memory and performance of mice, was isolated in 1999. Mice that have the smart gene are better able to navigate mazes and remember things.

Scientists at Princeton University such as Joseph Tsien have created a strain of genetically altered mice with an extra gene called NR2B that helps to trigger the production of the neurotransmitter N-methyl-D-aspartate (NMDA) in the forebrain of mice. The creators of the smart mice have christened them Doogie mice (after the TV character Doogie Howser, MD).

These smart mice outperformed normal mice on a variety of tests. If a mouse is placed in a vat of milky water, it must find a platform hidden just beneath the surface where it can rest. Normal mice forget where this platform is and swim randomly around the vat, while smart mice make a beeline to it on the first try. If the mice are shown two objects, one an old one and one a new one, the normal mice do not pay attention to the new object. But the smart mice immediately recognize the presence of this new object.

What is most important is that scientists understand how these smart mice genes work: they regulate the synapses of the brain. If you think of the brain as a vast collection of freeways, then the synapse would be equivalent to a toll booth. If the toll is too high, then cars cannot pass through the gate: a message is stopped within the brain. But if the toll is low, then cars can pass and the message is transmitted through the brain. Neurotransmitters like NMDA lower the toll at the synapse, making it possible for messages to pass freely. The smart mice have two copies of the NR2B gene, which in turn helps to produce the NMDA neurotransmitter.

These smart mice verify Hebb's rule: learning takes place when certain neural pathways are reinforced. Specifically, these pathways could be reinforced by regulating the synapses that connect two nerve fibers, making it easier for signals to cross a synapse.

This result may help to explain certain peculiarities about learning. It's been known that aging animals have a reduced ability to learn. Scientists see this throughout the animal kingdom. This might be explained because the NR2B gene becomes less active with age.

Also, as we saw earlier with Hebb's rule, memories might be created when neurons form a strong connection. This might be true, since activating the NMDA receptor creates a strong connection.

MIGHTY MOUSE GENE

In addition, the "mighty mouse gene" has been isolated, which increases the muscle mass so that the mouse appears to be musclebound. It was first found in mice with unusually large muscles. Scientists now realize that the key lies in the myostatin gene, which helps to keep muscle growth in check. But in 1997, scientists found that when the myostatin gene is silenced in mice, muscle growth expands enormously.

Another breakthrough was made soon afterward in Germany, when scientists examined a newborn boy who had unusual muscles in his upper legs and arms. Ultrasound analysis showed that this boy's muscles were twice as large as normal. By sequencing the genes of this baby and of his mother (who was a professional sprinter), they found a similar genetic pattern. In fact, an analysis of the boy's blood showed no myostatin whatsoever.

Scientists at the Johns Hopkins Medical School were at first eager to make contact with patients suffering from degenerative muscle disorders who might benefit from this result, but they were disappointed to find that half the telephone calls to their office came from bodybuilders who wanted the gene to bulk themselves up, regardless of the consequences. Perhaps these bodybuilders were recalling the phenomenal success of Arnold Schwarzenegger, who has admitted to using steroids to jump-start his meteoric career. Because of the intense interest in the myostatin gene and ways to suppress it, even the Olympic Committee was forced to set up a special commission to look into it. Unlike steroids, which are relatively easy to detect via chemical tests, this new method, because it involves genes and the proteins they create, is much more difficult to detect.

Studies done on identical twins who have been separated at birth show that there is a wide variety of behavioral traits influenced by genetics. In fact, these studies show that roughly 50 percent of a twin's behavior is influenced by genes, the other 50 percent by environment. These traits include memory, verbal reasoning, spatial reasoning, processing speed, extroversion, and thrill seeking.

Even behaviors once thought to be complex are now revealing their genetic roots. For example, prairie voles are monogamous. Laboratory mice are promiscuous. Larry Young at Emory University shocked the world of biotechnology by showing that the transfer of one gene from prairie voles could create mice that exhibited monogamous characteristics. Each animal has a different version of a certain receptor for a brain peptide associated with social behavior and mating. Young inserted the vole gene for this receptor into the mice and found that the mice then exhibited behaviors more like the monogamous voles.

Young said, "Although many genes are likely to be involved in the evolution of complex social behaviors such as monogamy . . . changes in the expression of a single gene can have an impact on the expression of components of these behaviors, such as affiliation."

Depression and happiness may also have genetic roots. It has long been known that there are people who are happy even though they may have suffered tragic accidents. They always see the brighter side of things, even in the face of setbacks that may devastate another individual. These people also tend to be healthier than normal. Harvard psychologist Daniel Gilbert told me that there is a theory that might explain this. Perhaps we are born with a "happiness set point." Day by day we may oscillate around this set point, but its level is fixed at birth. In the future, via drugs or gene therapy, one may be able to shift this set point, especially for those who are chronically depressed.

SIDE EFFECTS OF THE BIOTECH REVOLUTION

By midcentury, scientists will be able to isolate and alter many of the single genes that control a variety of human characteristics. But this does not mean humanity will immediately benefit from them. There is also the long, hard work of ironing out side effects and unwanted consequences, which will take decades.

For example, Achilles was invincible in combat, leading the victorious Greeks in their epic battle with the Trojans. However, his power had a fatal flaw. When he was a baby, his mother dipped him into the magic river Styx in order to make him invincible. Unfortunately, she had to hold him by the heel when she placed him into the river, leaving that one crucial point of vulnerability. Later, he would die during the Trojan War after being hit in the heel by an arrow.

Today, scientists are wondering if the new strains of creatures emerging from their laboratories also have a hidden Achilles' heel. For example, today there are about thirty-three different "smart mouse" strains that have enhanced memory and performance. However, there is an unexpected side effect of having enhanced memory; smart mice are sometimes paralyzed by fear. If they are exposed to an extremely mild electric shock, for example, they will shiver in terror. "It's as if they remember too much," says Alcino Silva of UCLA, who developed his own strain of smart mice. Scientists now realize that forgetting may be as important as remembering in making sense of this world and organizing our knowledge. Perhaps we have to throw out a lot of files in order to organize our knowledge.

This is reminiscent of a case from the 1920s, documented by Russian neurologist A. R. Luria, of a man who had a photographic memory. After just a single reading of Dante's *Divine Comedy,* he had memorized every word. This was helpful in his work as a newspaper reporter, but he was incapable of understanding figures of speech. Luria observed, "The obstacles to his understanding were overwhelming: each expression gave rise to an image; this, in turn, would conflict with another image that had been evoked."

In fact, scientists believe that there has to be a balance between forgetting and remembering. If you forget too much, you may be able to forget the pain of previous mistakes, but you also forget key facts and skills. If you remember too much, you may be able to remember important details, but you might be paralyzed by the memory of every hurt and setback. Only a trade-off between these two may yield optimal understanding.

Bodybuilders are already flocking to different drugs and therapies that promise them fame and glory. The hormone erythropoietin (EPO) works by making more oxygen-containing red blood cells, which means increased endurance. Because EPO thickens the blood, it also has been linked to strokes and heart attacks. Insulin-like growth factors (IGF) are useful because they help proteins to bulk up muscles, but they have been linked to tumor growth.

Even if laws are passed banning genetic enhancements, they will be difficult to stop. For example, parents are genetically hardwired by evolution to want to give every advantage to their children. On the one hand, this might mean giving them violin, ballet, and sports lessons. But on the other hand, this might mean giving them genetic enhancements to improve their memory, attention span, athletic ability, and perhaps even their looks. If parents find out that their child is competing with a neighbor's child who is rumored to have been genetically enhanced, there will be enormous pressure to give the same benefit to their child.

As Gregory Benford has said, "We all know that good-looking people do well. What parents could resist the argument that they were giving the child a powerful leg up (maybe literally) in a brave new competitive world?"

By midcentury, genetic enhancements may become commonplace. In fact, genetic enhancements may even be indispensable if we are to explore the solar system and live on inhospitable planets.

Some say that we should use designer genes to make us healthier and happier. Others say that we should allow for cosmetic enhancements. The big question will be how far this will go. In any event, it may become increasingly difficult to control the spread of "designer genes" that enhance looks and performance. We don't want the human race to split into different genetic factions, the enhanced and the unenhanced, but society will have to democratically decide how far to push this technology.

Personally, I believe that laws will be passed to regulate this powerful technology, possibly to allow gene therapy when it cures disease and allows us to lead productive lives, but to restrict gene therapy for purely cosmetic reasons. This means that a black market might eventually develop to skirt these laws, so we might have to adjust to a society in which a small fraction of the population is genetically enhanced.

For the most part, this might not be a disaster. Already, one can use plastic surgery to improve appearance, so using genetic engineering to do this may be unnecessary. But the danger may arise when one tries to genetically change one's personality. There are probably many genes that influence behavior, and they interact in complex ways, so tampering with behavioral genes may create unintended side effects. It may take decades to sort through all these side effects.

But what about the greatest gene enhancement of all, extending the human life span?

FAR FUTURE (2070 TO 2100)

REVERSING AGING

Throughout history, kings and warlords had the power to command entire empires, but there was one thing that was forever beyond their control: aging. Hence, the search for immortality has been one of the oldest quests in human history.

In the Bible, God banishes Adam and Even from the Garden of Eden for disobeying his orders concerning the apple of knowledge. God's fear was that Adam and Eve might use this knowledge to unlock the secret of immortality and become gods themselves. In Genesis 3:22, the Bible reads, "Behold, the man is become as one of us, to know good and evil: and now, lest he put forth his hand, and take also of the tree of life, and eat, and live for ever."

Besides the Bible, one of the oldest and greatest tales in human civilization, dating back to the twenty-seventh century BC, is *The Epic of Gilgamesh*, about the great warrior of Mesopotamia. When his lifelong, loyal companion suddenly died, Gilgamesh decided to embark upon a journey to find the secret of immortality. He heard rumors that a wise man and his wife had been granted the gift of immortality by the gods, and were, in fact, the only ones in their land to have survived the Great Flood. After an epic quest, Gilgamesh finally found the secret of immortality, only to see a serpent snatch it away at the last minute.

Because *The Epic of Gilgamesh* is one of the oldest pieces of literature, historians believe that this search for immortality was the inspiration for the Greek writer Homer to write the *Odyssey*, and also for Noah's flood mentioned in the Bible.

Many early kings—like Emperor Qin, who unified China around 200 BC—sent huge fleets of ships to find the Fountain of Youth, but all failed. (According to mythology, Emperor Qin gave instructions to his fleet not to come back if they failed to find the Fountain of Youth. Unable to find the fountain, but too afraid to return, they founded Japan instead.)

For decades, most scientists believed that life span was fixed and immutable, beyond the reach of science. Within the last few years, this view has crumbled under the onslaught of a stunning series of experimental results that have revolutionized the field. Gerontology, once a sleepy, backwater area of science, has now become one of the hottest fields, attracting hundreds of millions of dollars in research funds and even raising the possibility of commercial development.

The secrets of the aging process are now being unraveled, and genetics will play a vital role in this process. Looking at the animal kingdom, we see a vast variety of life spans. For example, our DNA differs from that of our nearest genetic relative, the chimpanzee, by only 1.5 percent, yet we live 50 percent longer. By analyzing the handful of genes separating us from the chimpanzees, we may be able to determine why we live so much longer than our genetic relative.

This, in turn, has given us a "unified theory of aging" that brings the various strands of research into a single, coherent tapestry. Scientists now know what aging is. It is the accumulation of errors at the genetic and cellular level. These errors can build up in various ways. For example, metabolism creates free radicals and oxidation, which damage the delicate molecular machinery of our cells, causing them to age; errors can build up in the form of "junk" molecular debris accumulating inside and outside the cells.

The buildup of these genetic errors is a by-product of the second law of thermodynamics: total entropy (that is, chaos) always increases. This is why rusting, rotting, decaying, etc., are universal features of life. The second law is inescapable. Everything, from the flowers in the field to our bodies and even the universe itself, is doomed to wither and die.

But there is a small but important loophole in the second law that states *total* entropy always increases. This means that you can actually reduce entropy in one place and reverse aging, as long as you increase entropy somewhere else. So it's possible to get younger, at the expense of wreaking havoc elsewhere. (This was alluded to in Oscar Wilde's famous novel *The Picture of Dorian Gray*. Mr. Gray was mysteriously eternally young. But his secret was the painting of himself that aged horribly. So the total amount of aging still increased.) The principle of entropy can also be seen by looking behind a refrigerator. Inside the refrigerator, entropy decreases as the temperature drops. But to lower the entropy, you have to have a motor, which increases the heat generated behind the refrigerator, increasing the entropy outside the machine. That is why refrigerators are always hot in the back.

As Nobel laureate Richard Feynman once said, "There is nothing in biology yet found that indicates the inevitability of death. This suggests to me that it is not at all inevitable and that it is only a matter of time before biologists discover what it is that is causing us the trouble and that this terrible universal disease or temporariness of the human's body will be cured."

The second law can also be seen by the action of the female sex hormone estrogen, which keeps women young and vibrant until they hit menopause, when aging accelerates and the death rate increases. Estrogen is like putting high-octane fuel into a sports car. The car performs beautifully but at the price of causing more wear and tear on the engine. For women, this cellular wear and tear might be manifested in breast cancer. In fact, injections of estrogen are known to accelerate the growth of breast cancer. So the price women pay for youth and vigor before menopause is possibly an increase in total entropy, in this case, breast cancer. (There have been scores of theories proposed to explain the recent rise in breast cancer rates, which are still quite controversial. One theory says that this is in part related to the total number of menstrual cycles a woman has. Throughout ancient history, after puberty women were more or less constantly pregnant until they hit menopause, and then they died soon afterward. This meant they had few menstrual cycles, low levels of estrogen, and hence, possibly, a relatively low level of breast cancer. Today, young girls reach puberty earlier, have many menstrual cycles, bear an average of only 1.5 children, live past menopause, and hence have considerably more exposure to estrogen, leading to a possible rise in the occurrence of breast cancer.)

Recently, a series of tantalizing clues has been discovered about genes and aging. First, researchers have shown that it is possible to breed generations of animals that live longer than normal. In particular, yeast cells, nematode worms, and fruit flies can be bred in the laboratory to live longer than normal. The scientific world was stunned when Michael Rose of the University of California at Irvine announced that he was able to increase the life span of fruit flies by 70 percent by selective breeding. His "superflies," or Methuselah flies, were found to have higher quantities of the antioxidant superoxide dismutase (SOD), which can slow down the damage caused by free radicals. In 1991, Thomas Johnson of the University of Colorado at Boulder isolated a gene, which he dubbed age-1, that seems to be responsible for aging in nematodes and increases their life spans by 110 percent. "If something like age-1 exists in humans, we might really be able to do something spectacular," he noted.

Scientists have now isolated a number of genes (age-1, age-2, daf-2) that control and regulate the aging process in lower organisms, but these genes have counterparts in humans as well. In fact, one scientist remarked that changing the life span of yeast cells was almost like flicking on a light switch. When one activated a certain gene, the cells lived longer. When you deactivated it, they lived shorter lives.

Breeding yeast cells to live longer is simple compared to the onerous task of breeding humans, who live so long that testing is almost impossible. But isolating the genes responsible for aging could accelerate in the future, especially when all of us have our genomes on a CD-ROM. By then, scientists will have a tremendous database of billions of genes that can be analyzed by computers. Scientists will be able to scan millions of genomes of two groups of people, the young and the old. By comparing the two sets, one can then identify where aging takes place at the genetic level. A preliminary scan of these genes has already isolated about sixty genes on which aging seems to be concentrated.

For example, scientists know that longevity tends to run somewhat in families. People who live long tend to have parents who also lived long. The effect is not dramatic, but it can be measured. Scientists who analyze identical twins who were separated at birth can also see this at the genetic level. But our life expectancy is not 100 percent determined by our genes. Scientists who have studied this believe that our life expectancy is only 35 percent determined by our genes. So in the future, when everyone has their own $100 personal genome, one may be able to scan the genomes of millions of people by computer to isolate the genes that partially control our life span.

Furthermore, these computer studies may be able to locate precisely where aging primarily takes place. In a car, we know that aging takes place mainly in the engine, where gasoline is oxidized and burned. Likewise, genetic analysis shows that aging is concentrated in the "engine" of the cell, the mitochondria, or the cell's power plant. This has allowed scientists to narrow the search for "age genes" and look for ways to accelerate the gene repair inside the mitochondria to reverse the effects of aging.

By 2050, it might be possible to slow down the aging process via a variety of therapies, for example, stem cells, the human body shop, and gene therapy to fix aging genes. We could live to be 150 or older. By 2100, it might be possible to reverse the effects of aging by accelerating cell repair mechanisms to live well beyond that.

CALORIC RESTRICTION

This theory may also explain the strange fact that caloric restriction (that is, lowering the calories we eat by 30 percent or more) increases the life span by 30 percent. Every organism studied so far—from yeast cells, spiders, and insects to rabbits, dogs, and now monkeys—exhibits this strange phenomenon. Animals given this restricted diet have fewer tumors, less heart disease, a lower incidence of diabetes, and fewer diseases related to aging. In fact, caloric restriction is the *only* known mechanism guaranteed to increase the life span that has been tested repeatedly, over almost the entire animal kingdom, and it works every time. Until recently, the only major species that still eluded researchers of caloric restriction were the primates, of which humans are a member, because they live so long.

Scientists were especially anxious to see the results of caloric restriction on rhesus monkeys. Finally, in 2009, the long-awaited results came in. The University of Wisconsin study showed that, after twenty years of caloric restriction, monkeys on the restricted diet suffered less disease across the board: less diabetes, cancer, heart disease. In general, these monkeys were in better health than their cousins who were fed a normal diet.

There is a theory that might explain this: Nature gives animals two "choices" concerning how they use their energy. During times of plenty, energy is used to reproduce. During times of famine, the body shuts down reproduction, conserves energy, and tries to ride out the famine. In the animal kingdom, the state of near starvation is a common one, and hence animals frequently make the "choice" of shutting down reproduction, slowing metabolism, living longer, and hoping for better days in the future.

The Holy Grail of aging research is to somehow preserve the benefits of caloric restriction without the downside (starving yourself). The natural tendency of humans apparently is to gain weight, not lose it. In fact, living on a calorically restricted diet is no fun; you are fed a diet that would make a hermit gag. Also, animals fed a particularly severe, restricted diet become lethargic, sluggish, and lose all interest in sex. What motivates scientists is the search for a gene that controls this mechanism, whereby we can reap the benefits of caloric restriction without the downside.

An important clue to this was found in 1991 by MIT researcher Leonard P. Guarente and others, who were looking for a gene that might lengthen the life span of yeast cells. Guarente, David Sinclair of Harvard, and coworkers discovered the gene SIR2, which is involved in bringing on the effects of caloric restriction. This gene is responsible for detecting the energy reserves of a cell. When the energy reserves are low, as during a famine, the gene is activated. This is precisely what you might expect in a gene that controls the effects of caloric restriction. They also found that the SIR2 gene has a counterpart in mice and in people, called the SIRT genes, which produce proteins called sirtuins. They then looked for chemicals that activate the sirtuins, and found the chemical resveratrol.

This was intriguing, because scientists also believe that resveratrol may be responsible for the benefits of red wine and may explain the "French paradox." French cooking is famous for its rich sauces, which are high in fats and oils, yet the French seem to have a normal life span. Perhaps this mystery can be explained because the French consume so much red wine, which contains resveratrol.

Scientists have found that sirtuin activators can protect mice from an impressive variety of diseases, including lung and colon cancer, melanoma, lymphoma, type 2 diabetes, cardiovascular disease, and Alzheimer's disease, according to Sinclair. If even a fraction of these diseases can be treated in humans via sirtuins, it would revolutionize all medicine.

Recently, a theory has been proposed to explain all the remarkable properties of resveratrol. According to Sinclair, the main purpose of sirtuin is to prevent certain genes from being activated. A single cell's chromosomes, for example, if fully stretched, would extend six feet, making an astronomically long molecule. At any time, only a portion of the genes along this six feet of chromosomes are necessary; all the rest must be inactive. The cell gags most of the genes when they are not needed by wrapping the chromosome tightly with chromatin, which is maintained by sirtuin.

Sometimes, however, there are catastrophic disruptions of these delicate chromosomes, like a total break in one of the strands. Then the sirtuins spring into action, helping to repair the broken chromosome. But when the sirtuins temporarily leave their posts to come to the rescue, they must abandon their primary job of silencing the genes. Hence, genes get activated, causing genetic chaos. This breakdown, Sinclair proposes, is one of the chief mechanisms for aging.

If this is true, then perhaps sirtuins can not only halt the advance of aging but also reverse it. DNA damage to our cells is difficult to repair and reverse. But Sinclair believes that much of our aging is caused by sirtuins that have been diverted from their primary task, allowing cells to degenerate. The diversion of these sirtuins can be easily reversed, he claims.

FOUNTAIN OF YOUTH?

One unwanted by-product of this discovery, however, has been the media circus that it sparked. Suddenly, *60 Minutes* and *The Oprah Winfrey Show* featured resveratrol, creating a stampede on the Internet, with fly-by-night companies springing up overnight, promising the elixir of life. It seems as if every snake oil salesman and charlatan wanted to jump on the resveratrol bandwagon.

(I had a chance to interview Guarente, the man who started this media stampede, in his laboratory. He was cautious in his statements, realizing the media impact that his results may have and the misconceptions that may develop. In particular, he was incensed that so many Internet sites are now advertising resveratrol as some sort of fountain of youth. It was appalling, he noted, that people were trying to cash in on the sudden fame that resveratrol has gotten, although most of the results are still tentative. However, he wouldn't rule out the possibility that one day, if the fountain of youth is ever found, assuming it even exists, then SIR2 may play a part. His colleague Sinclair, in fact, admits that he takes large quantities of resveratrol every day.)

Interest in aging research is so intense within the scientific community that Harvard Medical School sponsored a conference in 2009 that drew some of the major researchers in the field. In the audience were many who were personally undergoing caloric restriction. Looking gaunt and frail, they were putting their scientific philosophy to the test by restricting their diets. There were also members of the 120 Club, who intend to live to the age of 120. In particular, interest was focused on Sirtris Pharmaceuticals, cofounded by David Sinclair and Christoph Westphal, which is now putting some of their resveratrol substitutes through clinical trials. Westphal says flatly, "In five or six or seven years, there will be drugs that prolong longevity."

Chemicals that did not even exist a few years ago are the subject of intense interest as they go through trials. SRT501 is being tested against multiple myeloma and colon cancer. SRT2104 is being tested against type 2 diabetes. Not only sirtuins but also a host of other genes, proteins, and chemicals (including IGF-1, TOR, and rapamycin) are being closely analyzed by various groups.

Only time will tell if these clinical trials will be successful. The history of medicine is riddled with tales of deception, chicanery, and fraud when it comes to the aging process. But science, not superstition, is based on reproducible, testable, and falsifiable data. As the National Institute on Aging sets up programs to test various substances for their effects on aging, then we will see if these intriguing studies on animals carry over to humans.

DO WE HAVE TO DIE?

William Haseltine, a biotech pioneer, once told me, "The nature of life is not mortality. It's immortality. DNA is an immortal molecule. That molecule first appeared perhaps 3.5 billion years ago. That selfsame molecule, through duplication, is around today. . . . It's true that we run down, but we've talked about projecting way into the future the ability to alter that. First to extend our lives two- or threefold. And perhaps, if we understand the brain well enough, to extend both our body and our brain indefinitely. And I don't think that will be an unnatural process."

Evolutionary biologists point out that evolutionary pressure is placed on animals during their reproductive years. After an animal is past its reproductive years, it may in fact become a burden on the group and hence perhaps evolution has programmed it to die of old age. So perhaps we are programmed to die. But maybe we can reprogram ourselves to live longer.

Actually, if we look at mammals, for example, we find that the larger the mammal, the lower its metabolism rate, and the longer it lives. Mice, for example, burn up an enormous amount of food for their body weight, and live for only about four years. Elephants have a much slower metabolism rate and live to seventy. If metabolism corresponds to the buildup of errors, then this apparently agrees with the concept that you live longer if your metabolism rate is lower. (This may explain the expression "burning the candle at both ends." I once read a short story about a genie who offered to grant a man any wish he wanted. He promptly asked to live 1,000 years. The genie granted him his wish and turned him into a tree.)

Evolutionary biologists try to explain life span in terms of how longevity may help a species survive in the wild. To them, a specific life span is determined genetically because it helps the species to survive and flourish. Mice live so briefly, in their view, because they are constantly being hunted by a variety of predators and often freeze to death in winter. The mice that pass on their genes to the next generation are the ones that have the most offspring, not the ones who live longer. (If this theory is correct, then we expect that mice that can somehow fly away from predators would live longer. Indeed, bats, which are the same size as mice, live 3.5 times longer.)

But one anomaly comes from the reptiles. Apparently, certain reptiles have no known life span. They might even live forever. Alligators and crocodiles simply get larger and larger, but remain as vigorous and energetic as ever. (Textbooks often claim that alligators live to be only seventy years of age. But this is perhaps because the zookeeper died at age seventy. Other textbooks are more honest and simply say that the life span of these creatures is greater than seventy but has never been carefully measured under laboratory conditions.) In reality, these animals are not immortal, because they die of accidents, starvation, disease, etc. But if left in a zoo, they have enormous life spans, almost seeming to live forever.

BIOLOGICAL CLOCK

Another intriguing clue comes from the telomeres of a cell, which act like a "biological clock." Like the plastic tips at the ends of shoelaces, the telomeres are found at the ends of a chromosome. After every reproduction cycle, they get shorter and shorter. Eventually, after sixty or so reproductions (for skin cells), the telomeres unravel. The cell then enters senescence and ceases to perform properly. So the telomeres are like the fuse on a stick of dynamite. If the fuse gets shorter after each reproduction cycle, eventually the fuse disappears and the cell stops reproducing.

This is called the Hayflick limit, which seems to put an upper limit on the life cycle of certain cells. Cancer cells, for example, have no Hayflick limit and produce an enzyme called telomerase that prevents the telomeres from getting shorter and shorter.

The enzyme telomerase can be synthesized. When applied to skin cells, they apparently reproduce without limit. They become immortal.

However, there is a danger here. Cancer cells are also immortal, dividing without limit inside a tumor. In fact, that is why cancer cells are so lethal, because they reproduce without limit, until the body can no longer function. So the enzyme telomerase has to be analyzed carefully. Any therapy using telomerase to rewind the biological clock must be checked to make sure it does not cause cancer.

IMMORTALITY PLUS YOUTH

The prospect of extending the human life span is a source of joy for some and a horror for others, as we contemplate a population explosion and a society of decrepit elderly who will bankrupt the country.

A combination of biological, mechanical, and nanotechnological therapies may in fact not only increase our life span but also preserve our youth in the process. Robert A. Freitas Jr., who applies nanotechnology to medicine, has said, "Such interventions may become commonplace a few decades from today. Using annual checkups and cleanouts, and some occasional major repairs, your biological age could be restored once a year to the more or less constant physiological age that you select. You might still eventually die of accidental causes, but you'll live at least ten times longer than you do now."

In the future, extending the life span will not be a matter of drinking of the fabled Fountain of Youth. More likely, it will be a combination of several methods:

1. growing new organs as they wear out or become diseased, via tissue engineering and stem cells
2. ingesting a cocktail of proteins and enzymes that are designed to increase cell repair mechanisms, regulate metabolism, reset the biological clock, and reduce oxidation

3. using gene therapy to alter genes that may slow down the aging process
4. maintaining a healthy lifestyle (exercise and a good diet)
5. using nanosensors to detect diseases like cancer years before they become a problem

POPULATION, FOOD, AND POLLUTION

But one nagging question is: If life expectancy can be increased, then will we suffer from overpopulation? No one knows.

Delaying the aging process brings up a host of social implications. If we live longer, won't we overpopulate the earth? But some point out that the bulk of life extension has already happened, with life expectancy exploding from forty-five to seventy to eighty in just one century. Instead of creating a population explosion, it has arguably done the reverse. As people are living longer, they are pursuing careers and delaying childbearing. In fact, the native European population is actually decreasing dramatically. So if people live longer and richer lives, they might space out their children accordingly, and have fewer of them. With many more decades to live, people will reset their time frames accordingly, and hence space out or delay their children.

Others claim that people will reject this technology because it is unnatural and may violate their religious beliefs. Indeed, informal polls taken of the general population show that most people think that death is quite natural and helps to give life meaning. (However, most of the people interviewed in these polls are young to middle-aged. If you go to a nursing home, where people are wasting away, living with constant pain, and waiting to die and ask the same question, you might get an entirely different answer.)

As UCLA's Greg Stock says, "Gradually, our agonizing about playing God and our worries about longer life spans would give way to a new chorus: 'When can I get a pill?' "

In 2002, with the best demographic data, scientists estimated that 6 percent of all humans who have ever walked the face of the earth are still alive today. This is because the human population hovered at around 1 million for most of human history. Foraging for meager supplies of food kept the human population down. Even during the height of the Roman Empire, its population was estimated to be only 55 million.

But within the last 300 years, there has been a dramatic spike in world population coincident with the rise of modern medicine and the Industrial Revolution, which produced a bounty of food and supplies. And in the twentieth century, the world population soared to new heights, more than doubling from 1950 to 1992: from 2.5 billion to 5.5 billion. It now stands at 6.7 billion. Every year, 79 million people join the human race, which is more than the entire population of France.

As a result, many predictions of doomsday have been made, yet so far humanity has been able to dodge the bullet. Back in 1798, Thomas Malthus warned us what would happen when the population exceeded the food supply. Famines, food riots, the collapse of governments, and mass starvation could ensue until a new equilibrium is found between population and resources. Since the food supply expands only linearly with time, while the population grows exponentially, it seemed inevitable that at some point the world would hit the breaking point. Malthus predicted mass famines by the mid-1800s.

But in the 1800s, the world population was only in the early stages of major expansion, and because of the discovery of new land, the founding of colonies, increases in the food supply, etc., the disasters Malthus predicted never took place.

In the 1960s, another Malthusian prediction was made, stating that a population bomb would soon hit the earth, with global collapse by the year 2000. The prediction was wrong. The green revolution successfully expanded the food supply. The data show that the increase in food supply exceeded the growth in the world population, thereby temporarily defeating the logic of Malthus. From 1950 to 1984, grain production increased by more than 250 percent, mainly due to new fertilizers and new farming technologies.

Once again, we were able to dodge the bullet. But now the population expansion is in full swing, and some say we are reaching the limit of the earth's ability to create food supplies.

Ominously, food production is beginning to flatten out, both in world grain production and in food harvested from the oceans. The UK government's chief scientist warned of a perfect storm of exploding population and falling food and energy supplies by 2030. The world will have to produce 70 percent more food by 2050 to feed an extra 2.3 billion people, the UN's Food and Agriculture Organization has said, or else face disaster.

These projections may underestimate the true scope of the problem. With hundreds of millions of people from China and India entering the middle class, they will want to enjoy the same luxuries that they have seen in Hollywood movies—such as two cars, spacious suburban homes, hamburgers and French fries, etc.—and may strain the world's resources. In fact, Lester Brown, one of the world's leading environmentalists and founder of the World Watch Institute in Washington, D.C., confided to me that the world may not be able to handle the strain of providing a middle-class lifestyle to so many hundreds of millions of people.

SOME HOPE FOR WORLD POPULATION

There are some glimmers of hope, however. Birth control, once a taboo topic, has taken hold in the developed world and is making inroads in the developing world.

In Europe and Japan, we see the implosion, not the explosion, of the population. The birthrate is as low as 1.2 to 1.4 children per family in some European nations, far below the replacement level of 2.1. Japan is being hit with a triple whammy. One, it has the fastest-aging population on earth. Japanese women, for example, have held the record for more than twenty years for having the longest life expectancy of any group. Two, Japan has a plunging birthrate. And three, the government keeps immigration extremely low. These three demographic forces are creating a train wreck in slow motion. And Europe is not far behind.

One lesson here is that the world's greatest contraceptive is prosperity. In the past, peasants without retirement plans or social security tried to have as many children as possible to toil in the fields and care for them when they got old, doing a simple calculation: each new child in the family means more hands to work, more income, and more people to nurse you in old age. But when a peasant enters the middle class, complete with retirement benefits and a comfortable lifestyle, the equation flips the other way: each child reduces income and quality of life.

In the third world, you have the opposite problem—a rapidly expanding population, where much of the population is below the age of twenty. Even where the population explosion is expected to be the largest, in Asia and sub-Saharan Africa, the birthrate has been falling, for several reasons.

First, you have the rapid urbanization of the peasant population, as farmers leave their ancestral lands to try their luck in the megacities. In 1800, only 3 percent of the population lived in cities. By the end of the twentieth century, that figure rose to 47 percent, and it is expected to soar above that in the coming decades. The expense of child rearing in the city drastically reduces the number of children in a family. With rents, food, and expenses being so high, workers in the slums of the megacities perform the same calculus and conclude that each child reduces their wealth.

Second, as countries industrialize, as in China and India, this creates a middle class that wants fewer children, as in the industrialized West. And third, the education of women, even in poor countries like Bangladesh, has created a class of women who want fewer children. Because of an extensive educational plan, the birthrate in Bangladesh has gone down from 7 to 2.7, even without large-scale urbanization or industrialization.

Given all these factors, the UN has continually revised its figures about future population growth. Estimates still vary, but the world population may hit 9 billion by 2040. Although the population will continue to increase, the rate of growth will eventually slow down and level off. Optimistically, it may even stabilize at around 11 billion by 2100.

Normally, one might consider this to be beyond the carrying capacity of the planet. But it depends on how one defines carrying capacity, because there might be another green revolution in the making.

One possible solution to some of these problems is biotechnology. In Europe, bioengineered foods have earned a bad reputation that may last for an entire generation. The biotech industry simultaneously marketed herbicides to farmers as well as herbicide-resistant crops. To the biotech industry, this meant more sales, but to the consumer, this meant more poisons in their food, and the market quickly imploded.

In the future, however, grains such as "super-rice" may enter the market, that is, crops specifically engineered to thrive in dry, hostile, and barren environments. On moral grounds, it would be difficult to oppose the introduction of crops that are safe and can feed hundreds of millions of people.

RESURRECTING EXTINCT LIFE-FORMS

But other scientists are not just interested in extending human life span and cheating death. They are interested in bringing back creatures from the dead.

In the movie *Jurassic Park,* scientists extract DNA from the dinosaurs, insert it into the eggs of reptiles, and bring dinosaurs back to life. Although usable DNA from dinosaurs has so far never been found, there are tantalizing hints that this dream is not totally far-fetched. By the end of this century, our zoos may be populated by creatures that ceased walking the surface of the earth thousands of years ago.

As we mentioned earlier, Robert Lanza took the first major step by cloning banteng, an endangered species. It would be a shame, he feels, if this rare ox dies out. So he is considering another possibility: creating a new cloned animal, but of the opposite sex. In mammals, the sex of an organism is determined by the X and Y chromosomes. By tinkering with these chromosomes, he is confident he can clone another animal from this carcass, except of the opposite sex. In this way, zoos around the world could enjoy watching animals from long-dead species have babies.

I once had dinner with Richard Dawkins of Oxford University and author of *The Selfish Gene,* who takes this a step further. He speculates that one day we might be able to resurrect a variety of life-forms that are not just endangered but also have been long extinct. He first notes that every twenty-seven months, the number of genes that have been sequenced doubles. Then he calculates that in the coming decades it will cost only $160 to fully sequence anyone's genome. He envisions a time when biologists will carry a small kit with them and then, within minutes, be able to sequence the entire genome of any life-form they encounter.

But he goes further and theorizes that, by 2050, we will be able to construct the entire organism from its genome alone. He writes, "I believe that by 2050, we shall be able to read the language [of life]. We shall feed the genome of an unknown animal into a computer which will reconstruct not only the form of the animal but the detailed world in which its ancestors . . . lived, including their predators or prey, parasites or hosts, nesting sites, and even hopes and fears." Quoting from the work of Sydney Brenner, Dawkins believes that we can reconstruct the genome of the "missing link" between humans and the apes.

This would be a truly remarkable breakthrough. Judging from the fossil and DNA evidence, we separated from the apes about 6 million years ago.

Since our DNA differs from that of chimpanzees by only 1.5 percent, in the future a computer program should be able to analyze our DNA and the chimpanzee's DNA and then mathematically approximate the DNA of the common ancestor who gave birth to both species. Once the hypothetical genome of our common ancestor is mathematically reconstructed, a computer program will then give us a visual reconstruction of what it looked like, as well as its characteristics. He calls this the Lucy Genome Project, named after the celebrated fossil of an *Australopithecus.*

He even theorizes that once the genome of the missing link has been mathematically re-created by a computer program, it might be possible to actually create the DNA of this organism, implant it into a human egg, and then insert the egg into a woman, who will then give birth to our ancestor.

Although this scenario would have been dismissed as preposterous just a few years ago, several developments indicate that it is not such a far-fetched dream.

First, the handful of key genes that separate us from the chimpanzees are now being analyzed in detail. One interesting candidate is the ASPM gene, which is responsible for controlling brain size. The human brain increased in size several million years ago, for reasons that are not understood. When this gene is mutated, it causes microcephaly, in which the skull is small and the brain reduced by 70 percent, about the size of our ancient ancestors' millions of years ago. Intriguingly, it is possible using computers to analyze the history of this gene. Analyses show that it mutated fifteen times in the last 5 to 6 million years, since we separated from the chimpanzees, which coincides with the increase in our brain size. Compared to our primate cousins, humans have experienced the fastest rate of change in this key gene.

Even more interesting is the HAR1 region of the genome, which contains only 118 letters. In 2004, it was discovered that the crucial difference between chimps and humans in this region was just 18 letters, or nucleic acids. Chimps and chickens diverged 300 million years ago, yet their base pairs in the HAR1 region differ by only two letters. What this means is that the HAR1 region was remarkably stable throughout evolutionary history, until the coming of humans. So perhaps the genes that make us human are contained there.

But there is an even more spectacular development that makes Dawkins's proposal seem feasible. The entire genome of our nearest genetic neighbor, the long-extinct Neanderthal, has now been sequenced. Perhaps by computer analysis of the genome of humans, chimpanzees, and Neanderthals, one might use pure mathematics to reconstruct the genome of the missing link.

BRING BACK THE NEANDERTHAL?

Humans and the Neanderthals probably diverged about 300,000 years ago. But these creatures died out about 30,000 years ago in Europe. So it was long thought that it was impossible to extract usable DNA from long-dead Neanderthals.

But in 2009, it was announced that a team led by Svante Pääbo of the Max Planck Institute for Evolutionary Anthropology in Leipzig had produced a first draft of the entire Neanderthal genome, analyzing the DNA from six Neanderthals. This was a monumental achievement. The Neanderthal genome, as expected, was very similar to the human genome, both containing 3 billion base pairs, but also different in key respects.

Anthropologist Richard Klein of Stanford, commenting on this work of Pääbo and his colleagues, said that this reconstruction might answer long-standing questions about Neanderthal behavior, such as whether they could talk. Humans have two particular changes in the FOXP2 gene, which, in part, allow us to speak thousands of words. A close analysis shows that the Neanderthal had the same two genetic changes in its FOXP2 gene. So it is conceivable that the Neanderthal might have been able to vocalize in a way similar to us.

Since the Neanderthals were our closest genetic relative, they are a subject of intense interest among scientists. Some have raised the possibility of one day reconstructing the DNA of the Neanderthal and inserting it into an egg, which may one day become a living Neanderthal. Then, after thousands of years, the Neanderthal may one day walk the surface of the earth.

George Church of the Harvard Medical School even estimated that it would cost only $30 million to bring the Neanderthal back to life, and he even laid out a plan to do so. One could first divide the entire human genome into chunks, with 100,000 DNA pairs in each piece. Each one would be inserted into a bacterium and then altered genetically so the genome matched that of the Neanderthal. Each of these altered chunks of DNA would then be reassembled into the complete Neanderthal DNA.

This cell would then be reprogrammed to revert to its embryonic state and then inserted into the womb of a female chimp.

However, Klein of Stanford brought up some reasonable concerns when he asked, "Are you going to put them in Harvard or in a zoo?"

All this talk of resurrecting another long-extinct species like the Neanderthal "will doubtless raise ethical worries," cautions Dawkins. Will the Neanderthal have rights? What happens if he or she wants to mate? Who is responsible if he or she gets hurt or hurts someone else?

So if the Neanderthal can be brought back to life, can scientists eventually create a zoo for long-extinct animals, like the mammoth?

BRING BACK THE MAMMOTH?

The idea is not as crazy as it sounds. Already, scientists have been able to sequence much of the genome of the extinct Siberian mammoth. Previously, only tiny fragments of DNA had been extracted from woolly mammoths that were frozen in Siberia tens of thousands of years ago. Webb Miller and Stephan C. Schuster of Pennsylvania State University did the impossible: they extracted 3 billion base pairs of DNA from the frozen carcasses of the mammoths. Previously, the record for sequencing the DNA of an extinct species was only 13 million base pairs, less than 1 percent of the animal's genome. (This breakthrough was made possible by a new sequencing machine, called the high-throughput sequencing device, that allows one to scan thousands of genes at once, rather than individually.) Another trick was knowing where to look for ancient DNA. Miller and Schuster found that the hair follicle of the woolly mammoth, not the body itself, contained the best DNA.

The idea of resurrecting an extinct animal may now be biologically possible. "A year ago, I would have said this was science fiction," Schuster said. But now, with so much of the mammoth genome sequenced, this is no longer out of the question. He even sketched how this might be done. He estimated that perhaps only 400,000 changes in the DNA of an Asian elephant could create an animal that had all the essential features of a woolly mammoth. It might be possible to genetically alter the elephant's DNA to accommodate these changes, insert this into the nucleus of an elephant egg, and then implant the egg into a female elephant.

Already, the team is looking to sequence the DNA from yet another extinct animal, the thylacine, an Australian marsupial, closely related to the Tasmanian devil, that became extinct in 1936. There is also some talk of sequencing the dodo bird. "Dead as a dodo" is a common expression, but it may become obsolete if scientists can extract usable DNA from the soft tissue and bones of carcasses of dodos that exist in Oxford and elsewhere.

JURASSIC PARK?

This naturally leads to the original question: Can we resurrect the dinosaurs? In a word, perhaps no. A Jurassic Park depends on being able to retrieve the intact DNA of a life-form that died out more than 65 million years ago, and this may be impossible. Although soft tissue has been found within the thigh bones of dinosaur fossils, so far no DNA has been extracted in this way, only proteins. Although these proteins have chemically proven the close relationship between the *Tyrannosaurus rex* and the frog and chicken, this is a far cry from being able to reclaim the genome of a dinosaur.

Dawkins holds out the possibility, however, of being able to genetically compare the genome of various bird species with reptiles and then mathematically reconstruct the DNA sequence of a "generalized dinosaur." He notes that it is possible to induce chicken beaks to grow tooth buds (and induce snakes to grow legs). Hence, ancient characteristics, which have long vanished into the sands of time, might be lingering within genomes.

This is because biologists now realize that genes can be turned on and hence can also be turned off. This means that the genes for ancient characteristics may still exist but simply be dormant. By turning on these long-dormant genes, it might be possible to bring back these ancient traits.

For example, in the ancient past, chicken feet once had webbing. The gene for webbing did not disappear but was simply turned off. By turning this gene back on, one can in principle create chicken with webbed feet. Similarly, humans once were covered with fur. However, we lost our fur when we began to sweat, which is a very efficient way to regulate the temperature of the body. (Dogs don't have sweat glands, and so cool themselves off by panting.) The gene for human fur apparently still exists but has been turned off. Thus, by turning on this gene, it might be possible to have people with fur all over their bodies. (Some have speculated that this may be responsible for the werewolf legend.)

If we assume that some of the genes of the dinosaurs were in fact turned off for millions of years but still survive in the genome of birds, then it might be possible to reactivate these long-dormant genes and induce dinosaur characteristics in birds. So Dawkins's proposal is speculative but not out of the question.

CREATING NEW LIFE-FORMS

This raises the final question: Can we create life according to our wishes? Is it possible to create not just long-extinct animals but also animals that have never existed before? For example, could we make a pig with wings or an animal described in ancient mythology? Even by the end of this century, science will not be able to create animals to order. However, science will go a long way to being able to modify the animal kingdom.

So far, the limiting factor has been our ability to move genes around. Only single genes can be reliably modified. For example, it is possible to find a gene that causes certain animals to glow in the dark. This gene can be isolated, then placed in other animals so they glow in the dark. In fact, research is currently going on whereby family pets may be modified by the addition of single genes.

But creating an entirely new animal, like a chimera from Greek mythology (which is the combination of three different animals), requires the transposition of thousands of genes. To create a pig with wings, you would have to move the hundreds of genes that represent the wing and make sure all the muscles and blood vessels match up properly. This is far beyond anything that can be done today.

However, inroads have been made that might facilitate this futuristic possibility. Biologists were amazed to find that the genes that describe the layout of the body (from head to toe) were mirrored in the order in which they appear in the chromosomes. These are called the HOX genes, and they describe how the body is constructed. Nature, apparently, has taken a short-cut, mirroring the order of the organs of the body with the sequence found in the chromosomes themselves. This, in turn, has greatly accelerated the process by which the evolutionary history of these genes can be deciphered.

Furthermore, there are master genes that apparently govern the properties of many other genes. By manipulating a handful of these master genes, you can manipulate the properties of dozens of other genes.

In retrospect, we see that Mother Nature has decided to create the layout of the body in much the way an architect might create blueprints. The geometric layout of the blueprint is in the same order as the actual physical layout of the building. Also, the blueprints are modular, so that blocks of sub-blueprints are contained in a single master blueprint.

In addition to creating entirely new hybrid animals by exploiting the modularity of the genome, there is also the possibility of applying genetics to humans, using biotechnology to bring back historical figures. Lanza believes that as long as an intact cell can be extracted from a long-dead person, it will be possible to bring this person back to life. In Westminster Abbey, we have the carefully preserved bodies of long-dead kings and queens, as well as poets, religious figures, politicians, and even scientists like Isaac Newton. One day, Lanza confided to me, it may be possible to find intact DNA within their bodies and bring them back to life.

In the movie *The Boys from Brazil,* the plot revolves around bringing back Hitler. One should not believe, however, that one will be able to bring back the genius or notoriety of any of these historic figures. As one biologist noted, if you bring back Hitler, maybe all you get is a second-rate artist (which is what Hitler was before he led the Nazi movement).

BAN ALL DISEASES?

The prophetic movie *Things to Come* was based on a novel by H. G. Wells and predicted the future of civilization, where World War II unleashed a cycle of endless suffering and misery. Eventually, all the achievements of the human race are reduced to rubble, with gangs of warlords ruling over crushed, impoverished people. But at the end of the movie, a group of far-sighted scientists, armed with powerful superweapons, begin to restore order. Civilization finally rises again from the ashes. In one scene, a child is taught the brutal history of the twentieth century and learns about something called colds. What is a cold, she asks? She is told that colds were something that were cured a long time ago.

Maybe not.

Curing all diseases has been one of our most ancient goals. But even by 2100, scientists will not be able to cure all diseases, since diseases mutate faster than we can cure them, and there are too many of them. We sometimes forget that we live in an ocean of bacteria and viruses, which existed billions of years before humans walked the surface of the earth, and will exist billions of years after *Homo sapiens* is gone.

Many diseases originally came from animals. This is one of the prices we paid for the domestication of animals, which began roughly 10,000 years ago. So there is a vast reservoir of diseases lurking in animals that will probably outlast the human race. Normally, these diseases infect only a handful of individuals. But with the rise of large cities, these communicable diseases could spread rapidly among the human population, reaching critical mass and creating pandemics.

For example, when scientists analyzed the genetic sequence of the flu virus, they were surprised to find its origin: birds. Many birds can carry variations of the flu virus without any effects. But then pigs sometimes act as genetic mixing bowls, after eating bird droppings. And then farmers often live near both. Some speculate that this is the reason the flu virus often comes from Asia, because farmers there engage in polyfarming, i.e., living in close proximiy to both ducks and pigs.

The recent H1N1 flu epidemic is only the most recent wave of bird flu and pig flu mutations.

One problem is that humans are continually expanding into new environments, cutting down forests, building suburbs and factories, and in the process encountering ancient diseases lurking among the animals. Because the human population is continuing to expand, this means that we expect to find more surprises coming out of the forest.

For example, there is considerable genetic evidence that HIV began as simian immunodeficiency virus (SIV), which originally infected monkeys but then jumped to humans. Similarly, the hantavirus affected people in the Southwest as they encroached on the territory of prairie rodents. Lyme disease, which is spread largely by ticks, has invaded the suburbs of the Northeast because people now build houses close to the forests where the ticks live. The Ebola virus probably affected tribes of humans back in antiquity, but it was only with the coming of jet travel that it spread to a larger population and made the headlines. Even Legionnaires' disease is probably an ancient one that spawned in stagnant water, but it was the proliferation of air-conditioning units that spread this disease to the elderly on cruise ships.

This means that there will be plenty of surprises to come, with new waves of exotic diseases dominating the headlines of the future.

Unfortunately, cures for these diseases may be late in coming.

For example, even the common cold currently has no cure. The plethora of products found in any drugstore for it treats only the symptoms, rather than killing the virus itself. The problem is that there are probably more than 300 variations of the rhinovirus that causes the common cold, and it is simply too expensive to create a vaccine for all 300.

The situation for HIV is much worse, since there may be thousands of different strains. In fact, HIV mutates so rapidly that, even if you can develop a vaccine for one variety, the virus will soon mutate. Devising a vaccine for HIV is like trying to hit a moving target.

So while we will cure many diseases in the future, probably we will always have some disease that can evade our most advanced science.

BRAVE NEW WORLD

By 2100, when we will have control over our genetic destiny, we have to compare our fate with the dystopia laid out by Aldous Huxley in his prophetic novel *Brave New World,* which is set in the year 2540. The book caused universal shock and dismay when it was first published in 1932.

Yet more than seventy-five years later, many of his predictions have already come to pass. He scandalized British society when he wrote about test tube babies, when recreation and procreation would be separated, and when drugs became commonplace, yet today we live in a world where in vitro fertilization and birth control pills are taken for granted. (The only major prediction he made that has not come to pass is human cloning.) He envisioned a hierarchical world where doctors deliberately clone brain-damaged human embryos, which grow up to become servants of the ruling elite. Depending on the level of mental damage, they could be ranked into the Alphas, who are perfect and destined to rule, down to the Epsilons, who are little more than mentally retarded slaves. So technology, instead of liberating humanity from poverty, ignorance, and disease, has become a nightmare, enforcing an artificial and corrupt stability at the expense of enslaving an entire population.

Although the novel was accurate in many ways, Huxley did not anticipate genetic engineering. If he had known about this technology, then he might have worried about another problem: Will the human species split into fragments, with fickle parents and devious governments meddling with the genes of our children? Parents already dress their kids in outlandish outfits and make them compete in silly contests, so why not change the genes to fit the parents' whims? Indeed, parents are probably hardwired by evolution to give every benefit to their progeny, so why not tamper with their genes as well?

As an elementary example of what might go wrong, consider the lowly sonogram. Although doctors innocently introduced the sonogram to help with pregnancies, this has led to a massive epidemic of abortions of female fetuses, especially in the countrysides of China and India. One study in Bombay found that 7,997 out of 8,000 aborted fetuses were female. In South Korea 65 percent of all third-born children are male. The generation of children whose parents chose this gender-based abortion will soon be of marriageable age, and millions will find that there are no females to be found. This in turn could cause enormous social dislocation. Peasants who wanted only boys to carry on their name will find that they have no grandchildren.

And in the United States, there is rampant misuse of human growth hormone (HGH), which is often touted as a cure for aging. Originally, HGH was intended to correct hormone deficiencies in children who were too short. Instead, HGH has grown into a huge underground industry based on questionable data concerning aging. In effect, the Internet has created a huge population of human guinea pigs for specious therapies.

So, given the chance, people will often misuse technology and create an enormous amount of mischief. What happens if they get hold of genetic engineering?

In a worst-case scenario, we might have the nightmare imagined by H. G. Wells in his classic science fiction novella *The Time Machine*, when the human race, in the year 802,701 AD, splits into two distinct races. He wrote, "Gradually, the truth dawned on me: that Man had not remained one species, but had differentiated into two distinct animals: that my graceful children of the Upper World were not the sole descendants of our generation, but that this bleached, obscene, nocturnal Thing, which had flashed before me, was also heir to all the ages."

To see what variations of the human race are possible, simply look at the household dog. Although there are thousands of breeds of dogs, all originally descended from *Canis lupus,* the gray wolf, which was domesticated roughly 10,000 years ago at the end of the last Ice Age. Because of selective breeding by their human masters, dogs today come in a bewildering variety of sizes and shapes. Body shape, temperament, color, and abilities have all been radically altered by selective breeding.

Since dogs age roughly seven times faster than humans, we can estimate that about 1,000 generations of dogs have existed since they separated from wolves. If we apply this to humans, then systematic breeding of humans might split the human race into thousands of breeds in only 70,000 years, although they would be of the same species. With genetic engineering, this process could conceivably be vastly accelerated, to a single generation.

Fortunately, there are reasons to believe the speciation of the human race will not happen, at least not in the coming century. In evolution, a single species usually splits apart if it separates geographically into two separate breeding populations. This happened, for example, in Australia, where the physical separation of many animal species has led to the evolution of animals found nowhere else on earth, such as marsupials like the kangaroo. Human populations, by contrast, are highly mobile, without evolutionary bottlenecks, and are highly intermingled.

As Gregory Stock of UCLA has said, "Traditional Darwinian evolution now produces almost no change in humans and has little prospect of doing so in the foreseeable future. The human population is too large and entangled, and selective pressures are too localized and transitory."

There are also constraints coming from the Cave Man Principle.

As we mentioned earlier, people often reject the advances of technology (for example, the paperless office) when it contradicts human nature, which has remained relatively constant over the past 100,000 years. People may not want to create designer children who deviate from the norm and are considered freaks by their peers. This decreases their chances of success in society. Dressing one's children in silly clothing is one thing, but permanently changing their heredity is an entirely different thing. (In a free market, there probably will be a place for weird genes, but it will be small, since the market will be driven by consumer demand.) More than likely, by the end of the century, a couple will be given a library of genes to choose from, mostly those for eliminating genetic diseases but also some for genetic enhancement. However, there will be little market pressure to finance the study of bizarre genes because the demand for them will be so small.

The real danger will come not so much from consumer demand but from dictatorial governments that may want to use genetic engineering for their own purposes, such as breeding stronger but more obedient soldiers.

Another problem arises in the distant future, when we have space colonies on other planets whose gravity and climactic conditions are much different from the earth. At that point, perhaps in the next century, it becomes realistic to think of engineering a new breed of humans who can adjust to different gravity fields and atmospheric conditions. For example, a new breed of humans may be able to consume different amounts of oxygen, adjust to a different length of day, and have a different body weight and metabolism. But space travel will be expensive for a long time. By the end of the century, we may have a small outpost on Mars, but an overwhelming fraction of the human race will still be on the earth. For decades to centuries to come, space travel will be for astronauts, the wealthy, and maybe a handful of hardy space colonists.

So the splitting of the human race into different spacefaring species around the solar system and beyond will not happen in this century, or perhaps even the next. For the foreseeable future, unless there are dramatic breakthroughs in space technology, we are largely stuck on the earth.

Lastly, there is yet another threat that faces us before we reach 2100: that this technology may be deliberately turned against us, in the form of designer germ warfare.

GERM WARFARE

Germ warfare is as old as the Bible. Ancient warriors used to hurl diseased bodies over the walls of enemy cities or poison their wells with the bodies of diseased animals. Deliberately giving smallpox-infected clothing to an adversary is another way to destroy them. But with modern technology, germs can be genetically bred to wipe out millions of people.

In 1972, the United States and the former Soviet Union signed an historic treaty banning the use of germ warfare for offensive purposes. However, the technology of bioengineering is so advanced today that the treaty is meaningless.

First, there is no such thing as offensive and defensive technology when it comes to DNA research. The manipulation of genes can be used for either purpose.

Second, with genetic engineering, it is possible to create weaponized germs, those that have been deliberately modified to increase their lethality or their ability to spread into the environment. It was once believed that only the United States and Russia possessed the last vials containing smallpox, the greatest killer in the history of the human race. In 1992, a Soviet defector claimed that the Russians had weaponized smallpox and actually produced up to twenty tons of it. With the breakup of the Soviet Union, there is the nagging fear that one day a terrorist group may pay to gain access to weaponized smallpox.

In 2005, biologists successfully resurrected the Spanish flu virus of 1918, which killed more people than World War I. Remarkably, they were able to resurrect the virus by analyzing a woman who had died and was buried in the permafrost of Alaska, as well as samples taken from U.S. soldiers during the epidemic.

The scientists then proceeded to publish the entire genome of the virus on the Web, making it known to the entire world. Many scientists felt uneasy about this, since one day even a college student with access to a university laboratory might be able to resurrect one of the greatest killers in the history of the human race.

In the short term, the publication of the genome of the Spanish flu virus was a bonanza for scientists, who then could examine the genes to solve a long-standing puzzle: How did a tiny mutation cause such widespread damage to the human population? The answer was soon found. The Spanish flu virus, unlike other varieties, causes the body's immune system to overreact, releasing large amounts of fluid that eventually kills the patient. The person literally drowns in his own fluids. Once this was understood, the genes that cause this deadly effect could be compared to the genes of the H1N1 flu and other viruses. Fortunately, none of them possessed this lethal gene. Moreover, one could actually calculate how close a virus was to attaining this alarming capability, and the H1N1 flu was still far from achieving this ability.

But in the long term, there is a price to pay. Every year, it becomes easier and easier to manipulate the genes of living organisms. Costs keep plummeting, and the information is widely available on the Internet.

Within a few decades, some scientists believe that it will be possible to create a machine that will allow you to create any gene simply by typing the desired components. By typing in the A-T-C-G symbols making up a gene, the machine will then automatically splice and dice DNA to create that gene. If so, then it means that perhaps even high school students may one day do advanced manipulations of life-forms.

One nightmare scenario is airborne AIDS. Cold viruses, for example, possess a few genes that allow them to survive in droplets of aerosols, so that sneezing can infect others. At present, the AIDS virus is quite vulnerable when it is exposed to the environment. But if the cold virus genes are implanted into the AIDS virus, then it is conceivable that they might make it able to survive outside the human body. This could then cause the AIDS virus to spread like the common cold, thereby infecting a large portion of the human race. It is also known that viruses and bacteria do exchange genes, so there is also the possibility that the AIDS and common cold viruses can exchange genes naturally, although this is less likely.

In the future, a terrorist group or nation-state may be able to weaponize AIDS. The only thing preventing them from unleashing it would be the fact that they, too, would also perish if the virus were to be dispersed into the environment.

This threat became real right after the tragedy of 9/11. An unknown person mailed packets of a white powder containing anthrax spores to well-known politicians around the country. A careful, microscopic analysis of the white powder showed that the anthrax spores had been weaponized for maximum death and destruction. Suddenly, the entire country was gripped with fear that a terrorist group had access to advanced biological weapons. Although anthrax is found in the soil and throughout our environment, only a person with advanced training and maniacal intentions could have purified and weaponized the anthrax and pulled off this feat.

Even after one of the largest manhunts in U.S. history, the culprit was never found, even to this day (although a leading suspect recently committed suicide). The point here is that even a single individual with some advanced biological training can terrorize an entire nation.

One restraining factor that has kept germ warfare in check is simple self-interest. During World War I, the efficacy of poison gas on the battle-field was mixed. The wind conditions were often unpredictable, so the gas could blow back onto your own troops. Its military value was largely in ter-rorizing the enemy, rather than defeating him. Not a single decisive battle was won using poison gas. And even at the height of the Cold War, both sides knew that poison gas and biological weapons could have unpredict-able effects on the battlefield, and could easily escalate to a nuclear confron-tation.

All the arguments mentioned in this chapter, as we have seen, involved the manipulation of genes, proteins, and molecules. Then the next ques-tion naturally arises: How far can we manipulate individual atoms?

The principles of physics, as far as I can see, do not speak against the possibility of maneuvering things atom by atom.

—RICHARD FEYNMAN, NOBEL LAUREATE

Nanotechnology has given us the tools to play with the ultimate toy box of nature—atoms and molecules. Everything is made from these, and the possibilities to create new things appear limitless.

—HORST STORMER, NOBEL LAUREATE

The role of the infinitely small is infinitely large.

—LOUIS PASTEUR

4 NANOTECHNOLOGY *Everything from Nothing?*

The mastery of tools is a crowning achievement that distinguishes humanity from the animals. According to Greek and Roman mythology, this process began when Prometheus, taking pity on the plight of humans, stole the precious gift of fire from Vulcan's furnace. But this act of thievery enraged the gods. To punish humanity, Zeus devised a clever trick. He asked Vulcan to forge a box and a beautiful woman out of metal. Vulcan created this statue, called Pandora, and then magically brought her to life, and told her never to open the box. Out of curiosity, one day she did, and unleashed all the winds of chaos, misery, and suffering in the world, leaving only hope in the box.

So from Vulcan's divine furnace emerged both the dreams and the suffering of the human race. Today, we are designing revolutionary new machines that are the ultimate tools, forged from individual atoms. But will they unleash the fire of enlightenment and knowledge or the winds of chaos?

Throughout human history, the mastery of tools has determined our fate. When the bow and arrow were perfected thousands of years ago, it meant that we could fire projectiles much farther than our hands could throw them, increasing the efficiency of our hunting and increasing our food supply. When metallurgy was invented around 7,000 years ago, it meant that we could replace huts of mud and straw and eventually create great buildings that soared above the earth. Soon, empires began to rise from the forest and the desert, built by the tools forged from metals.

And now we are on the brink of mastering yet another type of tool, much more powerful than anything we have seen before. This time, we will be able to master the atoms themselves out of which everything is created. Within this century, we may possess the most important tool ever imagined—nanotechnology that will allow us to manipulate individual atoms. This could begin a second industrial revolution, as molecular manufacturing creates new materials we can only dream about today, which are superstrong, superlight, with amazing electrical and magnetic properties.

Nobel laureate Richard Smalley has said, "The grandest dream of nanotechnology is to be able to construct with the atom as the building block." Philip Kuekes of Hewlett-Packard said, "Eventually, the goal is not just to make computers the size of dust particles. The idea would be to make simple computers the size of bacteria. Then you could get something as powerful as what's now on your desktop into a dust particle."

This is not just the hope of starry-eyed visionaries. The U.S. government takes it seriously. In 2009, because of nanotechnology's immense potential for medical, industrial, aeronautical, and commercial applications, the National Nanotechnology Initiative allocated $1.5 billion for research. The government's National Science Foundation Nanotechnology Report states, "Nanotechnology has the potential to enhance human performance, to bring sustainable development for materials, water, energy, and foods, to protect against unknown bacteria and viruses. . . ."

Ultimately, the world economy and fate of nations may depend on this. Around 2020 or soon afterward, Moore's law will begin to falter and perhaps eventually collapse. The world economy could be thrown into disarray unless physicists can find a suitable replacement for silicon transistors to power our computers. The solution to the problem may come from nanotechnology.

Nanotechnology might also, perhaps by the end of this century, create a machine that only the gods can wield, a machine that can create anything out of almost nothing.

THE QUANTUM WORLD

The first to call attention to this new realm of physics was Nobel laureate Richard Feynman, who asked a deceptively simple question: How small can you make a machine? This was not an academic question. Computers were gradually becoming smaller, changing the face of industry, so it was becoming apparent that the answer to this question could have an enormous impact on society and the economy.

In his prophetic talk given in 1959 to the American Physical Society titled "There's Plenty of Room at the Bottom," Feynman said, "It is interesting that it would be, in principle, possible (I think) for a physicist to synthesize any chemical substance that the chemist writes down. Give the orders and the physicist synthesizes it. How? Put the atoms down where the chemist says, and so you make the substance." Feynman concluded that machines made out of individual atoms were possible, but that new laws of physics would make them difficult, but not impossible, to create.

So ultimately, the world economy and the fate of nations may depend on the bizarre and counterintuitive principles of the quantum theory. Normally, we think that the laws of physics remain the same if you go down to smaller scales. But this is not true. In movies like Disney's *Honey, I Shrunk the Kids* and *The Incredible Shrinking Man,* we get the mistaken impression that miniature people would experience the laws of physics the same way we do. For example, in one scene in the Disney movie, our shrunken heroes ride on an ant during a rainstorm. Raindrops fall onto the ground and make tiny puddles, just as in our world. But in reality, raindrops can be larger than ants. So when an ant encounters a raindrop, it would see a huge hemisphere of water. The hemisphere of water does not collapse because surface tension acts like a net that holds the droplet together. In our world, surface tension of water is quite small, so we don't notice it. But on the scale of an ant, surface tension is proportionately huge, so rain beads up into droplets.

(Furthermore, if you tried to scale up the ant so that it was the size of a house, you have another problem: its legs would break. As you increase the size of the ant, its weight grows much faster than the strength of its legs. If you increase the size of an ant by a factor of 10, its volume and hence its weight is 10 × 10 × 10 = 1,000 times heavier. But its strength is related to the thickness of its muscles, which is only 10 × 10 = 100 times stronger. Hence, the giant ant is 10 times weaker, relatively speaking, than an ordinary ant. This also means that King Kong, instead of terrorizing New York City, would crumble if he tried to climb the Empire State Building.)

Feynman noted that other forces also dominate at the atomic scale, such as hydrogen bonding and the van der Waals force, caused by tiny electrical forces that exist between atoms and molecules. Many of the physical properties of substances are determined by these forces.

(To visualize this, consider the simple problem of why the Northeast has so many potholes in its highways. Every winter, water seeps into tiny cracks in the asphalt; the water expands as it freezes, causing the asphalt to crumble and gouging out a pothole. But it violates common sense to think that water expands when it freezes. Water does expand because of hydrogen bonding. The water molecule is shaped like a V, with the oxygen atom at the base. The water molecule has a slight negative charge at the bottom and a positive charge at the top. Hence, when you freeze water and stack water molecules, they expand, forming a regular lattice of ice with plenty of spaces between the molecules. The water molecules are arranged like hexagons. So water expands as it freezes since there is more space between the atoms in a hexagon. This is also the reason snowflakes have six sides, and explains why ice floats on water, when by rights it should sink.)

WALKING THROUGH WALLS

In addition to surface tension, hydrogen bonding, and van der Waals forces, there are also bizarre quantum effects at the atomic scale. Normally, we don't see quantum forces at work in everyday life. But quantum forces are everywhere. For example, by rights, since atoms are largely empty, we should be able to walk through walls. Between the nucleus at the center of the atom and the electron shells, there is only a vacuum. If the atom were the size of a football stadium, then the stadium would be empty, since the nucleus would be roughly the size of a grain of sand.

(We sometimes amaze our students with a simple demonstration. We take a Geiger counter, place it in front of a student, and put a harmless radioactive pellet in back. The student is startled that some particles pass right through his body and trigger the Geiger counter, as if he is largely empty, which he is.)

But if we are largely empty, then why can't we walk through walls? In the movie *Ghost*, Patrick Swayze's character is killed by a rival and turns into a ghost. He is frustrated every time he tries to touch his former fiancée, played by Demi Moore. His hands pass through ordinary matter; he finds that he has no material substance and simply floats through solid objects. In one scene, he sticks his head into a moving subway car. The train races by with his head sticking inside, yet he doesn't feel a thing. (The movie does not explain why gravity does not pull him through the floor so he falls to the center of the earth. Ghosts, apparently, can pass through anything except floors.)

So why can't we pass through solid objects like ghosts? The answer resides in a curious quantum phenomenon. The Pauli exclusion principle states that no two electrons can exist in the same quantum state. Hence when two nearly identical electrons get too close, they repel each other. This is the reason objects appear to be solid, which is an illusion. The reality is that matter is basically empty.

When we sit in a chair, we think we are touching it. Actually, we are hovering above the chair, floating less than a nanometer above it, repelled by the chair's electrical and quantum forces. This means that whenever we "touch" something, we are not making direct contact at all but are separated by these tiny atomic forces. (This also means that if we could somehow neutralize the exclusion principle, then we might be able to pass through walls. However, no one knows how to do this.)

Not only does the quantum theory keep atoms from crashing through one another, it also binds them together into molecules. Imagine for the moment that an atom is like a tiny solar system, with planets revolving around a sun. Now, if two such solar systems collided, then the planets would either crash into one another or fly out in all directions, causing the solar system to collapse. Solar systems are never stable when they collide with another solar system, so by rights, atoms should collapse when they bump into one another.

In reality, when two atoms get very close, they either bounce off each other or they combine to form a stable molecule. The reason atoms can form stable molecules is because electrons can be shared between two atoms. Normally, the idea of an electron being shared between two atoms is preposterous. It is impossible if the electron obeyed the commonsense laws of Newton. But because of the Heisenberg uncertainty principle, you don't know precisely where the electron is. Instead, it's smeared out between two atoms, which holds them together.

In other words, if you turn off the quantum theory, then your molecules fall apart when they bump into one another and you would dissolve into a gas of particles. So the quantum theory explains why atoms can bind to form solid matter, rather than disintegrate.

(This is also the reason you cannot have worlds within worlds. Some people imagine that our solar system or galaxy might be an atom in someone else's gigantic universe. This was, in fact, the final scene in the movie *Men in Black,* where the entire known universe was in fact just an atom in some alien's ball game. But according to physics, this is impossible, since the laws of physics change as we go from scale to scale. The rules governing atoms are quite different from the rules governing galaxies.)

Some of the mind-bending principles of the quantum theory are:

- you cannot know the exact velocity and location of any particle—there is always uncertainty
- particles can in some sense be in two places at the same time
- all particles exist as mixtures of different states simultaneously; for example, spinning particles can be mixtures of particles whose axes spin both up and down simultaneously
- you can disappear and reappear somewhere else

All these statements sound ridiculous. In fact, Einstein once said, "the more successful the quantum theory is, the sillier it looks." No one knows where these bizarre laws come from. They are simply postulates, with no explanation. The quantum theory has only one thing going for it: it is correct. Its accuracy has been measured to one part in ten billion, making it the most successful physical theory of all time.

The reason we don't see these incredible phenomena in daily life is because we are composed of trillions upon trillions of atoms, and these effects, in some sense, average out.

MOVING INDIVIDUAL ATOMS

Richard Feynman dreamed of the day when a physicist could manufacture any molecule, atom for atom. That seemed impossible back in 1959, but part of that dream is now a reality.

I had a chance to witness this up close, when I visited the IBM Almaden Research Center in San Jose, California. I came to observe a remarkable instrument, the scanning tunneling microscope, which allows scientists to view and manipulate individual atoms. This device was invented by Gerd Binnig and Heinrich Rohrer of IBM, for which they won the Nobel Prize in 1986. (I remember, as a child, my teacher telling us that we would never be able to see atoms. They are just too small, he said. By then, I had already decided to become an atomic scientist. I realized that I would spend the rest of my life studying something I would never be able to observe directly. But today, not only can we see atoms, but we can play with them, with atomic tweezers.)

The scanning tunneling microscope is actually not a microscope at all. It resembles an old phonograph. A fine needle (with a tip that is only a single atom across) passes slowly over the material being analyzed. A small electrical current travels from the needle, through the material, to the base of the instrument. As the needle passes over the object, the electrical current changes slightly every time it passes over an atom. After multiple passes, the machine prints out the stunning outline of the atom itself. Using an identical needle, the microscope is then capable not just of recording these atoms but also of moving them around. In this way, one can spell out the letters, such as the initials IBM, and in fact even design primitive machines built out of atoms.

(Another recent invention is the atomic force microscope, which can give us stunning 3-D pictures of arrays of atoms. The atomic force microscope also uses the needle with a very small point, but it shines a laser onto it. As the needle passes over the material being studied, the needle jiggles, and this motion is recorded by the laser beam image.)

I found that moving individual atoms around was quite simple. I sat in front of a computer screen, looking at a series of white spheres, each resembling a Ping-Pong ball about an inch across. Actually, each ball was an individual atom. I placed the cursor over an atom and then moved the cursor to another position. I pushed a button that then activated the needle to move the atom. The microscope rescanned the substance. The screen changed, showing that the ball had moved to precisely where I wanted it.

The whole process took only a minute to move each atom to any position I wanted. In fact, in about thirty minutes, I found that I could actually spell out some letters on the screen, made of individual atoms. In an hour, I could make rather complex patterns involving ten or so atoms.

I had to recover from the shock that I had actually moved individual atoms, something that was once thought to be impossible.

MEMS AND NANOPARTICLES

Although nanotechnology is still in its infancy, it has already generated a booming commercial industry in chemical coatings. By spraying thin layers of chemicals only a few molecules thick onto a commercial product, one can make it more resistant to rust or change its optical properties. Other commercial applications today are stain-resistant clothing, enhanced computer screens, stronger metal-cutting tools, and scratch-resistant coatings. In the coming years, more and more novel commercial products will be marketed that have microcoatings to improve their performance.

For the most part, nanotechnology is still a very young science. But one aspect of nanotechnology is now beginning to affect the lives of everyone and has already blossomed into a lucrative $40 billion worldwide industry—microelectromechanical systems (MEMS)—that includes everything from ink-jet cartridges, air bag sensors, and displays to gyroscopes for cars and airplanes. MEMS are tiny machines so small they can easily fit on the tip of a needle. They are created using the same etching technology used in the computer business. Instead of etching transistors, engineers etch tiny mechanical components, creating machine parts so small you need a microscope to see them.

Scientists have made an atomic version of the abacus, the venerable Asian calculating device, that consists of several vertical columns of wires containing wooden beads. In 2000, scientists at the IBM Zurich Research Laboratory made an atomic version of the abacus by manipulating individual atoms with a scanning microscope. Instead of wooden beads that move up and down the vertical wires, the atomic abacus used buckyballs, which are carbon atoms arranged to form a molecule shaped like a soccer ball, 5,000 times smaller than the width of a human hair.

At Cornell, scientists have even created an atomic guitar. It has six strings, each string just 100 atoms wide. Laid end to end, twenty of these guitars would fit inside a human hair. The guitar is real, with real strings that can be plucked (although the frequency of this atomic guitar is much too high to be heard by the human ear).

But the most widespread practical application of this technology is in air bags, which contain tiny MEM accelerometers that can detect the sudden braking of your car. The MEM accelerometer consists of a microscopic ball attached to a spring or lever. When you slam on the brakes, the sudden deceleration jolts the ball, whose movement creates a tiny electrical charge. This charge then triggers a chemical explosion that releases large amounts of nitrogen gas within 1/25 of a second. Already, this technology has saved thousands of lives.

NEAR TERM (PRESENT TO 2030)

NANOMACHINES IN OUR BODIES

In the near future, we should expect a new variety of nanodevices that may revolutionize medicine, such as nanomachines coursing throughout the bloodstream. In the movie *Fantastic Voyage,* a crew of scientists and their ship are miniaturized to the size of a red blood cell. They then embark on a voyage through the bloodstream and brain of a patient, encountering a series of harrowing dangers within the body. One goal of nanotechnology is to create molecular hunters that will zoom in on cancer cells and destroy them cleanly, leaving normal cells intact. Science fiction writers have long dreamed about molecular search-and-destroy craft floating in the blood, constantly on the lookout for cancer cells. But critics once considered this to be impossible, an idle dream of fiction writers.

Part of this dream is being realized today. In 1992, Jerome Schentag of the University at Buffalo invented the smart pill, which we mentioned earlier, a tiny instrument the size of a pill that you swallow and that can be tracked electronically. It can then be instructed to deliver medicines to the proper location. Smart pills have been built that contain TV cameras to photograph your insides as they go down your stomach and intestines. Magnets can be used to guide them. In this way, the device can search for tumors and polyps. In the future, it may be possible to perform minor surgery via these smart pills, removing any abnormalities and doing biopsies from the inside, without cutting the skin.

A much smaller device is the nanoparticle, a molecule that can deliver cancer-fighting drugs to a specific target, which might revolutionize the treatment of cancer. These nanoparticles can be compared to a molecular smart bomb, designed to hit a specific target with a chemical payload, vastly reducing collateral damage in the process. While a dumb bomb hits everything, including healthy cells, smart bombs are selective and home in on just the cancer cells.

Anyone who has experienced the horrific side effects of chemotherapy will understand the vast potential of these nanoparticles to reduce human suffering. Chemotherapy works by bathing the entire body with deadly toxins, killing cancer cells slightly more efficiently than ordinary cells. The collateral damage from chemotherapy is widespread. The side effects—including nausea, loss of hair, loss of strength, etc.—are so severe that some cancer patients would rather die of cancer than subject themselves to this torture.

Nanoparticles may change all this. Medicines, such as chemotherapy drugs, will be placed inside a molecule shaped like a capsule. The nanoparticle is then allowed to circulate in the bloodstream, until it finds a particular destination, where it releases its medicine.

The key to these nanoparticles is their size: between 10 to 100 nanometers, too big to penetrate a blood cell. So nanoparticles harmlessly bounce off normal blood cells. But cancer cells are different; their cell walls are riddled with large, irregular pores. The nanoparticles can enter freely into the cancer cells and deliver their medicine but leave healthy tissue untouched. So doctors do not need complicated guidance systems to steer these nanoparticles to their target. They will naturally accumulate in certain types of cancerous tumors.

The beauty of this is that it does not require complicated and dangerous methods, which might have serious side effects. These nanoparticles are simply the right size: too big to attack normal cells but just right to penetrate a cancer cell.

Another example is the nanoparticles created by the scientists at BIND

Biosciences in Cambridge, Massachusetts. Its nanoparticles are made of polylactic acid and copolylactic acid/glycolic acid, which can hold drugs inside a molecular mesh. This creates the payload of the nanoparticle. The guidance system of the nanoparticle is the peptides that coat the particle and specifically bind to the target cell.

What is especially appealing about this work is that these nanoparticles form by themselves, without complicated factories and chemical plants. The various chemicals are mixed together slowly, in proper sequence, under very controlled conditions, and the nanoparticles self-assemble.

"Because the self-assembly doesn't require multiple complicated chemical steps, the particles are very easy to manufacture. . . . And we can make them on a kilogram scale, which no one else has done," says BIND's Omid Farokhzad, a physician at the Harvard Medical School. Already, these nanoparticles have proven their worth against prostate, breast, and lung cancer tumors in rats. By using colored dyes, one can show that these nanoparticles are accumulating in the organ in question, releasing their payload in the desired way. Clinical trials on human patients start in a few years.

ZAPPING CANCER CELLS

Not only can these nanoparticles seek out cancer cells and deliver chemicals to kill them, they might actually be able to kill them on the spot. The principle behind this is simple. These nanoparticles can absorb light of a certain frequency. By focusing laser light on them, they heat up, or vibrate, destroying any cancer cells in the vicinity by rupturing their cell walls. The key, therefore, is to get these nanoparticles close enough to cancer cells.

Several groups have already developed prototypes. Scientists at the Argonne National Laboratory and the University of Chicago have created titanium dioxide nanoparticles (titanium dioxide is a common chemical found in sunscreen). This group found that they could bind these nanoparticles to an antibody that naturally seeks out certain cancer cells called glioblastoma multiforme (GBM). So these nanoparticles, by hitching a ride on this antibody, are carried to the cancer cells. Then a white light is illuminated for five minutes, heating and eventually killing the cancer cells. Studies have shown that 80 percent of the cancer cells can be destroyed in this way.

These scientists have also devised a second way to kill cancer cells. They created tiny magnetic disks that can vibrate violently. Once these disks are led to the cancer cells, a small external magnetic field can be passed over them, causing them to shake and tear apart the cell walls of the cancer. In tests, 90 percent of the cancer cells were killed after just 10 minutes of shaking.

This result is not a fluke. Scientists at the University of California at Santa Cruz have devised a similar system using gold nanoparticles. These particles are only 20 to 70 nanometers across and only a few atoms thick, arranged in the shape of a sphere. Scientists used a certain peptide that is known to be attracted to skin cancer cells. This peptide was made to connect with the gold nanoparticles, which then were carried to the skin cancer cells in mice. By shining an infrared laser, these gold particles could destroy the tumor cells by heating them up. "It's basically like putting a cancer cell in hot water and boiling it to death. The more heat the metal nanospheres generate, the better," says Jin Zhang, one of the researchers.

So in the future, nanotechnology will detect cancer colonies years to decades before they can form a tumor, and nanoparticles circulating in our blood might be used to destroy these cells. The basic science is being done today.

NANOCARS IN OUR BLOOD

One step beyond the nanoparticle is the nanocar, a device that can actually be guided in its travels inside the body. While the nanoparticle is allowed to circulate freely in the bloodstream, these nanocars are like remote-controlled drones that can be steered and piloted.

James Tour and his colleagues at Rice University have made such a nanocar. Instead of wheels, it has four buckyballs. One future goal of this research is to design a molecular car that can push a tiny robot around the bloodstream, zapping cancer cells along the way or delivering lifesaving drugs to precise locations in the body.

But one problem with the molecular car is that it has no engine. Scientists have created more and more sophisticated molecular machines, but creating a molecular power source has been one of the main roadblocks. Mother Nature has solved this problem by using the molecule adenosine triphosphate (ATP) as her energy source. The energy of ATP makes life possible; it energizes every second of our muscles' motions. This energy of ATP is stored within an atomic bond between its atoms. But creating a synthetic alternative has proven difficult.

Thomas Mallouk and Ayusman Sen of Pennsylvania State University have found a potential solution to this problem. They have created a nano-car that can actually move tens of microns per second, which is the speed of most bacteria. (They first created a nanorod, made of gold and platinum, the size of a bacterium. The nanorod was placed into a mixture of water and hydrogen peroxide. This created a chemical reaction at either end of the nanorod that caused protons to move from one end of the rod to the other. Since the protons push against the electrical charges of the water molecule, this propels the nanorod forward. The rod continues to move forward as long as there is hydrogen peroxide in the water.)

Steering these nanorods is also possible using magnetism. Scientists have embedded nickel disks inside these nanorods, so they act like compass needles. By moving an ordinary refrigerator magnet next to these nano-rods, you can steer them in any direction you want.

Yet another way to steer a molecular machine is to use a flashlight. Light can break up the molecules into positive and negative ions. These two types of ions diffuse through the medium at different speeds, which sets up an electric field. The molecular machines are then attracted by these electric fields. So by pointing the flashlight one can steer the molecular machines in that direction.

I had a demonstration of this when I visited the laboratory of Sylvain Martel of the Polytechnic Montréal in Canada. His idea was to use the tails of ordinary bacteria to propel a tiny chip forward in the bloodstream. So far, scientists have been unable to manufacture an atomic motor, like the one found in the tails of bacteria. Martel asked himself: If nanotechnology could not make these tiny tails, why not use the tails of living bacteria?

He first created a computer chip smaller than the period at the end of this sentence. Then he grew a batch of bacteria. He was able to place about eighty of these bacteria behind the chip, so that they acted like a propeller that pushed the chip forward. Since these bacteria were slightly magnetic, Martel could use external magnets to steer them anywhere he wanted.

I had a chance to steer these bacteria-driven chips myself. I looked in a microscope, and I could see a tiny computer chip that was being pushed by several bacteria. When I pressed a button, a magnet turned on, and the chip moved to the right. When I released the button, the chip stopped and then moved randomly. In this way, I could actually steer the chip. While doing this, I realized that one day, a doctor may be pushing a similar button, but this time directing a nanorobot in the veins of a patient.

One can imagine a future where surgery is completely replaced by molecular machines moving through the bloodstream, guided by magnets, homing in on a diseased organ, and then releasing medicines or performing surgery. This could make cutting the skin totally obsolete. Or, magnets

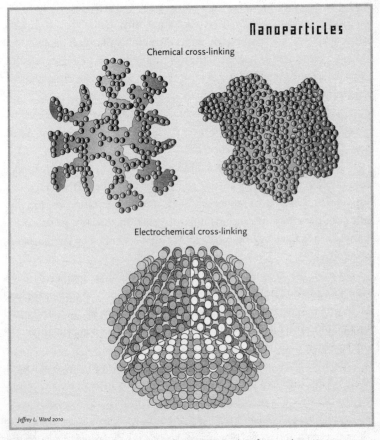

Nanoparticles

Chemical cross-linking

Electrochemical cross-linking

Jeffrey L. Ward 2010

Molecular robots will be patrolling our bloodstreams, identifying and zapping cancer cells and pathogens. They could revolutionize medicine.

could guide these nanomachines to the heart in order to remove a blockage of the arteries.

DNA CHIPS

As we mentioned in Chapter 3, in the future we will have tiny sensors in our clothes, body, and bathroom, constantly monitoring our health and detecting diseases like cancer years before they become a danger. The key to this is the DNA chip, which promises a "laboratory on a chip." Like the tricorder of *Star Trek*, these tiny sensors will give us a medical analysis within minutes.

Today, screening for cancer is a long, costly, and laborious process, often taking weeks. This severely limits the number of cancer analyses that can be performed. However, computer technology is changing all this. Already, scientists are creating devices that can rapidly and cheaply detect cancer, by looking for certain biomarkers produced by cancer cells.

Using the very same etching technology used in computer chips, it is possible to etch a chip on which there are microscopic sites that can detect specific DNA sequences or cancer cells.

Using transistor etching technology, DNA fragments are embedded into the chip. When fluids are passed over the chip, these DNA fragments can bind to specific gene sequences. Then, using a laser beam, one can rapidly scan the entire site and identify the genes. In this way, genes do not have to be read one by one as before, but can be scanned by the thousands all at once.

In 1997, the Affymetrix company released the first commercial DNA chip that could rapidly analyze 50,000 DNA sequences. By 2000, 400,000 DNA probes were available for a few thousand dollars. By 2002, prices had dropped to $200 for even more powerful chips. Prices continue to plunge due to Moore's law, down to a few dollars.

Shana Kelley, a professor at the University of Toronto's medical school, said, "Today, it takes a room filled with computers to evaluate a clinically relevant sample of cancer biomarkers and the results aren't quickly available. Our team was able to measure biomolecules on an electronic chip the size of your fingertip." She also envisions the day when all the equipment to analyze this chip will be shrunk to the size of a cell phone. This lab on a chip will mean that we can shrink a chemical laboratory found in a hospital or university down to a single chip that we can use in our own bathrooms.

Doctors at Massachusetts General Hospital have created their own custom-made biochip that is 100 times more powerful than anything on the market today. Normally, circulating tumor cells (CTCs) make up fewer than one in a million cells in our blood, but these CTCs eventually kill us if they proliferate. The new biochip is sensitive enough to find one in a billion CTCs circulating in our blood. As a result, this chip has been proven to detect lung, prostate, pancreatic, breast, and colorectal cancer cells by analyzing as little as a teaspoon of blood.

Standard etching technology carves out chips containing 78,000 microscopic pegs (each 100 microns tall). Under an electron microscope, they resemble a forest of round pegs. Each peg is coated with an antibody for the epithelial cell adhesion molecule (EpCAM), which is found in many types of cancer cells but is absent in ordinary cells. EpCAM is vital for cancer cells to communicate with one another as they form a tumor. If blood is passed through the chip, the CTC cells stick to the round pegs. In clinical trials, the chip successfully detected cancers in 115 out of 116 patients.

The proliferation of these labs on a chip will also radically affect the cost of diagnosing disease. At present, it may cost several hundred dollars to have a biopsy or chemical analysis, which might take a few weeks. In the future, it may cost a few pennies and take a few minutes. This could revolutionize the speed and accessibility of cancer diagnoses. Every time we brush our teeth, we will have a thorough checkup for a variety of diseases, including cancer.

Leroy Hood and his colleagues at the University of Washington created a chip, about 4 centimeters wide, that can test for specific proteins from a single drop of blood. Proteins are the building blocks of life. Our muscles, skin, hair, hormones, and enzymes are all made of proteins. Detecting proteins from diseases like cancer could lead to an early warning system for the body. At present, the chip costs only ten cents and can identify a specific protein within ten minutes, so it is several million times more efficient than the previous system. Hood envisions a day when a chip will be able to rapidly analyze hundreds of thousands of proteins, alerting us to a wide variety of diseases years before they become serious.

CARBON NANOTUBES

One preview of the power of nanotechnology is carbon nanotubes. In principle, carbon nanotubes are stronger than steel and can also conduct electricity, so carbon-based computers are a possibility. Although they are enormously strong, one problem is that they must be in pure form, and the longest pure carbon fiber is only a few centimeters long. But one day, entire computers may be made of carbon nanotubes and other molecular structures.

Carbon nanotubes are made of individual carbon atoms bonded to form a tube. Imagine chicken wire, where every joint is a carbon atom. Now roll up the chicken wire into a tube, and you have the geometry of a carbon nanotube. Carbon nanotubes are formed every time ordinary soot is created, but scientists never realized that carbon atoms could bond in such a novel way.

The near-miraculous properties of carbon nanotubes owe their power to their atomic structure. Usually, when you analyze a solid piece of matter, like a rock or wood, you are actually analyzing a huge composite of many overlapping structures. It is easy to create tiny fractures within this composite, which cause it to break. So the strength of a material depends on imperfections in its molecular structure. For example, graphite is made of pure carbon, but it is extremely soft because it is made of layers that can slide past each other. Each layer consists of carbon atoms, each of which is bonded with three other carbon atoms.

Diamonds are also made of pure carbon, but they are the strongest naturally occurring mineral. The carbon atoms in diamonds are arranged in a tight, interlocking crystal structure, giving them their phenomenal strength. Similarly, carbon nanotubes owe their amazing properties to their regular atomic structure.

Already, carbon nanotubes are finding their way into industry. Because of their conductivity, they can be used to create cables to carry large amounts of electrical power. Because of their strength, they can be used to create substances tougher than Kevlar.

But perhaps the most important application of carbon will be in the computer business. Carbon is one of several candidates that may eventually succeed silicon as the basis of computer technology. The future of the world economy may eventually depend on this question: What will replace silicon?

POST-SILICON ERA

As we mentioned earlier, Moore's law, one of the foundations of the information revolution, cannot last forever. The future of the world economy and the destiny of nations may ultimately hinge on which nation develops a suitable replacement for silicon.

The question—When will Moore's law collapse?—sends shudders throughout the world economy. Gordon Moore himself was asked in 2007 if he thought the celebrated law named after him could last forever. Of course not, he said, and predicted that it would end in ten to fifteen years.

This rough assessment agreed with a previous estimate made by Paolo Gargini, an Intel Fellow, who is responsible for all external research at Intel. Since the Intel Corporation sets the pace for the entire semiconductor industry, his words were carefully analyzed. At the annual Semicon West conference in 2004, he said, "We see that for at least the next fifteen to twenty years, we can continue staying on Moore's law."

The current revolution in silicon-based computers has been driven by one overriding fact: the ability of UV light to etch smaller and smaller transistors onto a wafer of silicon. Today, a Pentium chip may have several hundred million transistors on a wafer the size of your thumbnail. Because the wavelength of UV light can be as small as 10 nanometers, it is possible to use etching techniques to carve out components that are only thirty atoms across. But this process cannot continue forever. Sooner or later, it collapses, for several reasons.

First, the heat generated by powerful chips will eventually melt them. One naive solution is to stack the wafers on top of one another, creating a cubical chip. This would increase the processing power of the chip but at the expense of creating more heat. The heat from these cubical chips is so intense you could fry an egg on top of them. The problem is simple: there is not enough surface area on a cubical chip to cool it down. In general, if you pass cool water or air across a hot chip, the cooling effect is greater if you have more surface contact with the chip. But if you have a cubical chip, the surface area is not enough. For example, if you could double the size of a cubical chip, the heat it generates goes up by a factor of eight (since the cube contains eight times more electrical components), but its surface area increases only by a factor of four. This means that the heat generated in a cubical chip rises faster than the ability to cool it down. The larger the cubical chip, the more difficult it is to cool. So cubical chips will provide only a partial, temporary solution to the problem.

Some have suggested that we simply use X-rays instead of UV light to etch the circuits. In principle, this might work, since X-rays can have a wavelength 100 times smaller than UV light. But there is a trade-off. As you move from UV light to X-rays, you also increase the energy of the beam by a factor of 100 or so. This means that etching with X-rays may destroy the wafer you are trying to etch. X-ray lithography can be compared to an artist trying to use a blowtorch to create a delicate sculpture. X-ray lithography has to be very carefully controlled, so X-ray lithography is only a short-term solution.

Second, there is a fundamental problem posed by the quantum theory: the uncertainty principle, which says that you cannot know for certain the location and velocity of any atom or particle. Today's Pentium chip may have a layer about thirty atoms thick. By 2020, that layer could be five atoms across, so that the electron's position is uncertain, and it begins to leak through the layer, causing a short circuit. Thus, there is a quantum limit to how small a silicon transistor can be.

As I mentioned earlier, I once keynoted a major conference of 3,000 of Microsoft's top engineers in their headquarters in Seattle, where I highlighted the problem of the slowing down of Moore's law. These top software engineers confided to me that they are now taking this problem very seriously, and parallel processing is one of their top answers to increase computer processing power. The easiest way to solve this problem is to string a series of chips in parallel, so that a computer problem is broken down into pieces and then reassembled at the end.

Parallel processing is one of the keys to how our own brain works. If you do an MRI scan of the brain as it thinks, you find that various regions of the brain light up simultaneously, meaning that the brain breaks up a task into small pieces and processes each piece simultaneously. This explains why neurons (which carry electrical messages at the excruciatingly slow pace of 200 miles per hour) can outperform a supercomputer, in which messages travel at nearly the speed of light. What our brain lacks in speed, it more than makes up for by doing billions of small calculations simultaneously and then adding them all up.

The difficulty with parallel processing is that every problem has to be broken into several pieces. Each piece is then processed by different chips, and the problem is reassembled at the end. The coordination of this breakup can be exceedingly complicated, and it depends specifically on each problem, making a general procedure very difficult to find. The human brain does this effortlessly, but Mother Nature has had millions of years to solve this problem. Software engineers have had only a decade or so.

ATOMIC TRANSISTORS

One possible replacement for silicon chips is transistors made of individual atoms. If silicon transistors fail because wires and layers in a chip are going down in size to the atomic scale, then why not start all over again and compute on atoms?

One way of realizing this is with molecular transistors. A transistor is a switch that allows you to control the flow of electricity down a wire. It's possible to replace a silicon transistor with a single molecule, made of chemicals like rotaxane and benzenethiol. When you see a molecule of benzenethiol, it looks like a long tube, with a "knob," or valve, made of atoms in the middle. Normally, electricity is free to flow down the tube, making it conductive. But it is also possible to twist the "knob," which shuts off the flow of electricity. In this way, the entire molecule acts like a switch that can control the flow of electricity. In one position, the knob allows electricity to flow, which can represent the number "1." If the knob is turned, then the electric flow is stopped, which represents the number "0." Thus, digital messages can be sent by using molecules.

Molecular transistors already exist. Several corporations have announced that they have created transistors made of individual molecules. But before they can be commercially viable, one must be able to wire them up correctly and mass-produce them.

One promising candidate for the molecular transistor comes from a substance called graphene, which was first isolated from graphite in 2004 by Andre Geim and Kostya Novoselov of the University of Manchester, who won a Nobel Prize for their work. It is like a single layer of graphite. Unlike carbon nanotubes, which are sheets of carbon atoms rolled up into long, narrow tubes, graphene is a single sheet of carbon, no more than one atom thick. Like carbon nanotubes, graphene represents a new state of matter, so scientists are teasing apart its remarkable properties, including conducting electricity. "From the point of view of physics, graphene is a goldmine. You can study it for ages," remarks Novoselov. (Graphene is also the strongest material ever tested in science. If you placed an elephant on a pencil, and balanced the pencil on a sheet of graphene, the graphene would not tear.)

Novoselov's group has employed standard techniques used in the computer industry to carve out some of the smallest transistors ever made. Narrow beams of electrons can carve out channels in graphene, making the world's smallest transistor: one atom thick and ten atoms across. (At present, the smallest molecular transistors are about 30 nanometers in size. Novoselov's smallest transistors are thirty times smaller than that.)

These transistors of graphene are so small, in fact, they may represent the ultimate limit for molecular transistors. Any smaller, and the uncertainty principle takes over and electrons leak out of the transistor, destroying its properties. "It's about the smallest you can get," says Novoselov.

Although there are several promising candidates for molecular transistors, the real problem is more mundane: how to wire them up and assemble them into a commercially viable product. Creating a single molecular transistor is not enough. Molecular transistors are notoriously hard to manipulate, since they can be thousands of times thinner than a human hair. It is a nightmare thinking of ways to mass-produce them. At present, the technology is not yet in place.

For example, graphene is such a new material that scientists do not know how to produce large quantities of it. Scientists can produce only about .1 millimeter of pure graphene, much too small for commercial use. One hope is that a process can be found that self-assembles the molecular transistor. In nature, we sometimes find arrays of molecules that condense into a precise pattern, as if by magic. So far, no one has been able to reliably re-create this magic.

QUANTUM COMPUTERS

The most ambitious proposal is to use quantum computers, which actually compute on individual atoms themselves. Some claim that quantum computers are the ultimate computer, since the atom is the smallest unit that one can calculate on.

An atom is like a spinning top. Normally, you can store digital information on spinning tops by assigning the number "0" if the top is spinning upward, or "1" if the top is spinning down. If you flip over a spinning top, then you have converted a 0 into a 1 and have done a calculation.

But in the bizarre world of the quantum, an atom is in some sense spinning up and down simultaneously. (In the quantum world, being several places at the same time is commonplace.) An atom can therefore contain much more information than a 0 or a 1. It can describe a mixture of 0 and 1. So quantum computers use "qubits" rather than bits. For example, it can be 25 percent spinning up and 75 percent spinning down. In this way, a spinning atom can store vastly more information than a single bit.

Quantum computers are so powerful that the CIA has looked into their code-breaking potentials. When the CIA tries to break the code of another nation, it searches for the key. Nations have devised ingenious ways of constructing the key that encodes their messages. For example, the key may be based on factorizing a large number. It's easy to factorize the number 21 as the product of 3 and 7. Now let's say that you have an integer of 100 digits, and you ask a digital computer to rewrite it as the product of two other integers. It might take a digital computer a century to be able to factorize this number. A quantum computer, however, is so powerful that in principle it can effortlessly crack any such code. A quantum computer quickly outperforms a standard computer on these huge tasks.

Quantum computers are not science fiction but actually exist today. In fact, I had a chance to see a quantum computer for myself when I visited the MIT laboratory of Seth Lloyd, one of the pioneers in the field. His laboratory is full of computers, vacuum pumps, and sensors, but the heart of his experiment is a machine that resembles a standard MRI machine, except much smaller. Like the MRI machine, his device has two large coils of wire that create a uniform magnetic field in the space between them. In this uniform magnetic field, he places his sample material. The atoms inside the sample align, like spinning tops. If the atom points up, it corresponds to a 0. If it points down, it corresponds to a 1. Then he sends an electromagnetic pulse into the sample, which changes the alignment of the atoms. Some of the atoms flip over, so a 1 becomes a 0. In this way, the machine has performed a calculation.

So why don't we have quantum computers sitting on our desks, solving the mysteries of the universe? Lloyd admitted to me the real problem that has stymied research in quantum computers is the disturbances from the outside world that destroy the delicate properties of these atoms.

When atoms are "coherent" and vibrating in phase with one another, the tiniest disturbances from the outside world can ruin this delicate balance and make the atoms "decohere," so they no longer vibrate in unison. Even the passing of a cosmic ray or the rumble of a truck outside the lab can destroy the delicate spinning alignment of these atoms and destroy the computation.

The decoherence problem is the single most difficult barrier to creating quantum computers. Anyone who can solve the problem of decoherence will not only win a Nobel Prize but also become the richest man on earth.

As you can imagine, creating quantum computers out of individual coherent atoms is an arduous process, because these atoms quickly decohere and fall out of phase. So far, the world's most complex calculation done on a quantum computer is $3 \times 5 = 15$. Although this might not seem much, remember that this calculation was done on individual atoms.

In addition, there is another bizarre complication coming from the quantum theory, again based on the uncertainty principle. All calculations done on a quantum computer are uncertain, so you have to repeat the experiment many times. So $2 + 2 = 4$, at least sometimes. If you repeat the calculation of $2 + 2$ a number of times, the final answer averages out to 4. So even arithmetic becomes fuzzy on a quantum computer.

No one knows when one might solve this problem of decoherence. Vint Cerf, one of the original creators of the Internet, predicts, "By 2050, we will surely have found ways to achieve room-temperature quantum computation."

We should also point out that the stakes are so high that a variety of computer designs have been explored by scientists. Some of these competing designs include:

- **optical computers:** These computers calculate on light beams rather than electrons. Since light beams can pass through each other, optical computers have the advantage that they can be cubical, without wires. Also, lasers can be fabricated using the same lithographic techniques as ordinary transistors, so you can in theory pack millions of lasers onto a chip.

- **quantum dot computers:** Semiconductors used in chips can be etched into tiny dots so small they consist of a collection of perhaps 100 atoms. At that point, these atoms can begin to vibrate in unison. In 2009, the world's smallest quantum dot was built out of a single electron. These quantum dots have already proven their worth with light-emitting diodes and computer displays. In the future, if these quantum dots are arranged properly, they might even create a quantum computer.

- **DNA computers:** In 1994, the first computer made of DNA molecules was created at the University of Southern California. Since a strand of DNA encodes information on amino acids represented by the letters A,T,C,G instead of 0s and 1s, DNA can be viewed as ordinary computer tape, except it can store more information. In the same way that a large digital number can be manipulated and rearranged by a computer, one can also perform analogous manipulations by mixing tubes of fluids containing DNA, which can be cut and spliced in various ways. Although the process is slow, there are so many trillions of DNA molecules acting simultaneously that a DNA computer can solve certain calculations more conveniently than a digital computer. Although a digital computer is quite convenient and can be placed inside your cell phone, DNA computers are more awkward, involving mixing tubes of liquid containing DNA.

MIDCENTURY (2030 TO 2070)

SHAPE-SHIFTING

In the movie *Terminator 2: Judgment Day,* Arnold Schwarzenegger is attacked by an advanced robot from the future, a T-1000, which is made of liquid metal. Resembling a quivering mass of mercury, it can change shape and slither its way through any obstacle. It can seep through the tiniest cracks and fashion deadly weapons by reshaping its hands and feet. And then it can suddenly re-form into its original shape to carry on its murderous rampage. The T-1000 appeared to be unstoppable, the perfect killing machine.

All this was science fiction, of course. The technology of today does not allow you to change a solid object at will. Yet by midcentury a form of this shape-shifting technology may become commonplace. In fact, one of the main companies driving this technology is Intel.

Ironically, by 2050, most of the fruits of nanotechnology will be everywhere, but hidden from view. Almost every product will be enhanced via molecular manufacturing techniques, so they will become superstrong, resistant, conductive, and flexible. Nanotechnology will also give us sensors that constantly protect and help us, distributed in the environment, hidden away, beneath the surface of our consciousness. We will walk down the street and everything will appear to be the same, so we will never know how nanotechnology has changed the world around us.

But there is one consequence of nanotechnology that will be obvious.

The *Terminator* T-1000 killer robot is perhaps the most dramatic example of an object from the field called programmable matter, which may allow us one day to change the shape, color, and physical form of an object with the push of a button. On a primitive level, even a neon sign is a form of programmable matter, since you can flick a light switch and send electricity through a tube of gas. The electricity excites the gas atoms, which then decay back to their normal state, releasing light in the process. A more sophisticated version of this is the LCD display found on computer screens everywhere. The LCD contains a liquid crystal that becomes opaque when a small electrical current is applied. Thus, by regulating the electrical current flowing inside a liquid crystal, one can create colors and shapes on a screen with the push of a button.

The scientists at Intel are much more ambitious. They visualize using programmable matter to actually change the shape of a solid object, just like in science fiction. The idea is simple: create a computer chip in the shape of a tiny grain of sand. These smart grains of sand allow you to change the static electric charge on the surface, so that these grains can attract and repel each other. With one set of charges, these grains can line up to form a certain array. But you can reprogram these grains so that their electrical charges change. Instantly, these grains rearrange themselves, forming an entirely different arrangement. These grains are called "catoms" (short for claytronic atoms) since they can form a wide range of objects by simply changing their charges, much like atoms. (Programmable matter has much in common with the modular robots we saw in Chapter 2. While the modular robots contain smart blocks, about 2 inches in size, that can rearrange themselves, programmable matter shrinks these building blocks to submillimeter size and beyond.)

One of the promoters of this technology is Jason Campbell, a senior researcher at Intel. He says, "Think of a mobile device. My cell phone is too big to fit comfortably in my pocket and too small for my fingers. It's worse if I try to watch movies or do my e-mail. But if I had 200 to 300 milliliters of catoms, I could have it take on the shape of the device that I need at that moment." So one moment, I have a cell phone in my hand. The next moment, it morphs into something else. This way, I don't have to carry so many electronic gadgets.

In its laboratories, Intel has already created an array of catoms that are about an inch in size. The catom resembles a cube with scores of tiny electrodes spread evenly on its surfaces. What makes the catom unique is that you can change the charge on each of its electrodes, so that catoms bind to each other in different orientations. With one set of charges, these cubes might combine to create a large cube. Change the charges on each cube's electrode, and then the catoms disassemble and quickly rearrange themselves into an entirely different shape, such as a boat.

The point is to shrink each catom to the size of a grain of sand, or even smaller. If one day silicon-etching techniques allow us to create catoms that are as small as a cell, then we might be able to realistically change one shape into another, simply by pushing a button. Justin Rattner, a senior fellow at Intel, says, "Sometime over the next forty years, this will become everyday technology." One immediate application would be for automobile designers, airline engineers, artists, architects, and anyone who has to design three-dimensional models of their projects and then continually modify them. If one has a mold of a four-door sedan, for example, one can grab the mold, stretch it, and it suddenly morphs into a hatchback. Compress the mold a bit more and it turns into a sports car. This is far superior to molding clay, which has no memory or intelligence. Programmable matter has intelligence, can remember previous shapes, adapt to new ideas, and respond to the designers' wishes. Once the mold is finalized, the design can simply be e-mailed to thousands of other designers, who can then create exact copies.

This could have a profound effect on consumer products. Toys, for example, can be programmed to change shape by inserting new software instructions. So for Christmas, one need only download the software for a new toy, reprogram the old toy, and an entirely new toy appears. Children might celebrate Christmas not by opening presents under the tree but by downloading software for their favorite toy that Santa has e-mailed them, and the catoms making up last year's toy become the hottest thing on the market. This means that a wide array of consumer products may eventually be reduced to software programs sent over the Internet. Instead of hiring a truck to deliver your new furniture and appliances, you may simply download the software off the net and recycle your old products. Renovating homes and apartments won't be such a chore with programmable matter. In your kitchen, replacing the tiles, tabletops, appliances, and cabinets might simply involve pushing a button.

In addition, this could cut down on waste disposal. You don't have to throw out many of your unwanted things if you can simply reprogram them. If an appliance or piece of furniture breaks, you have only to reprogram it and it becomes new again.

Despite its enormous promise, there are also numerous problems facing the Intel team. One is how to orchestrate the movements of all these millions of catoms. There will be bandwidth problems when we try to upload all this information into the programmable matter. But there are also shortcuts one can take.

For example, in science fiction movies it is common to see "morphing," that is, one person suddenly changing into a monster. This used to be a very complex, tedious process to create on film, but can now be done easily by computer. First, you identify certain vectors that mark different key points on the face, such as the nose and eyes, for both the human and the monster. Each time a vector is moved, the face changes gradually. Then computers are programmed to move these vectors, from one face to the next, thereby slowly changing one face into another. In the same way, it might be possible to use shortcuts when shape-shifting a 3-D object.

Another problem is that the static electrical forces between the catoms are weak when compared to the tough interatomic forces that hold most solids together. As we have seen, quantum forces can be quite powerful, responsible for the tough properties of metals and the elastic properties of plastic. Duplicating these quantum forces with static electrical forces to ensure that these products remain stable is going to be an issue in the future.

I had a chance to witness firsthand the remarkable, rapid advances in programmable matter when I took a Science Channel film crew to visit Seth Goldstein at Carnegie Mellon University. In his laboratory you could see large stacks of cubes scattered all over a table in various sizes, each with chips inside. I saw two of these cubes bound tightly together by electrical forces, and he asked me to try to rip them apart by hand. Surprisingly, I couldn't. I found that the electrical forces binding these two cubes were quite powerful. Then he pointed out that these electrical forces would be correspondingly greater if you miniaturized the cubes. He took me to another lab, where he showed me just how small these catoms can become. By employing the same techniques used to carve out millions of transistors on silicon wafers, he could carve out microscopic catoms that were only millimeters across. In fact, they were so small that I had to look at them under a microscope to see them clearly. He hopes that eventually, by controlling their electrical forces, he can get them to arrange in any shape with a push of a button, almost like a sorcerer conjuring up anything he wants.

Then I asked him, How can you give detailed instructions to billions upon billions of catoms, so that a refrigerator, say, might suddenly transform into an oven? It seems like a programming nightmare, I said. But he replied that it wasn't necessary to give detailed instructions to every single catom. Each catom has to know only which neighbors it must attach to. If each catom is instructed to bind with only a tiny set of neighboring catoms, then the catoms would magically rearrange themselves into complex structures (much like the neurons of a baby's brain need to know only how to attach themselves to neighboring neurons as the brain develops).

Assuming that the problem of programming and stability can be solved, then by late century there is the possibility that entire buildings or even cities may rise at the push of a button. One need only lay out the location of the buildings, dig their foundations, and allow trillions of catoms to create entire cities rising from the desert or forest.

However, these Intel engineers envision the day when the catoms may even take human form. "Why not? It's an interesting thing to speculate on," says Rattner. (Then perhaps the T-1000 robot may become a reality.)

FAR FUTURE (2070 TO 2100)

HOLY GRAIL: THE REPLICATOR

By 2100, advocates of nanotechnology envision an even more powerful machine: a molecular assembler, or "replicator," capable of creating anything. It would consist of a machine perhaps the size of a washing machine. You would put the basic raw materials into the machine and then push a button. Trillions upon trillions of nanobots would then converge on the raw materials, each one programmed to take them apart molecule by molecule and then reassemble them into an entirely new product. This machine would be able to manufacture anything. The replicator would be the crowning achievement of engineering and science, the ultimate culmination of our struggles ever since we picked up the first tool back in prehistory.

One problem with the replicator is the sheer number of atoms that must be rearranged in order to copy an object. The human body, for example, has over 50 trillion cells and in excess of 10^{26} atoms. That is a staggering number, requiring a colossal amount of memory space just to store the locations of all these atoms.

But one way to overcome this problem is to create a nanobot, a still-hypothetical molecular robot. These nanobots have several key properties. First, they can reproduce themselves. If they can reproduce once, then they can, in principle, create an unlimited number of copies of themselves. So the trick is to create just the first nanobot. Second, they are capable of identifying molecules and cutting them up at precise points. Third, by following a master code, they are capable of reassembling these atoms into different arrangements. So the task of rearranging 10^{26} atoms is reduced to making a similar number of nanobots, each one designed to manipulate individual atoms. In this way, the sheer number of atoms of the body is no longer such a daunting obstacle. The real problem is creating just the first one of these mythical nanobots and letting it reproduce by itself.

However, the scientific community is split on the question of whether the full-blown dream of a nanofabricator is physically possible. A few, like Eric Drexler, a pioneer in nanotechnology and author of *The Engines of*

Creation, envision a future where all products are manufactured at the molecular level, creating a cornucopia of goods that we can only dream of today. Every aspect of society would be turned upside down by the creation of a machine that can create anything you want. Other scientists, however, are skeptical.

The late Nobel laureate Richard Smalley, for example, raised the problem of "sticky fingers" and "fat fingers" in an article in *Scientific American* in 2001. The key question is: Can a molecular nanobot be built that is nimble enough to rearrange molecules at will? He said the answer was no.

This debate spilled open when Smalley squared off with Drexler in a series of letters, reprinted in the pages of *Chemical and Engineering News* in 2003 to 2004. The repercussions of that debate are being felt even today. Smalley's position was that the "fingers" of a molecular machine would not be able to perform this delicate task for two reasons.

First, the "fingers" would face tiny attractive forces that would make them stick to other molecules. Atoms stick to each other, in part, because of tiny electrical forces, like the van der Waals force, that exist between their electrons. Think of trying to repair a watch when your tweezers are covered with honey. Assembling anything as delicate as watch components would be impossible. Now imagine assembling something even more complicated than a watch, like a molecule, that constantly sticks to your fingers.

Second, these fingers might be too "fat" to manipulate atoms. Think of trying to repair that watch wearing thick cotton gloves. Since the "fingers" are made of individual atoms, as are the objects being manipulated, the fingers may simply be too thick to perform the delicate operations needed.

Smalley concluded, "Much like you can't make a boy and a girl fall in love with each other simply by pushing them together, you cannot make precise chemistry occur as desired between two molecular objects with simple mechanical motion. . . . Chemistry, like love, is more subtle than that."

This debate goes to the very heart of whether a replicator will one day revolutionize society or be treated as a curiosity and relegated to the trash bin of technology. As we have seen, the laws of physics in our world do not easily translate to the physics of the nanoworld. Effects that we can ignore, such as van der Waals forces, surface tension, the uncertainty principle, the Pauli exclusion principle, etc., become dominant in the nanoworld.

To appreciate this problem, imagine that the atom is the size of a marble and that you have a swimming pool full of these atoms. If you fell into the swimming pool, it would be quite different from falling into a swimming pool of water. These "marbles" would be constantly vibrating and hitting you from all directions, because of Brownian motion. Trying to swim in this pool would be almost impossible, since it would be like trying to swim in molasses. Every time you tried to grab one of the marbles, it would either move away from you or stick to your fingers, due to a complex combination of forces.

In the end, both scientists agreed to disagree. Although Smalley was unable to throw a knockout punch against the molecular replicator, several things became clear after the dust settled. First, both agreed that the naive idea of a nanobot armed with molecular tweezers cutting and pasting molecules had to be modified. New quantum forces become dominant at the atomic scale.

Second, although this replicator, or universal fabricator, is science fiction today, a version of it already exists. Mother Nature, for example, can take hamburgers and vegetables and turn them into a baby in just nine months. This process is carried out by DNA molecules (which encode the blueprint for the baby) that guide the actions of ribosomes (which cut and splice the molecules into correct order) using the proteins and amino acids present in your food.

And third, a molecular assembler might work, but in a more sophisticated version. For example, as Smalley pointed out, bringing two atoms together does not guarantee a reaction. Mother Nature often gets around this problem by employing a third party, an enzyme in a water solution, to facilitate a chemical reaction. Smalley pointed out that many chemicals found in computers and the electronics industry cannot be dissolved in water. But Drexler countered by saying that not all chemical reactions involve water or enzymes.

One possibility, for example, is called self-assembly, or the bottom-up approach. Since antiquity, humans have used the top-down approach to building. With tools like a hammer and saw, one begins to cut wood and then piece together boards to create larger structures like a house according to a plan. You have to carefully guide this process from above at every step of the way.

In the bottom-up approach, things assemble by themselves. In nature, for example, beautiful snowflakes crystallize all by themselves in a thunderstorm. Trillions upon trillions of atoms rearrange to create novel forms. No one has to design each snowflake. This often occurs in biological systems as well. Bacterial ribosomes, which are complex molecular systems containing at least fifty-five different protein molecules and several RNA molecules, can spontaneously self-assemble in a test tube.

Self-assembly is also used in the semiconductor industry. Components used in transistors sometimes assemble by themselves. By applying various complex techniques and processes in a precise sequence (such as quenching, crystallization, polymerization, vapor deposition, solidification, etc.) one can produce a variety of commercially valuable computer components. As we saw earlier, a certain type of nanoparticle used against cancer cells can be produced using this method.

However, most things do not create themselves. In general, only a tiny fraction of nanomaterials have been shown to self-assemble properly. You cannot order a nanomachine using self-assembly like you can order from a menu. So progress in creating nanomachines this way will be steady but slow.

In sum, molecular assemblers apparently violate no law of physics, but they will be exceedingly difficult to build. Nanobots do not exist now, and will not in the near future, but once (and if) the first nanobot is successfully produced, it might alter society as we know it.

BUILDING A REPLICATOR

What might a replicator look like? No one knows exactly, since we are decades to a century away from actually building one, but I got a taste of how a replicator might appear when I had my head examined (literally). For a Science Channel special, they created a realistic 3-D copy of my face out of plastic by scanning a laser beam horizontally across my face. As the beam bounced off my skin, the reflection was recorded by a sensor that fed the image into a computer. Then the beam made the next pass across my face, but slightly lower. Eventually, it scanned my entire face, dividing it up into many horizontal slices. By looking at a computer screen, you could see a 3-D image of the surface of my face emerge, to an accuracy of perhaps a tenth of a millimeter, consisting of these horizontal slices.

Then this information was fed into a large device, about the size of a refrigerator, that can create a plastic 3-D image of almost anything. The device has a tiny nozzle that moves horizontally, making many passes. On each pass, it sprays out a tiny amount of molten plastic, duplicating the original laser image of my face. After about ten minutes and numerous passes, the mold emerged from this machine, bearing an eerie resemblance to my face.

The commercial applications of this technology are enormous, since you can create a realistic copy of any 3-D object, such as complicated machine parts, within a matter of a few minutes. However, one can imagine a device that, decades to centuries from now, may be able to create a 3-D copy of a real object, down to the cellular and atomic level.

At the next level, it is possible to use this 3-D scanner to create living organs of the human body. At Wake Forest University, scientists have pioneered a novel way to create living heart tissue, with an ink-jet printer. First, they have to carefully write a software program that successively sprays out living heart cells as the nozzle makes each pass. For this, they use an ordinary ink-jet printer but one whose ink cartridge is filled with a mixture of fluids containing living heart cells. In this way, they have control over the precise 3-D placement of every cell. After multiple passes, they can actually create the layers of heart tissue.

There is another instrument that might one day record the location of every atom of our body: the MRI. As we observed earlier, the accuracy of the MRI scan is about a tenth of a millimeter. This means that every pixel of a sensitive MRI scan may contain thousands of cells. But if you examine the physics behind the MRI, you find that the accuracy of the image is related to the uniformity of the magnetic field within the machine. Thus, by making the magnetic field increasingly uniform, one can even go below a tenth of a millimeter.

Already, scientists are envisioning an MRI-type machine with a resolution down to the size of a cell, and even smaller, one that can scan down to the individual molecules and atoms.

In summary, a replicator does not violate the laws of physics, but it would be difficult to create using self-assembly. By late in this century, when the techniques of self-assembly are finally mastered, we can think about commercial applications of replicators.

GRAY GOO?

Some people, including Bill Joy, a founder of Sun Microsystems, have expressed reservations about nanotechnology, writing that it's only a matter of time before the technology runs wild, devours all the minerals of the earth, and spits out useless "gray goo" instead. Even Prince Charles of England has spoken out against nanotechnology and the gray-goo scenario.

The danger lies in the key property of these nanobots: they can reproduce themselves. Like a virus, they cannot be recalled once they are let loose into the environment. Eventually, they could proliferate wildly, taking over the environment and destroying the earth.

My own belief is that there are many decades to centuries before this technology is mature enough to create a replicator, so concerns about the gray goo are premature. As the decades pass, there will be plenty of time to design safeguards against nanobots that run amok. For example, one can design a fail-safe system so that, by pressing a panic button, all the nanobots are rendered useless. Or one could design "killer bots," specifically designed to seek out and destroy nanobots that have run out of control.

Another way to deal with this is to study Mother Nature, who has had billions of years of experience with this problem. Our world is full of self-replicating molecular life-forms, called viruses and bacteria, that can proliferate out of control and mutate as well. However, our body has also created "nanobots" of its own, antibodies and white blood cells in our immune system that seek out and destroy alien life-forms. The system is certainly not perfect, but it provides a model for dealing with this out-of-control-nanobot problem.

SOCIAL IMPACT OF REPLICATORS

For a BBC/Discovery Channel special I once hosted, Joel Garreau, author of *Radical Evolution*, said, "If a self-assembler ever does become possible, that's going to be one of history's great 'holy s—!' moments. Then you are really talking about changing the world into something we've never recognized before."

There is an old saying, Be careful what you wish for, because it may come true. The holy grail of nanotechnology is to create the molecular assembler, or replicator, but once it is invented, it could alter the very foundation of society itself. All philosophies and social systems are ultimately based on scarcity and poverty. Throughout human history, this has been the dominant theme running through society, shaping our culture, philosophy, and religion. In some religions, prosperity is viewed as a divine reward and poverty as just punishment. Buddhism, by contrast, is based on the universal nature of suffering and how we cope with it. In Christianity, the New Testament reads: "It is easier for a camel to go through the eye of a needle than for a rich man to enter into the kingdom of God."

The distribution of wealth also defines the society itself. Feudalism is based on preserving the wealth of a handful of aristocrats against the poverty of the peasants. Capitalism is based on the idea that energetic, productive people are rewarded for their labors by starting companies and getting rich. But if lazy, nonproductive individuals can get as much as they want almost for free by pushing a button, then capitalism no longer works. A replicator upsets the entire apple cart, turning human relations upside down. The distinctions between the haves and have-nots may disappear, and along with it the notion of status and political power.

This conundrum was explored in an episode in *Star Trek: The Next Generation,* in which a capsule from the twentieth century is found floating in outer space. Inside the capsule are the frozen bodies of people who suffered from incurable diseases of that primitive time period, hoping to be revived in the future. The doctors of the starship *Enterprise* quickly cure these individuals of their diseases and revive them. These fortunate individuals are surprised that their gamble paid off, but one of them is a shrewd capitalist. The first thing he asks is: What time period is this? When he finds out that he is now alive in the twenty-fourth century, he quickly realizes that his investments must today be worth a fortune. He immediately demands to contact his banker back on earth. But the crew of the *Enterprise* is bewildered. Money? Investments? These do not exist in the future. In the twenty-fourth century, you simply ask for something, and it is given to you.

This also calls into question the search for the perfect society, or utopia, a word coined in the novel written by Sir Thomas More in 1516 titled *Utopia.* Appalled by the suffering and squalor he saw around him, he envisioned a paradise on a fictional island in the Atlantic Ocean. In the nineteenth century, there were many social movements in Europe that searched for various forms of utopia, and many of them eventually found sanctuary by escaping to the United States, where we see evidence of their settlements even today.

On one hand, a replicator could give us the utopia that was once envisioned by nineteenth-century visionaries. Previous experiments in utopia failed because of scarcity, which led to inequalities, then bickering, and ultimately collapse. But if replicators solve the problem of scarcity, then perhaps utopia is within reach. Art, music, and poetry will flourish, and people will be free to explore their fondest dreams and wishes.

On the other hand, without the motivating factor of scarcity and money, it could lead to a self-indulgent, degenerate society that sinks to the lowest level. Only a tiny handful, the most artistically motivated, will strive to write poetry. The rest of us, the critics claim, will become good-for-nothing loafers and slackers.

Even the definitions used by the utopians are called into question. The mantra for socialism, for example, is: "From each according to his ability, to each according to his contribution." The mantra for communism, the highest stage of socialism, is: "From each according to his ability, to each according to his need."

But if replicators are possible, then the mantra simply becomes: "To each according to his desire."

There is, however, a third way of looking at this question. According to the Cave Man Principle, people's basic personalities have not changed much in the past 100,000 years. Back then, there was no such thing as a job. Anthropologists say that primitive societies were largely communal, sharing goods and hardships equally. Daily rhythms were not governed by a job and pay, since neither of them existed.

Yet people back then did not become loafers, for several reasons. First, they would starve to death. People who did not do their share of the work were simply thrown out of the tribe, and they soon perished. Second, people became proud of their work, and even found meaning in their tasks. Third, there was enormous social pressure to remain a productive member of society. Productive individuals could marry to pass their genes onto the next generation, while the genes of loafers usually died with them.

So why will people live productive lives when replicators are invented and everyone can have anything they want? First of all, replicators would guarantee that no one starves. But second, most people will probably still continue to work because they are proud of their skills and find meaning in their labor. But the third reason, social pressure, is harder to maintain without infringing on personal liberties. Instead of social pressure there would probably have to be a major shift in education to change people's attitudes toward work and reward, so that the replicator is not abused.

Fortunately, since progress will be slow and the replicator is a century or so away, society will have plenty of time to debate the merits and implications of this technology and adjust to this new reality so that society does not disintegrate.

More than likely, the first replicators will be expensive. As MIT robotics expert Rodney Brooks says, "Nanotechnology will thrive, much as photolithography thrives—in very expensive, controlled situations rather than as a freestanding mass-market technology." The problem of unlimited free goods will not be so much a problem. Given the sophistication of these machines, it may take many decades after they are first created to bring down the cost.

I once had an interesting conversation with Jamais Cascio, a leading futurist with a long career of thoughtfully contemplating the outlines of tomorrow. First, he told me that he doubted the singularity theory mentioned in Chapter 2, observing that human nature and social dynamics are much too messy, complicated, and unpredictable to be fit into a simple neat theory. But he also admitted that remarkable advances in nanotechnology might eventually create a society in which there was an overabundance of goods, especially with replicators and robots. So I asked him: How will society behave when goods are nearly for free, when society is finally so rich that there is no necessity to work?

Two things would happen, he said. First, he thought there would be enough wealth to guarantee a decent, minimum income for everyone, even if they did not work. So there probably would be a fraction of the population who become permanent slackers. He foresaw a permanent safety net for society. This might be undesirable, but it is unavoidable, especially if replicators and robots meet all our material needs. Second, this would be compensated for, he thought, by unleashing a revolution in the entrepreneurial spirit. Freed from the fear of plunging into poverty and ruin, the more industrious individuals would have more initiative and take on additional risks to create new industries and new opportunities for others. He foresaw a new renaissance of society, as the creative spirit was unleashed from the fear of bankruptcy.

In my own field, physics, I see that most of us engage in physics not for the money but for the sheer joy of discovery and innovation. Often, we passed up lucrative jobs in other fields because we wanted to pursue a dream, not the dollar. The artists and intellectuals I know also feel the same way—that their goal is not to amass as big a bank account as possible but to be creative and ennoble the human spirit.

Personally, if by 2100 society becomes so rich that we are surrounded by material wealth, I feel that society may react in a similar way. A fraction of the population will form a permanent class of people who simply refuse to work. Others may be liberated from the constraints of poverty and pursue creative scientific and artistic achievement. For them, the sheer joy of being creative, innovative, and artistic will outweigh the lure of a materialistic world. But the majority will continue to work and be useful simply because it is part of our genetic heritage, the Cave Man Principle within us.

But there is one problem that even replicators cannot solve. And this is the problem of energy. All these miraculous technologies need vast amounts of energy to drive them. Where will this energy come from?

The Stone Age did not end for lack of stone. And the Oil Age will end long before the world runs out of oil.

—JAMES CANTON

In my mind, (fusion) ranks with the original gift of fire, back in the mists of prehistory.

—BEN BOVA

5 FUTURE OF ENERGY *Energy from the Stars*

The stars were the energy source of the gods. When Apollo rode across the sky in a chariot drawn by fire-breathing horses, he illuminated the heavens and the earth with the infinite power of the sun. His power was rivaled only by that of Zeus himself. Once, when Semele, one of Zeus's numerous mortal lovers, begged to see him in his true form, he reluctantly obliged. The resulting burst of blinding, cosmic energy burned her to a crisp.

In this century, we will harness the power of the stars, the energy source of the gods. In the short term, this means ushering in an era of solar/hydrogen power to replace fossil fuels. But in the long term, it means harnessing the power of fusion and even solar energy from outer space. Further advances in physics could usher in the age of magnetism, whereby cars, trains, and even skateboards will float through the air on a cushion of magnetism. Our energy consumption could be drastically reduced, since almost all the energy used in cars and trains is simply to overcome the friction of the road.

END OF OIL?

Today our planet is thoroughly wedded to fossil fuels in the form of oil, natural gas, and coal. Altogether, the world consumes about 14 trillion watts of power, of which 33 percent comes from oil, 25 percent from coal, 20 percent from gas, 7 percent from nuclear, 15 percent from biomass and hydroelectric, and a paltry .5 percent from solar and renewables.

Without fossil fuels, the world economy would come to a grinding halt.

One man who clearly saw the end of the age of oil was M. King Hubbert, a Shell Oil petroleum engineer. In 1956, Hubbert presented a far-reaching talk to the American Petroleum Institute, making a disturbing prediction that was universally derided by his colleagues at the time. He predicted that U.S. oil reserves were being depleted so rapidly that soon 50 percent of the oil would be taken out of the ground, triggering an irreversible era of decline that would set in between 1965 and 1971. He saw that the total amount of oil in the United States could be plotted as a bell-shaped curve, and that we were then near the top of that curve. From then on, things could only go downhill, he predicted. This meant that oil would become increasingly difficult to extract, hence the unthinkable would happen: the United States would begin importing oil.

His prediction seemed rash, even outlandish and irresponsible, since the United States was still pumping an enormous amount of oil from Texas and elsewhere in this country. But oil engineers are not laughing anymore. Hubbert's prediction was right on the button. By 1970, U.S. oil production peaked at 10.2 million barrels a day and then fell. It has never recovered. Today, the United States imports 59 percent of its oil. In fact, if you compare a graph of Hubbert's estimates made decades ago with a graph of actual U.S. oil production through 2005, the two curves are almost identical.

Now the fundamental question facing oil engineers is: Are we at the top of Hubbert's peak in world oil reserves? Back in 1956, Hubbert also predicted that global oil production would peak in about fifty years. He could be right again. When our children look back at this era, will they view fossil fuels the same way we view whale oil today, as an unfortunate relic of the distant past?

I have lectured many times in Saudi Arabia and throughout the Middle East, speaking about science, energy, and the future. On one hand, Saudi Arabia has 267 billion barrels of oil, so this country seems to be floating on a huge underground lake of crude oil. Traveling throughout Saudi Arabia and the Persian Gulf states, I could see an exorbitant waste of energy, with huge fountains gushing in the middle of the desert, creating mammoth artificial ponds and lakes. In Dubai, there is even an indoor ski slope with thousands of tons of artificial snow, in utter defiance of the sweltering heat outside.

But now the oil ministers are worried. Behind all the rhetoric of "proven oil reserves," which are supposed to reassure us that we will have plenty of oil for decades to come, there is the realization that many of these authoritative oil figures are a deceptive form of make-believe. "Proven oil reserves" sounds soothingly authoritative and definitive, until you realize that the reserves are often the creation of a local oil minister's wishful thinking and political pressure.

Speaking to the experts in energy, I could see that a rough consensus is emerging: we are either at the top of Hubbert's peak for world oil production, or are perhaps a decade away from that fateful point. This means that in the near future, we may be entering a period of irreversible decline.

Of course, we will never totally run out of oil. New pockets are being found all the time. But the cost of extracting and refining these will gradually skyrocket. For example, Canada has huge tar sands deposits, enough to supply the world's oil for decades to come, but it is not cost-effective to extract and refine it. The United States probably has enough coal reserves to last 300 years, but there are legal restrictions, and the cost of extracting all the particulate and gaseous pollutants is onerous.

Furthermore, oil continues to be found in politically volatile regions of the world, contributing to foreign instability. Oil prices, when graphed over the decades, are like a roller-coaster ride, peaking at an astonishing $140 per barrel in 2008 (and more than $4 per gallon at the gas pump) and then plunging due to the great recession. Although there are wild swings, due to political unrest, speculation, rumors, etc., one thing is clear: the average price of oil will continue to rise over the long term.

This will have profound implications for the world economy. The rapid rise of modern civilization in the twentieth century has been fueled by two things: cheap oil and Moore's law. With energy prices rising, this puts pressure on the world's food supply as well as on the control of pollution. As novelist Jerry Pournelle has said, "Food and pollution are not primary problems: they are energy problems. Given sufficient energy we can produce as much food as we like, if need be, by high-intensity means such as hydroponics and greenhouses. Pollution is similar: given enough energy, pollutants can be transformed into manageable products; if need be, disassembled into their constituent products."

We also face another issue: the rise of a middle class in China and India, one of the great demographic changes of the postwar era, which has created enormous pressure on oil and commodity prices. Seeing McDonald's hamburgers and two-car garages in Hollywood movies, they also want to live the American dream of wasteful energy consumption.

NEAR FUTURE (PRESENT TO 2030)

SOLAR/HYDROGEN ECONOMY

In this regard, history seems to be repeating itself. Back in the 1900s, Henry Ford and Thomas Edison, two longtime friends, made a bet as to which form of energy could fuel the future. Henry Ford bet on oil replacing coal, with the internal combustion engine replacing steam engines. Thomas Edison bet on the electric car. It was a fateful bet, whose outcome would have a profound effect on world history. For a while, it appeared that Edison would win the bet, since whale oil was extremely hard to get. But the rapid discovery of cheap oil deposits in the Middle East and elsewhere soon had Ford emerging victorious. The world has never been the same since. Batteries could not keep up with the phenomenal success of gasoline. (Even today, pound for pound, gasoline contains roughly forty times more energy than a battery.)

But now the tide is slowly turning. Perhaps Edison will win yet, a century after the bet was made.

The question being asked in the halls of government and industry is: What will replace oil? There is no clear answer. In the near term, there is no immediate replacement for fossil fuels, and there most likely will be an energy mix, with no one form of energy dominating the others.

But the most promising successor is solar/hydrogen power (based on renewable technologies like solar power, wind power, hydroelectric power, and hydrogen).

At the present time, the cost of electricity produced from solar cells is several times the price of electricity produced from coal. But the cost of solar/hydrogen keeps plunging due to steady technological advances, while the cost of fossil fuels continues its slow rise. It is estimated that within ten to fifteen years or so, the two curves will cross. Then market forces will do the rest.

WIND POWER

In the short term, renewables like wind power are a big winner. Worldwide, generating capacity from wind grew from 17 billion watts in 2000 to 121 billion watts in 2008. Wind power, once considered a minor player, is becoming increasingly prominent. Recent advances in wind turbine technology have increased the efficiency and productivity of wind farms, which are one of the fastest-growing sectors of the energy market.

The wind farms of today are a far cry from the old windmills that used to power farms and mills in the late 1800s. Nonpolluting and safe, a single wind power generator can produce 5 megawatts of power, enough for a small village. A wind turbine has huge, sleek blades, about 100 feet long, that turn with almost no friction. Wind turbines create electricity in the same way as hydroelectric dams and bicycle generators. The rotating motion spins a magnet inside a coil. The spinning magnetic field pushes electrons inside the coil, creating a net current of electricity. A large wind farm, consisting of 100 windmills, can produce 500 megawatts, comparable to the 1,000 megawatts produced by a single coal-burning or nuclear power plant.

Over the past few decades, Europe has been the world's leader in wind technology. But recently, the United States overtook Europe in generating electricity from wind. In 2009, the United States produced just 28 billion watts from wind power. But Texas alone produces 8 billion watts from wind power and has 1 billion watts in construction, and even more in development. If all goes as planned, Texas will generate 50 billion watts of electrical power from wind, more than enough to satisfy the state's 24 million people.

China will soon surpass the United States in wind power. Its Wind Base program will create six wind farms with a generating capacity of 127 billion watts.

Although wind power looks increasingly attractive and will undoubtedly grow in the future, it cannot supply the bulk of energy for the world. At best, it will be an integral part of a larger energy mix. Wind power faces several problems. Wind power is generated only intermittently, when the wind blows, and only in a few key regions of the world. Also, because of losses in the transmission of electricity, wind farms have to be close to cities, which further limits their usefulness.

HERE COMES THE SUN

Ultimately, all energy comes from the sun. Even oil and coal are, in some sense, concentrated sunlight, representing the energy that fell on plants and animals millions of years ago. As a consequence, the amount of concentrated sunlight energy stored within a gallon of gasoline is much larger than the energy we can store in a battery. That was the fundamental problem facing Edison in the last century, and it is the same problem today.

Solar cells operate by converting sunlight directly into electricity. (This process was explained by Einstein in 1905. When a particle of light, or a photon, hits a metal, it kicks out an electron, thereby creating a current.)

Solar cells, however, are not efficient. Even after decades of hard work by engineers and scientists, solar cell efficiency hovers around 15 percent. So research has gone in two directions. The first is to increase the efficiency of solar cells, which is a very difficult technical problem. The other is to reduce the cost of the manufacture, installation, and construction of solar parks.

For example, one might be able to supply the electrical needs of the United States by covering the entire state of Arizona with solar cells, which is impractical. However, land rights to large chunks of Saharan real estate have suddenly become a hot topic, and investors are already creating massive solar parks in this desert to meet the needs of European consumers.

Or in cities, one might be able to reduce the cost of solar power by covering homes and buildings with solar cells. This has several advantages, including eliminating the losses that occur during the transmission of power from a central power plant. The problem is one of reducing costs. A quick calculation shows that you would have to squeeze every possible dollar to make these ventures profitable.

Although solar power still has not lived up to its promise, the recent instability in oil prices has spurred efforts to finally bring solar power to the marketplace. The tide could be turning. Records are being broken every few months. Solar voltaic production is growing by 45 percent per year, almost doubling every two years. Worldwide, photovoltaic installation is now 15 billion watts, growing by 5.6 billion watts in 2008 alone.

In 2008, Florida Power & Light announced the largest solar plant project in the United States. The contract was given by SunPower, which plans to generate 25 megawatts of power. (The current record holder in the United States is the Nellis Air Force Base in Nevada, with a solar plant that generates 15 megawatts of solar power.)

In 2009, BrightSource Energy, based in Oakland, California, announced plans to beat that record by building fourteen solar plants, generating 2.6 billion watts, across California, Nevada, and Arizona.

One of BrightSource's projects is the Ivanpah solar plant, consisting of three solar thermal plants to be based in Southern California, which will produce 440 megawatts of power. In a joint project with Pacific Gas and Electric, BrightSource plans to build a 1.3 billion watt plant in the Mojave Desert.

In 2009, First Solar, the world's largest manufacturer of solar cells, announced that it will create the world's largest solar plant just north of the Great Wall of China. The ten-year contract, whose details are still being hammered out, envisions a huge solar complex containing 27 million thin-film solar panels that will generate 2 billion watts of power, or the equivalent of two coal-fired plants, producing enough energy to supply 3 million homes. The plant, which will cover twenty-five square miles, will be built in Inner Mongolia and is actually part of a much larger energy park. Chinese officials state that solar power is just one component of this facility, which will eventually supply 12 billion watts of power from wind, solar, biomass, and hydroelectric.

It remains to be seen whether these ambitious projects will finally negotiate the gauntlet of environmental inspections and cost overruns, but the point is that solar economics are gradually undergoing a sea change, with large solar companies seriously viewing solar power as being competitive with fossil fuel plants.

ELECTRIC CAR

Since about half the world's oil is used in cars, trucks, trains, and planes, there is enormous interest in reforming that sector of the economy. There is now a race to see who will dominate the automotive future, as nations make the historic transition from fossil fuels to electricity. There are several stages in this transition. The first is the hybrid car, already on the market, which uses a combination of electricity from a battery and gasoline. This design uses a small internal combustion engine to solve the long-standing problems with batteries: it is difficult to create a battery that can operate for long distances as well as provide instantaneous acceleration.

But the hybrid is the first step. The plug-in hybrid car, for example, has a battery powerful enough to run the car on electrical power for the first fifty miles or so before the car has to switch to its gasoline engine. Since most people do their commuting and shopping within fifty miles, it means that these cars are powered only by electricity during that time.

One major entry into the plug-in hybrid race is the Chevy Volt, made by General Motors. It has a range of 40 miles (using only a lithium-ion battery) and a range of 300 miles using the small gasoline engine.

And then there is the Tesla Roadster, which has no gasoline engine at all. It is made by Tesla Motors, a Silicon Valley company that is the only one in North America selling fully electric cars in series production. The Roadster is a sleek sports car that can go head-to-head with any gasoline-fired car, putting to rest the idea that electric lithium-ion batteries cannot compete against gasoline engines.

I had a chance to drive a two-seat Tesla, owned by John Hendricks, founder of Discovery Communications, the parent company of the Discovery Channel. As I sat in the driver's seat, Mr. Hendricks urged me to hit the accelerator with all my might to test his car. Taking his advice, I floored the accelerator. Immediately, I could feel the sudden surge in power. My body sank into the seat as I hit 60 miles per hour in just 3.9 seconds. It is one thing to hear an engineer boast about the performance of fully electric cars; it is another thing to hit the accelerator and feel it for yourself.

The successful marketing of the Tesla has forced mainstream automakers to play catch-up, after decades of putting down the electric car. Robert Lutz, when he was vice chairman of General Motors, said, "All the geniuses here at General Motors kept saying lithium-ion technology is ten years away, and Toyota agreed with us—and boom, along comes Tesla. So I said, 'How come some tiny little California startup, run by guys who know nothing about the car business, can do this and we can't?'"

Nissan Motors is leading the charge to introduce the fully electric car to the average consumer. It is called the Leaf, has a range of 100 miles, a top speed of up to ninety miles per hour, and is fully electric.

After the fully electric car, another car that will eventually hit the showrooms is the fuel cell car, sometimes called the car of the future. In June 2008, Honda Motor Company announced the debut of the world's first commercially available fuel cell car, the FCX Clarity. It has a range of 240 miles, has a top speed of 100 miles per hour, and has all the amenities of a standard four-door sedan. Using only hydrogen as fuel, it needs no gasoline and no electric charge. However, because the infrastructure for hydrogen does not yet exist, it is available for leasing in the United States only in Southern California. Honda is also advertising a sports car version of its fuel cell car, called the FC Sport.

Then in 2009, GM, emerging from bankruptcy after its old management was summarily fired, announced that its fuel cell car, the Chevy Equinox, had passed the million-mile mark in terms of testing. For the past twenty-five months 5,000 people have been testing 100 of these fuel cell cars. Detroit, chronically lagging behind Japan in introducing small car technology and hybrids, is trying to get a foothold in the future.

On the surface, the fuel cell car is the perfect car. It runs by combining hydrogen and oxygen, which then turns into electrical energy, leaving only water as the waste product. It creates not an ounce of smog. It's almost eerie looking at the tailpipe of a fuel cell car. Instead of choking on the toxic fumes billowing from the back, all you see are colorless, odorless droplets of water.

"You put your hand over the exhaust pipe and the only thing coming out is water. That was such a cool feeling," observed Mike Schwabl, who test-drove the Equinox for ten days.

Fuel cell technology is nothing new. The basic principle was demonstrated as far back as 1839. NASA has used fuel cells to power its instruments in space for decades. What is new is the determination of car manufacturers to increase production and bring down costs.

Another problem facing the fuel cell car is the same problem that dogged Henry Ford when he marketed the Model T. Critics claimed that gasoline was dangerous, that people would die in horrible car accidents, being burned alive in a crash. Also, you would have to have a gasoline pump on nearly every block. On all these points, the critics were right. People do die by the thousands every year in gruesome car accidents, and we see gasoline stations everywhere. But the convenience and utility of the car are so great that people ignore these facts.

Now the same objections are being raised against fuel cell cars. Hydrogen fuel is volatile and explosive, and hydrogen pumps would have to be built every few blocks. Most likely, the critics are right again. But once the hydrogen infrastructure is in place, people will find pollution-free fuel cell cars to be so convenient that they will overlook these facts. Today, there are only seventy refueling stations for fuel cell cars in the entire United States. Since fuel cell cars have a range of about 170 miles per fill-up, it means you have to watch the fuel meter carefully as you drive. But this will change gradually, especially if the price of the fuel car begins to drop with mass production and advances in technology.

But the main problem with the electric car is that the electric battery does not create energy from nothing. You have to charge the battery in the first place, and that electricity usually comes from a coal-burning plant. So even though the electric car is pollution free, ultimately the energy source for it is fossil fuels.

Hydrogen is not a net producer of energy. Rather, it is a carrier of energy. You have to create hydrogen gas in the first place. For example, you have to use electricity to separate water into hydrogen and oxygen. So although electric and fuel cell cars give us the promise of a smog-free future, there is still the problem that the energy they use comes largely from burning coal. Ultimately, we bump up against the first law of thermodynamics: the total amount of matter and energy cannot be destroyed or created out of nothing. You can't get something for nothing.

This means that, as we make the transition from gasoline to electricity, we need to replace the coal-burning plants with an entirely new form of energy.

NUCLEAR FISSION

One possibility to create energy, rather than just transmit energy, is by splitting the uranium atom. The advantage is that nuclear energy does not produce copious quantities of greenhouse gases, like coal- and oil-burning plants, but technical and political problems have tied nuclear power in knots for decades. The last nuclear power plant in the United States began construction in 1977, before the fateful 1979 accident at Three Mile Island, which crippled the future of commercial nuclear energy. The devastating 1986 accident at Chernobyl sealed the fate of nuclear power for a generation. Nuclear power projects dried up in the United States and Europe, and were kept on life support in France, Japan, and Russia only through generous subsidies from the government.

The problem with nuclear energy is that when you split the uranium atom, you produce enormous quantities of nuclear waste, which is radioactive for thousands to tens of millions of years. A typical 1,000-megawatt reactor produces about thirty tons of high-level nuclear waste after one year. It is so radioactive that it literally glows in the dark, and has to be stored in special cooling ponds. With about 100 commercial reactors in the United States, this amounts to thousands of tons of high-level waste being produced per year.

This nuclear waste causes problems for two reasons. First, it remains hot even after the reactor has been turned off. If the cooling water is accidentally shut off, as in Three Mile Island, then the core starts to melt. If this molten metal comes into contact with water, it can cause a steam explosion that can blow the reactor apart, spewing tons of high-level radioactive debris into the air. In a worst-case class-9 nuclear accident, you would have to immediately evacuate perhaps millions of people out to 10 to 50 miles from the reactor. The Indian Point reactor is just 24 miles north of New York City. One government study estimated that an accident at Indian Point could conceivably cost hundreds of billions of dollars in property damages. At Three Mile Island, the reactor came within minutes of a major catastrophe that would have crippled the Northeast. Disaster was narrowly averted when workers successfully reintroduced cooling water into the core barely thirty minutes before the core would have reached the melting point of uranium dioxide.

At Chernobyl, outside Kiev, the situation was much worse. The safety mechanism (the control rods) were manually disabled by the workers. A small power surge occurred, which sent the reactor out of control. When cold water suddenly hit molten metal, it created a steam explosion that blew off the entire top of the reactor, releasing a large fraction of the core into the air. Many of the workers sent in to control the accident eventually died horribly of radiation burns. With the reactor fire burning out of control, eventually the Red Air Force had to be called in. Helicopters with special shielding were sent in to spray borated water onto the flaming reactor. Finally, the core had to be encased in solid concrete. Even today, the core is still unstable and continues to generate heat and radiation.

In addition to the problems of meltdowns and explosions, there is also the problem of waste disposal. Where do we put it? Embarrassingly, fifty years into the atomic age, there is still no answer. In the past, there has been a string of costly errors with regard to the permanent disposal of the waste. Originally, some waste was simply dumped into the oceans by the United States and Russia, or buried in shallow pits. In the Ural Mountains one plutonium waste dump even exploded catastrophically in 1957, requiring a massive evacuation and causing radiological damage to a 400-square-mile area between Sverdlovsk and Chelyabinsk.

Originally, in the 1970s the United States tried to bury the high-level waste in Lyons, Kansas, in salt mines. But later, it was discovered that the salt mines were unusable, as they already were riddled with numerous holes drilled by oil and gas explorers. The United States was forced to close the Lyons site, an embarrassing setback.

Over the next twenty-five years, the United States spent $9 billion studying and building the giant Yucca Mountain waste-disposal center in Nevada, only to have it canceled by President Barack Obama in 2009. Geologists have testified that the Yucca Mountain site may be incapable of containing nuclear waste for 10,000 years. The Yucca Mountain site will never open, leaving commercial operators of nuclear power plants without a permanent waste-storage facility.

At present, the future of nuclear energy is unclear. Wall Street remains skittish about investing several billion dollars in each new nuclear power plant. But the industry claims that the latest generation of plants is safer than before. The Department of Energy, meanwhile, is keeping its options open concerning nuclear energy.

NUCLEAR PROLIFERATION

Yet with great power also comes great danger. In Norse mythology, for example, the Vikings worshipped Odin, who ruled Asgard with wisdom and justice. Odin presided over a legion of gods, including the heroic Thor, whose honor and valor were the most cherished qualities of any warrior.

However, there was also Loki, the god of mischief, who was consumed by jealousy and hate. He was always scheming and excelled in deception and deceit. Eventually, Loki conspired with the giants to bring on the final battle between darkness and light, the epic battle Ragnarok, the twilight of the gods.

The problem today is that jealousies and hatreds between nations could unleash a nuclear Ragnarok. History has shown that when a nation masters commercial technology, it can, if it has the desire and political will, make the transition to nuclear weapons. The danger is that nuclear weapons technology will proliferate into some of the most unstable regions of the world.

During World War II, only the greatest nations on earth had the resources, know-how, and capability to create an atomic bomb. However, in the future, the threshold could be dramatically lowered as the price of uranium enrichment plummets due to the introduction of new technologies. This is the danger we face: newer and cheaper technologies may place the atomic bomb into unstable hands.

The key to building the atomic bomb is to secure large quantities of uranium ore and then purify it. This means separating uranium 238 (which makes up 99.3 percent of naturally occurring uranium) from uranium 235, which is suitable for an atomic bomb but makes up only .7 percent. These two isotopes are chemically identical, so the only way to reliably separate the two is to exploit the fact that uranium 235 weighs about 1 percent less than its cousin.

During World War II, the only way of separating the two isotopes of uranium was the laborious process of gaseous diffusion: uranium was made into a gas (uranium hexafluoride) and then forced to travel down hundreds of miles of tubing and membranes. At the end of this long journey, the faster (that is, lighter) uranium 235 won the race, leaving the heavier uranium 238 behind. After the gas containing uranium 235 was extracted, the process was repeated, until the enrichment level of uranium 235 rose from .7 percent to 90 percent, which is bomb-grade uranium. But pushing the gas required vast amounts of electricity. During the war, a significant fraction of the total U.S. electrical supply was diverted to Oak Ridge National Laboratory for this purpose. The enrichment facility was gigantic, occupying 2 million square feet and employing 12,000 workers.

After the war, only the superpowers, the United States and the Soviet

Union, could amass huge stockpiles of nuclear weapons, up to 30,000 apiece, because they had mastered the art of gaseous diffusion. But today, only 33 percent of the world's enriched uranium comes from gaseous diffusion.

Second-generation enrichment plants use a more sophisticated, cheaper technology: ultracentrifuges, which have created a dramatic shift in world politics as a result. Ultracentrifuges can spin a capsule containing uranium to speeds of up to 100,000 revolutions per minute. This accentuates the 1 percent difference in mass between uranium 235 and uranium 238. Eventually, the uranium 238 sinks to the bottom. After many revolutions, one can remove the uranium 235 from the top of the tube.

Ultracentrifuges are fifty times more efficient in energy than gaseous diffusion. About 54 percent of the world's uranium is purified in this way.

With ultracentrifuge technology, it takes only 1,000 ultracentrifuges operating continuously for one year to produce one atomic bomb's worth of enriched uranium. Ultracentrifuge technology can easily be stolen. In one of the worst breeches of nuclear security in history, an obscure atomic engineer, A. Q. Khan, was able to steal blueprints for the ultracentrifuge and components of the atomic bomb and sell them for profit. In 1975, while working in Amsterdam for URENCO, which was established by the British, West Germany, and the Netherlands to supply European reactors with uranium, he gave these secret blueprints to the Pakistani government, which hailed him as a national hero, and he is also suspected of selling this classified information to Saddam Hussein and to the governments of Iran, North Korea, and Libya.

Using this stolen technology, Pakistan was able to create a small stockpile of nuclear weapons, which it began testing in 1998. The ensuing nuclear rivalry between Pakistan and India, with each exploding a series of atomic bombs, almost led to a nuclear confrontation between these two rival nations.

Perhaps because of the technology it purchased from A. Q. Khan, Iran reportedly accelerated its nuclear program, building 8,000 ultracentrifuges by 2010, with the intention of building 30,000 more. This put pressure on other Middle East states to create their own atomic bombs, furthering instability.

The second reason the geopolitics of the twenty-first century might be altered is because another generation of enrichment technology—laser enrichment—is coming online, one potentially even cheaper than ultra-centrifuges.

If you examine the electron shells of these two isotopes of uranium, they are apparently the same, since the nucleus has the same charge. But if you analyze the equations for the electron shells very carefully, you find that there is a tiny separation in energy between the electron shells of uranium 235 and uranium 238. By shining a laser beam that is extremely fine-tuned, you can knock out electrons from the shell of uranium 235 but not from that of uranium 238. Once the uranium 235 atoms are ionized, they can be easily separated from uranium 238 by an electric field.

But the difference in energy between the two isotopes is so small that many nations have tried to exploit this fact and have failed. In the 1980s and 1990s, the United States, France, Britain, Germany, South Africa, and Japan attempted to master this difficult technology and were unsuccessful. In the United States, one attempt actually involved 500 scientists and $2 billion.

But in 2006, Australian scientists announced that not only have they solved the problem, they intend to commercialize it. Since 30 percent of the cost of uranium fuel comes from the enrichment process, the Australian company Silex thinks there could be a market for this technology. Silex even signed a contract with General Electric to begin commercialization. Eventually, they hope to produce up to one-third of the world's uranium using this method. In 2008, GE Hitachi Nuclear Energy announced plans to build the first commercial laser enrichment plant in Wilmington, North Carolina, by 2012. The plant will occupy 200 acres of a 1,600-acre site.

For the nuclear power industry, this is good news, since it will drive down the cost of enriched uranium over the next few years. However, others are worried because it is only a matter of time before this technology proliferates into unstable regions of the world. In other words, we have a window of opportunity to sign treaties to restrict and regulate the flow of enriched uranium. Unless we control this technology, the bomb will continue to proliferate, perhaps even to terrorist groups.

One of my acquaintances was the late Theodore Taylor, who had the rare distinction of designing some of the biggest and smallest nuclear warheads for the Pentagon. One of his designs was the Davy Crockett, weighing only fifty pounds, but capable of hurling a small atomic bomb at the enemy. Taylor was such a gung ho advocate of nuclear bombs that he worked on the Orion project, which was to use nuclear bombs to propel a spaceship to the nearby stars. He calculated that by successively dropping nuclear bombs out the end, the resulting shock wave would propel such a spacecraft to near the speed of light.

I once asked him why he got disillusioned with designing nuclear bombs and switched to working on solar energy. He confided to me that he had a recurring nightmare. His work on nuclear weapons, he felt, was leading to one thing: producing third-generation atomic warheads. (First-generation warheads of the 1950s were huge and difficult to carry to their targets. Second-generation warheads of the 1970s were small, compact, and ten of them could fit into the nose cone of a missile. But third-generation bombs are "designer bombs," specifically tailored to work in various environments, such as the forest, the desert, even outer space.) One of these third-generation bombs is a miniature atomic bomb, so small that a terrorist could carry it in a suitcase and use it to destroy an entire city. The idea that his life's work could one day be used by a terrorist haunted him for the rest of his life.

MIDCENTURY (2030 TO 2070)

GLOBAL WARMING

By midcentury, the full impact of a fossil fuel economy should be in full swing: global warming. It is now indisputable that the earth is heating up. Within the last century, the earth's temperature rose 1.3° F, and the pace is accelerating. The signs are unmistakable everywhere we look:

- The thickness of Arctic ice has decreased by an astonishing 50 percent in just the past fifty years. Much of this Arctic ice is just below the freezing point, floating on water. Hence, it is acutely sensitive to small temperature variations of the oceans, acting as a canary in a mineshaft, an early warning system. Today, parts of the northern polar ice caps disappear during the summer months, and may disappear entirely during summer as early as 2015. The polar ice cap may vanish permanently by the end of the century, disrupting the world's weather by altering the flow of ocean and air currents around the planet.

- Greenland's ice shelves shrank by twenty-four square miles in 2007. This figure jumped to seventy-one square miles in 2008. (If all the Greenland ice were somehow to melt, sea levels would rise about twenty feet around the world.)

- Large chunks of Antarctica's ice, which have been stable for tens of thousands of years, are gradually breaking off. In 2000, a piece the size of Connecticut broke off, containing 4,200 square miles of ice. In 2002, a piece of ice the size of Rhode Island broke off the Thwaites Glacier. (If all Antarctica's ice were to melt, sea levels would rise about 180 feet around the world.)

- For every vertical foot that the ocean rises, the horizontal spread of the ocean is about 100 feet. Already, sea levels have risen 8 inches in the past century, mainly caused by the expansion of seawater as it heats up. According to the United Nations, sea levels could rise by 7 to 23 inches by 2100. Some scientists have said that the UN report was too cautious in interpreting the data. According to scientists at the University of Colorado's Institute of Arctic and Alpine Research, by 2100 sea levels could rise by 3 to 6 feet. So gradually the map of the earth's coastlines will change.

- Temperatures started to be reliably recorded in the late 1700s; 1995, 2005, and 2010 ranked among the hottest years ever recorded; 2000 to 2009 was the hottest decade. Likewise, levels of carbon dioxide are rising dramatically. They are at the highest levels in 100,000 years.

- As the earth heats up, tropical diseases are gradually migrating northward. The recent spread of the West Nile virus carried by mosquitoes may be a harbinger of things to come. UN officials are especially concerned about the spread of malaria northward. Usually, the eggs of many harmful insects die every winter when the soil freezes. But with the shortening of the winter season, it means the inexorable spread of dangerous insects northward.

CARBON DIOXIDE—GREENHOUSE GAS

According to the UN's Intergovernmental Panel on Climate Change, scientists have concluded with 90 percent confidence that global warming is driven by human activity, especially the production of carbon dioxide via the burning of oil and coal. Sunlight easily passes through carbon dioxide. But as sunlight heats up the earth, it creates infrared radiation, which does not pass back through carbon dioxide so easily. The energy from sunlight cannot escape back into space and is trapped.

We also see a somewhat similar effect in greenhouses or cars. The sunlight warms the air, which is prevented from escaping by the glass.

Ominously, the amount of carbon dioxide generated has grown explosively, especially in the last century. Before the Industrial Revolution, the carbon dioxide content of the air was 270 parts per million (ppm). Today, it has soared to 387 ppm. (In 1900, the world consumed 150 million barrels of oil. In 2000, it jumped to 28 billion barrels, a 185-fold jump. In 2008, 9.4 billion tons of carbon dioxide were sent into the air from fossil fuel burning and also deforestation, but only 5 billion tons were recycled into the oceans, soil, and vegetation. The remainder will stay in the air for decades to come, heating up the earth.)

VISIT TO ICELAND

The rise in temperature is not a fluke, as we can see by analyzing ice cores. By drilling deep into the ancient ice of the Arctic, scientists have been able to extract air bubbles that are thousands of years old. By chemically analyzing the air in these bubbles, scientists can reconstruct the temperature and carbon dioxide content of the atmosphere going back more than 600,000 years. Soon, they will be able to determine the weather conditions going back a million years.

I had a chance to see this firsthand. I once gave a lecture in Reykjavik, the capital of Iceland, and had the privilege of visiting the University of Iceland, where ice cores are being analyzed. When your airplane lands in Reykjavik, at first all you see is snow and jagged rock, resembling the bleak landscape of the moon. Although barren and forbidding, the terrain makes the Arctic an ideal place to analyze the climate of the earth hundreds of thousands of years ago.

When I visited their laboratory, which is kept at freezing temperatures, I had to pass through thick refrigerator doors. Once inside, I could see racks and racks containing long metal tubes, each about an inch and a half in diameter and about ten feet long. Each hollow tube had been drilled deep into the ice of a glacier. As the tube penetrated the ice, it captured samples from snows that had fallen thousands of years ago. When the tubes were removed, I could carefully examine the icy contents of each. At first, all I could see was a long column of white ice. But upon closer examination, I could see that the ice had stripes made of tiny bands of different colors.

Scientists have to use a variety of techniques to date them. Some of the ice layers contain markers indicating important events, such as the soot emitted from a volcanic eruption. Since the dates of these eruptions are known to great accuracy, one can use them to determine how old that layer is.

These ice cores were then cut in various slices so they could be examined. When I peered into one slice under a microscope, I saw tiny, microscopic bubbles. I shuddered to realize that I was seeing air bubbles that were deposited tens of thousands of years ago, even before the rise of human civilization.

The carbon dioxide content within each air bubble is easily measured. But calculating the temperature of the air when the ice was first deposited is more difficult. (To do this, scientists analyze the water in the bubble. Water molecules can contain different isotopes. As the temperature falls, heavier water isotopes condense faster than ordinary water molecules. Hence, by measuring the amount of the heavier isotopes, one can calculate the temperature at which the water molecule condensed.)

Finally, after painfully analyzing the contents of thousands of ice cores, these scientists have come to some important conclusions. They found that temperature and carbon dioxide levels have oscillated in parallel, like two roller coasters moving together, in synchronization over many thousands of years. When one curve rises or falls, so does the other.

Most important, they found a sudden spike in temperature and carbon dioxide content happening just within the last century. This is highly unusual, since most fluctuations occur slowly over millennia. This unusual spike is not part of this natural heating process, scientists claim, but is a direct indicator of human activity.

There are other ways to show that this sudden spike is caused by human activity, and not natural cycles. Computer simulations are now so advanced that we can simulate the temperature of the earth with and without the presence of human activity. Without civilization producing carbon dioxide, we find a relatively flat temperature curve. But with the addition of human activity, we can show that there should be a sudden spike in both temperature and carbon dioxide. The predicted spike fits the actual spike perfectly.

Lastly, one can measure the amount of sunlight that lands on every square foot of the earth's surface. Scientists can also calculate the amount of heat that is reflected into outer space from the earth. Normally, we expect these two amounts to be equal, with input equaling output. But in reality, we find the net amount of energy that is currently heating the earth. Then if we calculate the amount of energy being produced by human activity, we find a perfect match. Hence, human activity is causing the current heating of the earth.

Unfortunately, even if we were to suddenly stop producing any carbon dioxide, the gas that has already been released into the atmosphere is enough to continue global warming for decades to come.

As a result, by midcentury, the situation could be dire.

Scientists have created pictures of what our coastal cities will look like at midcentury and beyond if sea levels continue to rise. Coastal cities may disappear. Large parts of Manhattan may have to be evacuated, with Wall Street underwater. Governments will have to decide which of their great cities and capitals are worth saving and which are beyond hope. Some cities may be saved via a combination of sophisticated dikes and water gates. Other cities may be deemed hopeless and allowed to vanish under the ocean, creating mass migrations of people. Since most of the commercial and population centers of the world are next to the ocean, this could have a disastrous effect on the world economy.

Even if some cities can be salvaged, there is still the danger that large storms can send surges of water into a city, paralyzing its infrastructure. For example, in 1992 a huge storm surge flooded Manhattan, paralyzing the subway system and trains to New Jersey. With transportation flooded, the economy grinds to a halt.

FLOODING BANGLADESH AND VIETNAM

A report by the Intergovernmental Panel on Climate Change isolated three hot spots for potential disaster: Bangladesh, the Mekong Delta of Vietnam, and the Nile Delta in Egypt.

The worst situation is that of Bangladesh, a country regularly flooded by storms even without global warming. Most of the country is flat and at sea level. Although it has made significant gains in the last few decades, it is still one of the poorest nations on earth, with one of the highest population densities. (It has a population of 161 million, comparable to that of Russia, but with 1/120 of the land area.) About 50 percent of the land area will be permanently flooded if sea levels rise by three feet. Natural calamities occur there almost every year, but in September 1998, the world witnessed in horror a preview of what may become commonplace. Massive flooding submerged two-thirds of the nation, leaving 30 million people homeless almost overnight; 1,000 were killed, and 6,000 miles of roads were destroyed. This was one of the worst natural disasters in modern history.

Another country that would be devastated by a rise in sea level is Vietnam, where the Mekong Delta is particularly vulnerable. By midcentury, this country of 87 million people could face a collapse of its main food-growing area. Half the rice in Vietnam is grown in the Mekong Delta, home to 17 million people, and much of it will be flooded permanently by rising sea levels. According to the World Bank, 11 percent of the entire population would be displaced if sea levels rise by three feet by midcentury. The Mekong Delta will also be flooded with salt water, permanently destroying the fertile soil of the area. If millions are flooded out of their homes in Vietnam, many will flock to Ho Chi Minh City seeking refuge. But one-fourth of the city will also be underwater.

In 2003 the Pentagon commissioned a study, done by the Global Business Network, that showed that, in a worst-case scenario, chaos could spread around the world due to global warming. As millions of refugees cross national borders, governments could lose all authority and collapse, so countries could descend into the nightmare of looting, rioting, and chaos. In this desperate situation, nations, when faced with the prospect of the influx of millions of desperate people, may resort to nuclear weapons.

"Envision Pakistan, India, and China—all armed with nuclear weapons—skirmishing at their borders over refugees, access to shared rivers, and arable land," the report said. Peter Schwartz, founder of the Global Business Network and a principal author of the Pentagon study, confided to me the details of this scenario. He told me that the biggest hot spot would be the border between India and Bangladesh. In a major crisis in Bangladesh, up to 160 million people could be driven out of their homes, sparking one of the greatest migrations in human history. Tensions could rapidly rise as borders collapse, local governments are paralyzed, and mass rioting breaks out. Schwartz sees that nations may use nuclear weapons as a last resort.

In a worst-case scenario, we could have a greenhouse effect that feeds on itself. For example, the melting of the tundra in the Arctic regions may release millions of tons of methane gas from rotting vegetation. Tundra covers nearly 9 million square miles of land in the Northern Hemisphere, containing vegetation frozen since the last Ice Age tens of thousands of years ago. This tundra contains more carbon dioxide and methane than the atmosphere, and this poses an enormous threat to the world's weather. Methane gas, moreover, is a much deadlier greenhouse gas than carbon dioxide. It does not stay in the atmosphere as long, but it causes much more damage than carbon dioxide. The release of so much methane gas from the melting tundra could cause temperatures to rapidly rise, which will cause even more methane gas to be released, causing a runaway cycle of global warming.

TECHNICAL FIXES

The situation is dire, but we have not yet reached the point of no return. The problem of controlling greenhouse gases is actually largely economic and political, not technical. Carbon dioxide production coincides with economic activity, and hence wealth. For example, the United States generates roughly 25 percent of the world's carbon dioxide. This is because the United States has roughly 25 percent of the world's economic activity. And in 2009, China overtook the United States in creating greenhouse gases, mainly because of the explosive growth of its economy. This is the fundamental reason that nations are so reluctant to deal with global warming: it interferes with economic activity and prosperity.

Various schemes have been devised to deal with this global crisis, but ultimately, a quick fix may not be enough. Only a major shift in the way we consume energy will solve the problem. Some technical measures have been advocated by serious scientists, but none has won wide acceptance. The proposals include:

- **Launching pollutants into the atmosphere.** One proposal is to send rockets into the upper atmosphere, where they would release pollutants, such as sulfur dioxide, in order to reflect sunlight into space, thereby cooling the earth. In fact, Nobel laureate Paul Crutzen has advocated shooting pollution into space as a "doomsday device," providing one final escape route for humanity to stop global warming. This idea has its roots in 1991, when scientists carefully monitored the huge volcanic explosion of Mount Pinatubo in the Philippines, which lofted 10 billion metric tons of dirt and debris into the upper atmosphere. This darkened the skies and caused the average temperature around the earth to drop by 1° F. This made it possible to calculate how much pollutants would be necessary to reduce the world temperature. Although this is a serious proposal, some critics doubt that it can solve the problem by itself. Little is known about how a huge quantity of pollutants will affect the world temperature. Maybe the benefits will be short-lived, or the unintended side effects may be worse than the original problem. For example, there was a sudden drop in global precipitation after the Mount Pinatubo eruption; if the experiment goes awry, it could similarly cause massive droughts. Cost estimates show that $100 million would be required to conduct field tests. Since the effect of the sulfate aerosols is temporary, it would cost a minimum of $8 billion per year to regularly inject massive amounts of them into the atmosphere.

- **Creating algae blooms.** Another suggestion is to dump iron-based chemicals into the oceans. These mineral nutrients will cause algae to thrive in the ocean, which in turn will increase the amount of carbon dioxide that is absorbed by the algae. However, after Planktos, a corporation based in California, announced that it would unilaterally begin a private effort to fertilize part of the South Atlantic with iron—hoping to deliberately spawn plankton blooms that would absorb the carbon dioxide in the air—countries bound by the London Convention, which regulates dumping at sea, issued a "statement of concern" about this effort. Also, a United Nations group called for a temporary moratorium on such experiments. The experiment was ended when Planktos ran out of funds.

- **Carbon sequestration.** Yet another possibility is to use carbon sequestration, a process by which the carbon dioxide emitted from

coal-burning power plants is liquefied and then separated from
the environment, perhaps by being buried underground. Although
this might work in principle, it is a very expensive process, and it
cannot remove the carbon dioxide that has already been lofted into
the atmosphere. In 2009, engineers were carefully monitoring the
first major test of carbon sequestration. The huge Mountaineer
power plant, built in 1980 in West Virginia, was retrofitted to
separate carbon dioxide from the environment, making it the
United States' first electricity-generating coal-burning plant to
experiment with sequestration. The liquefied gas will be injected
7,800 feet underground, eventually into a layer of dolomite. The
liquid will eventually form a mass thirty to forty feet high and
hundreds of yards long. The plant's owner, American Electric
Power, plans to inject 100,000 tons of carbon dioxide annually
for two to five years. This is only 1.5 percent of the plant's yearly
emission, but eventually the system could capture up to 90 percent.
The initial costs are about $73 million. But if it's successful, then
this model could rapidly be disseminated to other sites such as
four nearby giant coal-burning plants generating 6 billion watts
of energy (so much that this area is dubbed Megawatt Valley).
There are large unknowns: it is not clear if the carbon dioxide will
eventually migrate or if the gas will combine with water, perhaps
creating carbonic acid that may poison groundwater. However,
if the project is a success, it may very well be part of a mix of
technologies used to deal with global warming.

- **Genetic engineering.** Another proposal is to use genetic engineering
to specifically create life-forms that can absorb large quantities of
carbon dioxide. One enthusiastic promoter of this approach is J.
Craig Venter, who gained fame and fortune pioneering high-speed
techniques that successfully led to sequencing the human genome
years ahead of schedule. "We view the genome as the software,
or even the operating system, of the cell," he says. His goal is to
rewrite that software, so that microbes can be genetically modified,
or even constructed almost from scratch, so that they absorb
the carbon dioxide from coal-burning plants and convert it into
useful substances, such as natural gas. He notes, "There are already
thousands, perhaps millions, of organisms on our planet that know

how to do this." The trick is to modify them so that they can increase their output and also flourish in a coal-fired plant. "We think this field has tremendous potential to replace the petrochemical industry, possibly within a decade," he said optimistically.

Princeton physicist Freeman Dyson has advocated another variation, creating a genetically engineered variety of trees that would be adept at absorbing carbon dioxide. He has stated that perhaps a trillion such trees might be enough to control the carbon dioxide in the air. In his paper "Can We Control the Carbon Dioxide in the Atmosphere?" he advocated creating a "carbon bank" of "fast-growing trees" to regulate carbon dioxide levels.

However, as with any plan to use genetic engineering on a large scale, one must be careful about side effects. One cannot recall a life-form in the same way that we can recall a defective car. Once it is released into the environment, the genetically engineered life-form may have unintended consequences for other life-forms, especially if it displaces local species of plants and upsets the balance of the food chain.

Sadly, there has been a conspicuous lack of interest among politicians to fund any of these plans. However, one day, global warming will become so painful and disruptive that politicians will be forced to implement some of them.

The critical period will be the next few decades. By midcentury, we should be in the hydrogen age, where a combination of fusion, solar power, and renewables should give us an economy that is much less dependent on fossil fuel consumption. A combination of market forces and advances in hydrogen technology should give us a long-term solution to global warming. The danger period is now, before a hydrogen economy is in place. In the short term, fossil fuels are still the cheapest way to generate power, and hence global warming will pose a danger for decades to come.

FUSION POWER

By midcentury, a new option arises that is a game changer: fusion. By that time, it should be the most viable of all technical fixes, perhaps giving us a permanent solution to the problem. While fission power relies on splitting the uranium atom, thereby creating energy (and a large amount of nuclear waste), fusion power relies on fusing hydrogen atoms with great heat, thereby releasing vastly more energy (with very little waste).

Unlike fission power, fusion power unleashes the nuclear energy of the sun. Buried deep inside the hydrogen atom is the energy source of the universe. Fusion power lights up the sun and the heavens. It is the secret of the stars. Anyone who can successfully master fusion power will have unleashed unlimited eternal energy. And the fuel for these fusion plants comes from ordinary seawater. Pound for pound, fusion releases 10 million times more energy than gasoline. An 8-ounce glass of water is equal to the energy content of 500,000 barrels of petroleum.

Fusion (not fission) is nature's preferred way to energize the universe. In star formation, a hydrogen-rich ball of gas is gradually compressed by gravity, until it starts to heat up to enormous temperatures. When the gas reaches around 50 million degrees or so (which varies depending on the specific conditions), the hydrogen nuclei inside the gas are slammed into one another, until they fuse to form helium. In the process, vast amounts of energy are released, which causes the gas to ignite. (More precisely, the compression must satisfy something called Lawson's criterion, which states that you have to compress hydrogen gas of a certain density to a certain temperature for a certain amount of time. If these three conditions involving density, temperature, and time are met, you have a fusion reaction, whether it is a hydrogen bomb, a star, or a fusion in a reactor.)

So that is the key: heating and compressing hydrogen gas until the nuclei fuse, releasing cosmic amounts of energy.

But previous attempts to harness this cosmic power have failed. It is a fiendishly difficult task to heat hydrogen gas to tens of millions of degrees, until the protons fuse to form helium gas and release vast amounts of energy.

Moreover, the public is cynical about these claims, since every twenty years scientists claim that fusion power is twenty years away. But after decades of overoptimistic claims, physicists are increasingly convinced that fusion power is finally arriving, perhaps as early as 2030. Sometime by mid-century, we may see fusion plants dotting the countryside.

The public has a right to be skeptical about fusion, since there have been so many hoaxes, frauds, and failures in the past. Back in 1951, when the United States and the Soviet Union were gripped in Cold War frenzy and were feverishly developing the first hydrogen bomb, President Juan Perón of Argentina announced, with huge fanfare and a media blitz, that his country's scientists had made a breakthrough in controlling the power of the sun. The story sparked a firestorm of publicity. It seemed unbelievable, yet it made the front page of the *New York Times*. Argentina, boasted Perón, had scored a major scientific breakthrough where the superpowers had failed. An unknown German-speaking scientist, Ronald Richter, had convinced Perón to fund his "thermotron," which promised unlimited energy and eternal glory for Argentina.

The American scientific community, which was still grappling with fusion in the fierce race with Russia to produce the H-bomb, declared that the claim was nonsense. Atomic scientist Ralph Lapp said, "I know what the other material is that the Argentines are using. It's baloney."

The press quickly dubbed it the Baloney Bomb. Atomic scientist David Lilienthal was asked if there was the "slightest chance" the Argentines could be correct. He shot back, "Less than that."

Under intense pressure, Perón simply dug in his heels, hinting that the superpowers were jealous that Argentina had scooped them. The moment of truth finally came the next year, when Perón's representatives visited Richter's lab. Under fire, Richter was acting increasingly erratic and bizarre. When inspectors arrived, he blew the laboratory door off using tanks of oxygen and then scribbled on a piece of paper the words "atomic energy." He ordered gunpowder to be injected into the reactor. The verdict was that he was probably insane. When inspectors placed a piece of radium next to Richter's "radiation counters," nothing happened, so clearly his equipment was fraudulent. Richter was later arrested.

But the most celebrated case was that of Stanley Pons and Martin Fleischmann, two well-respected chemists from the University of Utah who in 1989 claimed to have mastered "cold fusion," that is, fusion at room temperature. They claimed to have placed palladium metal in water, which then somehow magically compressed hydrogen atoms until they fused into helium, releasing the power of the sun on a tabletop.

The shock was immediate. Almost every newspaper in the world put this discovery on its front page. Overnight, journalists talked of ending the energy crisis and ushering in a new age of unlimited energy. A feeding frenzy hit the world media. The state of Utah immediately passed a $5 million bill to create a National Institute for Cold Fusion. Even Japanese car manufacturers began to donate millions of dollars to promote research in this hot new field. A cultlike following began to emerge based around cold fusion.

Unlike Richter, Pons and Fleischmann were well respected in the scientific community and were glad to share their results with others. They carefully laid out their equipment and their data for the world to see.

But then things got complicated. Since the apparatus was so simple, groups around the world tried to duplicate these astonishing results. Unfortunately, most groups failed to find any net release of energy, declaring cold fusion a dead end. However, the story was kept alive because there were sporadic claims that certain groups had successfully duplicated the experiment.

Finally, the physics community weighed in. They analyzed Pons and Fleischmann's equations, and found them deficient. First, if their claims were correct, a blistering barrage of neutrons would have radiated from the glass of water, killing Pons and Fleischmann. (In a typical fusion reaction, two hydrogen nuclei are slammed together and fuse, creating energy, a helium nuclei, and also a neutron.) So the fact that Pons and Fleischmann were still alive meant the experiment hadn't worked. If their experiments had produced cold fusion, they would be dying of radiation burns. Second, more than likely Pons and Fleischmann had found a chemical reaction rather than a thermonuclear reaction. And last, the physicists concluded, palladium metal cannot bind hydrogen atoms closely enough to cause the hydrogen to fuse into helium. It would violate the laws of the quantum theory.

But the controversy has not died down, even today. There are still occasional claims that someone has achieved cold fusion. The problem is that no one has been able to reliably attain cold fusion on demand. After all, what is the point of making an automobile engine if it works only occasionally? Science is based on reproducible, testable, and falsifiable results that work every time.

HOT FUSION

But the advantages of fusion power are so great that many scientists have heeded its siren call.

For example, fusion creates minimal pollution. It is relatively clean, and is nature's way of energizing the universe. One by-product of fusion is helium gas, which is actually commercially valuable. Another is the radioactive steel of the fusion chamber, which eventually has to be buried. It is mildly dangerous only for a few decades. But a fusion plant produces an insignificant amount of nuclear waste compared to a standard uranium fission plant (which produces thirty tons of high-level nuclear waste per year that lasts for thousands to tens of millions of years).

Also, fusion plants cannot suffer a catastrophic meltdown. Uranium fission plants, precisely because they contain tons of high-level nuclear waste in their core, produce volatile amounts of heat even after shutdown. It is this residual heat that can eventually melt the solid steel and enter the groundwater, creating a steam explosion and the nightmare of the China Syndrome accident.

Fusion plants are inherently safer. A "fusion meltdown" is a contradiction in terms. For example, if one were to shut down a fusion reactor's magnetic field, the hot plasma would hit the walls of the chamber and the fusion process would stop immediately. So a fusion plant, instead of undergoing a runaway chain reaction, spontaneously turns itself off in case of an accident.

"Even if the plant were flattened, the radiation level one kilometer outside the fence would be so small that evacuation would not be necessary," says Farrokh Najmabadi, who directs the Center for Energy Research at the University of California at San Diego.

Although commercial fusion power has all these marvelous advantages, there is still one small detail: it doesn't exist. No one has yet produced an operating fusion plant.

But physicists are cautiously optimistic. "A decade ago, some scientists questioned whether fusion was possible, even in the lab. We now know that fusion will work. The question is whether it is economically practical," says David E. Baldwin of General Atomics, who oversees one of the largest fusion reactors in the United States, the DIII-D.

NIF—FUSION BY LASER

All this could change rather dramatically in the next few years.

Several approaches are being tried simultaneously, and after decades of false starts, physicists are convinced that they will finally attain fusion. In France, there is the International Thermonuclear Experimental Reactor (ITER), backed by many European nations, the United States, Japan, and others. And in the United States, there is the National Ignition Facility (NIF).

I had a chance to visit the NIF laser fusion machine, and it is a colossal sight. Because of the close connection with hydrogen bombs, the NIF reactor is based at the Lawrence Livermore National Laboratory, where the military designs hydrogen warheads. I had to pass through many layers of security to finally gain access.

But when I reached the reactor, it was a truly awesome experience. I am used to seeing lasers in university laboratories (in fact, one of the largest laser laboratories in New York State is directly beneath my office at the City University of New York), but seeing the NIF facility was overwhelming. It is housed in a ten-story building the size of three football fields, with 192 giant laser beams being fired down a long tunnel. It is the largest laser system in the world, delivering sixty times more energy than any previous one.

After these laser beams are fired down this long tunnel, they eventually hit an array of mirrors that focus each beam onto a tiny pinhead-size target, consisting of deuterium and tritium (two isotopes of hydrogen). Incredibly, 500 trillion watts of laser power are focused onto a tiny pellet that is barely visible to the naked eye, scorching it to 100 million degrees, much hotter than the center of the sun. (The energy of that colossal pulse is equivalent to the output of half a million nuclear power plants in a brief instant.) The surface of this microscopic pellet is quickly vaporized, which unleashes a shock wave that collapses the pellet and unleashes the power of fusion.

It was completed in 2009, and is currently undergoing tests. If all goes well, it may be the first machine to create as much energy as it consumes. Although this machine is not designed to produce commercial electrical power, it is designed to show that laser beams can be focused to heat hydrogen-rich materials and produce net energy.

I talked to one of the directors of the NIF facility, Edward Moses, about his hopes and dreams for his project. Wearing a hard hat, he looked more like a construction worker than a top nuclear physicist in charge of the largest laser lab in the world. He admitted to me that in the past there have been numerous false starts. But this, he believed, was the real thing: he and his team were about to realize an important achievement, one that will enter the history books, the first to peacefully capture the power of the sun on earth. Talking to him, you realize how projects like NIF are kept alive by the passion and energy of their true believers. He savored the day, he told me, when he could invite the president of the United States to this laboratory to announce that history had just been made.

But from the beginning, NIF got off to a bad start. (Even strange things have happened, such as when the previous associate director of NIF, E. Michael Campbell, was forced to resign in 1999 when it was revealed that he lied about completing a Ph.D. at Princeton.) Then the completion date, originally set for 2003, began to slip. Costs ballooned, from $1 billion to $4 billion. It was finally finished in March 2009, six years late.

The devil, they say, is in the details. In laser fusion, for example, these 192 laser beams have to hit the surface of a tiny pellet with utmost precision, so that it implodes evenly. The beams must hit this tiny target to within 30 trillionths of a second of one another. The slightest misalignment of the laser beams or irregularity of the pellet means that the pellet will heat unsymmetrically, causing it to blow out to one side rather than implode spherically.

If the pellet is irregular by more than 50 nanometers (or about 150 atoms), the pellet will also fail to implode evenly. (That is like trying to throw a baseball within the strike zone from a distance of 350 miles.) So alignment of the laser beams and evenness of the pellet are the main problems facing laser fusion.

In addition to NIF, the European Union is backing its own version of laser fusion. The reactor will be built at the High Power Laser Energy Research Facility (HiPER), and it is smaller but perhaps more efficient than NIF. Construction for HiPER starts in 2011.

The hopes of many ride on NIF. However, if laser fusion does not work as expected, there is another, even more advanced proposal for controlled fusion: putting the sun in a bottle.

ITER—FUSION IN A MAGNETIC FIELD

Yet another design is being exploited in France. The International Thermonuclear Experimental Reactor (ITER) uses huge magnetic fields to contain hot hydrogen gas. Instead of using lasers to instantly collapse a tiny pellet of hydrogen-rich material, ITER uses a magnetic field to slowly compress hydrogen gas. The machine looks very much like a huge hollow doughnut made of steel, with magnetic coils surrounding the hole of the doughnut. The magnetic field keeps the hydrogen gas inside the doughnut-shaped chamber from escaping. Then an electrical current is sent surging through the gas, heating it. The combination of squeezing the gas with the magnetic field and sending a current surging through it causes the gas to heat up to many millions of degrees.

The idea of using a "magnetic bottle" to create fusion is not new. It goes back to the 1950s, in fact. But why has it taken so long, with so many delays, to commercialize fusion power?

The problem is that the magnetic field has to be precisely tuned so that the gas is compressed evenly without bulging or becoming irregular. Think of taking a balloon and trying to compress it with your hands so that the balloon is evenly compressed. You will find that the balloon bulges out from the gaps between your hands, making a uniform compression almost impossible. So the problem is instability and is not one of physics but of engineering.

This seems strange, because stars easily compress hydrogen gas, creating the trillions of stars we see in our universe. Nature, it seems, effortlessly creates stars in the heavens, so why can't we do it on earth? The answer speaks to a simple but profound difference between gravity and electromagnetism.

Gravity, as shown by Newton, is strictly attractive. So in a star, the gravity of the hydrogen gas compresses it evenly into a sphere. (That is why stars and planets are spherical and not cubical or triangular.) But electrical charges come in two types: positive and negative. If one collects a ball of negative charges, they repel each other and scatter in all directions. But if one brings a positive and negative charge together, you get what is called a "dipole," with a complicated set of electrical field lines resembling a spider web. Similarly, magnetic fields form a dipole; hence squeezing hot gas evenly inside a doughnut-shaped chamber is a fiendishly difficult task. It takes a supercomputer, in fact, to plot the magnetic and electric fields emanating from a simple configuration of electrons.

It all boils down to this. Gravity is attractive and can compress gas

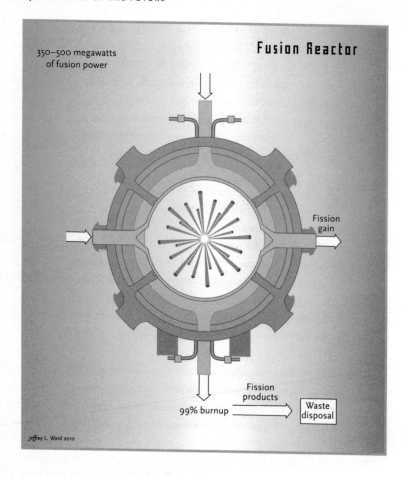

Fusion Reactor

350–500 megawatts of fusion power

Fission gain

Fission products

99% burnup

Waste disposal

Jeffrey L. Ward 2010

evenly into a sphere. Stars can form effortlessly. But electromagnetism is both attractive and repulsive, so gases bulge out in complex ways when compressed, making controlled fusion exceedingly difficult. This is the fundamental problem that has dogged physicists for fifty years.

Until now. Physicists now claim that the ITER has finally worked out the kinks in the stability problem with magnetic confinement.

The ITER is one of the largest international scientific projects ever attempted. The heart of the machine consists of a doughnut-shaped metal chamber. Altogether, it will weigh 23,000 tons, far surpassing the weight of the Eiffel Tower, which weighs only 7,300 tons.

Two types of fusion. On the left, lasers compress a pellet of hydrogen-rich materials. On the right, magnetic fields compress a gas containing hydrogen. By midcentury, the world may derive its energy from fusion.

The components are so heavy that the roads transporting the equipment have to be specially modified. A large convoy of trucks will transport the components, with the heaviest weighing 900 tons and the tallest being four stories high. The ITER building will be nineteen stories tall and sit on a huge platform the size of sixty soccer fields. It is projected to cost 10 billion euros, a cost shared by seven member states (the European Union, the United States, China, India, Japan, Korea, and Russia).

When it is finally fired up, it will heat hydrogen gas to 270 million degrees Fahrenheit, far surpassing the 27 million degrees Fahrenheit found in the center of the sun. If all goes well, it will generate 500 megawatts of energy, which is ten times the amount of energy originally going into the reactor. (The current record for fusion power is 16 megawatts, created by the European JET (Joint European Torus) reactor at the Culham Science Center, in Oxfordshire, UK.) After some delays, the target date for break-even is now set to be 2019.

The ITER is still just a science project. It is not designed to produce commercial power. But physicists already are laying the groundwork for the next step, taking fusion power to the marketplace. Farrokh Najmabadi, who leads a working group looking into commercial designs for fusion plants, has proposed ARIES-AT, a smaller machine than the ITER, which would produce a billion watts at roughly 5 cents per kilowatt-hour, making it competitive with fossil fuels. But even Najmabadi, who is optimistic about fusion, admits that fusion won't be ready for widespread commercialization until the middle of the century.

Another commercial design is the DEMO fusion reactor. While the ITER is designed to produce 500 megawatts for a minimum of 500 seconds, the DEMO will be designed to produce energy continually. The DEMO adds one extra step lacking in the ITER. When fusion takes place, an extra neutron is formed, which quickly escapes from the chamber. However, it is possible to surround the chamber with a special coating, called the blanket, specifically designed to absorb the energy of this neutron. The blanket then heats up. Pipes inside the blanket carry water, which then boils. This steam is sent against the blades of a turbine that generates electricity.

If all goes well, the DEMO will go online in 2033. It will be 15 percent larger than the ITER reactor. DEMO will produce twenty-five times more energy than it consumes. Altogether, DEMO is expected to produce 2 billion watts of power, making it comparable to a conventional power plant. If the DEMO plant is successful, it could lead to rapid commercialization of this technology.

But many uncertainties remain. The ITER reactor has already secured the funding necessary for construction. But since the DEMO reactor is still in its planning stages, delays are to be expected.

Fusion scientists believe that they have finally turned the corner. After decades of overstatements and failures, they believe that fusion is within grasp. Not one but two designs (NIF and ITER) may eventually bring fusion electricity into the living room. But since neither NIF nor ITER is yet delivering commercial fusion power, there is still room for the unexpected, such as tabletop fusion and bubble fusion.

TABLETOP FUSION

Because the stakes are so high, it is also important to acknowledge the possibility of solving the problem from an entirely different, unexpected direction. Because fusion is such a well-defined process, several proposals have been made that are outside the usual mainstream of large-scale funding but that still have some merit. In particular, some of them might one day achieve fusion on a tabletop.

In the final scene in the movie *Back to the Future*, Doc Brown, the crazy scientist, is seen scrambling to get fuel for his DeLorean time machine. Instead of fueling up with gasoline, he searches garbage cans for banana peels and trash and then dumps everything into a small canister called Mr. Fusion.

Given a hundred years, is it possible that some breakout design may reduce huge football field–size machines to the size of a coffeemaker, like in the movie?

One serious possibility for tabletop fusion is called sonoluminescence, which uses the sudden collapses of bubbles to unleash blistering temperatures. It is sometimes called sonic fusion or bubble fusion. This curious effect has been known for decades, going back to 1934, when scientists at the University of Cologne were experimenting with ultrasound and photographic film, hoping to speed up the development process. They noticed tiny dots in the film, caused by flashes of light generated by the ultrasound creating bubbles in the fluid. Later, the Nazis noticed that bubbles emitted from their propeller blades often glowed, indicating that high temperatures were somehow being produced inside the bubbles.

Later, it was shown that these bubbles were glowing brightly because they collapsed evenly, thereby compressing the air in the bubble to enormously high temperatures. Hot fusion, as we saw earlier, is plagued by the uneven compression of hydrogen, either because laser beams striking the pellet of fuel are misaligned or the gas is being squeezed unevenly. As a bubble shrinks, the motion of the molecules is so rapid that air pressure inside the bubble quickly becomes uniform along the bubble walls. In principle, if one can collapse a bubble under such perfect conditions, one might attain fusion.

Sonoluminescence experiments have successfully produced temperatures of tens of thousands of degrees. Using noble gases, one can increase the intensity of light emitted from these bubbles. But there is some controversy over whether it can achieve temperatures hot enough to produce nuclear fusion. The controversy stems from the work of Rusi Taleyarkhan, formerly of the Oak Ridge National Laboratory, who claimed in 2002 that he was able to achieve fusion with his sonic fusion device. He claimed to have detected neutrons from his experiment, a sure sign that nuclear fusion was taking place. However, after years of work by other researchers who have failed to reproduce his work, this result, for the moment, has been discredited.

Yet another wild card is the fusion machine of Philo Farnsworth, the unsung coinventor of TV. As a child, Farnsworth originally got the idea for TV by thinking of the way a farmer plows his fields, row after row. He even sketched the details of his prototype at the age of fourteen. He was the first to transfer this idea to a fully electronic device capable of capturing moving images on a screen. Unfortunately, he was unable to capitalize on his landmark invention and was mired in lengthy, messy patent fights with RCA. His legal battles even drove him crazy, and he voluntarily checked himself into an insane asylum. His pioneering work on TV went largely unnoticed.

Later in life, he turned his attention to the fusor, a small tabletop device that can actually generate neutrons via fusion. It consists of two large spheres, one inside the other, each made of a wire mesh. The outer mesh is positively charged, while the inner mesh is negatively charged, so protons injected through this mesh are repelled by the outer mesh and attracted to the inner mesh. The protons then smash into a hydrogen-rich pellet in the middle, creating fusion and a burst of neutrons.

The design is so simple that even high school students have done what Richter, Pons, and Fleischmann could not do: successfully generate neutrons. However, it is unlikely that this device will ever yield usable energy. The number of protons that are accelerated is extremely small, and hence the energy resulting from this device is very tiny.

In fact, it is also possible to produce fusion on a tabletop using a standard atom smasher or particle accelerator. An atom smasher is more complicated than a fusor, but it can also be used to accelerate protons so that they can slam into a hydrogen-rich target and create fusion. But again, the number of protons that are fused is so small that this is an impractical device. So both the fusor and atom smasher can attain fusion, but they are simply too inefficient and their beams are too thin to produce usable power.

Given the enormous stakes, no doubt other enterprising scientists and engineers will have their chance to turn their basement contraptions into the next mega invention.

FAR FUTURE (2070 TO 2100)

AGE OF MAGNETISM

The previous century was the age of electricity. Because electrons are so easily manipulated, this has opened up entirely new technologies, making possible radio, TV, computers, lasers, MRI scans, etc. But sometime in this century, it is likely that physicists will find their holy grail: room temperature superconductors. This will usher in an entirely new era, the age of magnetism.

Imagine riding in a magnetic car, hovering above the ground and traveling at several hundred miles per hour, using almost no fuel. Imagine trains and even people traveling in the air, floating on magnetism.

We forget that most of the gasoline we use in our cars goes to overcoming friction. In principle, it takes almost no energy to ride from San Francisco to New York City. The main reason this trip consumes hundreds of dollars of gasoline is because you have to overcome the friction of the wheels on the road and the friction of the air. But if you could somehow cover the road from San Francisco to New York with a layer of ice, you could simply coast most of the way almost for free. Likewise, our space probes can soar beyond Pluto with only a few quarts of fuel because they coast through the vacuum of space. In the same way, a magnetic car would float above the ground; you simply blow on the car, and the car begins to move.

The key to this technology is superconductors. It has been known since 1911 that mercury, when cooled to four degrees (Kelvin) above absolute zero, loses all electrical resistance. This means that superconducting wires have no energy loss whatsoever, since they lack any resistance. (This is because electrons moving through a wire lose energy as they collide with atoms. But at near absolute zero, these atoms are almost at rest, so the electrons can easily slip through them without losing energy.)

These superconductors have strange but marvelous properties, but one severe disadvantage is that you have to cool them to near absolute zero with liquid hydrogen, which is very expensive.

Therefore, physicists were in shock in 1986 when it was announced that a new class of superconductors had been found that did not need to be cooled to these fantastically low temperatures. Unlike previous materials like mercury or lead, these superconductors were ceramics, previously thought to be unlikely candidates for superconductors, and became superconductors at 92 degrees (Kelvin) above absolute zero. Embarrassingly, they became superconductors at a temperature that was thought to be theoretically impossible.

So far, the world record for these new ceramic superconductors is 138 degrees (Kelvin) above absolute zero (or −211° F). This is significant, since liquid nitrogen (which costs as little as milk) forms at 77° K (−321° F) and hence can be used to cool these ceramics. This fact alone has drastically cut the costs of superconductors. So these high-temperature superconductors have immediate practical applications.

But these ceramic superconductors have just whetted the appetite of physicists. They are a giant step in the right direction, but still they are not enough. First, although liquid nitrogen is relatively cheap, you still have to have some refrigeration equipment to cool the nitrogen. Second, these ceramics are difficult to mold into wires. Third, physicists are still bewildered by the nature of these ceramics. After several decades, physicists are not quite sure how they work. The quantum theory of these ceramics is too complicated to solve at the present time, so no one knows why they become superconductors. Physicists are clueless. There is a Nobel Prize waiting for the enterprising individual who can explain these high-temperature superconductors.

But every physicist knows the tremendous impact that a room temperature superconductor would have. It could set off another industrial revolution. Room temperature superconductors would not require any refrigeration equipment, so they could create permanent magnetic fields of enormous power.

For example, if electricity is flowing inside a copper loop, its energy dissipates within a fraction of a second because of the resistance of the wire. However, experiments have shown that electricity within a superconducting loop can remain constant for years at a time. The experimental evidence points to a lifetime of 100,000 years for currents inside a superconducting coil. Some theories maintain that the maximum limit for such an electrical current in a superconductor is the lifetime of the known universe itself.

At the very least, such superconductors could reduce the waste found in high-voltage electrical cables, thereby reducing the cost of electricity. One of the reasons an electrical plant has to be so close to a city is because of losses in the transmission lines. That is why nuclear power plants are so close to cities, which poses a health hazard, and why wind power plants cannot be placed in areas with the maximum wind.

Up to 30 percent of the electricity generated by an electrical plant can be wasted in the transmission. Room temperature superconducting wires could change all that, thereby saving significantly on electrical costs and pollution. This could also have a profound impact on global warming. Since the world's production of carbon dioxide is tightly connected to energy use, and since most of that energy is wasted to overcome friction, the age of magnetism could permanently reduce energy consumption and carbon dioxide production.

THE MAGNETIC CAR AND TRAIN

Without any extra input of energy, room temperature superconductors could produce supermagnets capable of lifting trains and cars so they hover above the ground.

One simple demonstration of this power can be done in any lab. I've done it several times myself for BBC-TV and the Science Channel. It's possible to order a small piece of ceramic high-temperature superconductor from a scientific supply company. It's a tough, gray ceramic about an inch in size. Then you can buy some liquid nitrogen from a dairy supply company. You place the ceramic in a plastic dish and gently pour the liquid nitrogen over it. The nitrogen starts to boil furiously as it hits the ceramic. Wait until the nitrogen stops boiling, then place a tiny magnet on top of the ceramic. Magically, the magnet floats in midair. If you tap the magnet, it starts to spin by itself. In that tiny dish, you may be staring at the future of transportation around the world.

The reason the magnet floats is simple. Magnetic lines of force cannot penetrate a superconductor. This is the Meissner effect. (When a magnetic field is applied to a superconductor, a small electric current forms on the surface and cancels it, so the magnetic field is expelled from the superconductor.) When you place the magnet on top of the ceramic, its field lines bunch up since they cannot pass through the ceramic. This creates a "cushion" of magnetic field lines, which are all squeezed together, thereby pushing the magnet away from the ceramic, making it float.

Room temperature superconductors may also usher in an era of supermagnets. MRI machines, as we have seen, are extremely useful but require large magnetic fields. Room temperature superconductors will allow scientists to create enormous magnetic fields cheaply. This will allow the future miniaturization of MRI machines. Already, using nonuniform magnetic fields, MRI machines about a foot tall can be created. With room temperature superconductors, it might be possible to reduce them to the size of buttons.

In the movie *Back to the Future Part III*, Michael J. Fox was filmed riding a hoverboard, a skateboard that floated in air. After the movie debut, stores were flooded with calls from kids asking to purchase the hoverboard. Unfortunately, hoverboards do not exist, but they might become possible with room temperature superconductors.

MAGLEV TRAINS AND CARS

One simple application of room temperature superconductors is to revolutionize transportation, introducing cars and trains that float above the ground and thus move without any friction.

Imagine riding in a car that uses room temperature superconductors. The roads would be made of superconductors instead of asphalt. The car would either contain a permanent magnet or generate a magnetic field via a superconductor of its own. The car would float. Even compressed air would be enough to get the car going. Once in motion, it would coast almost forever if the road were flat. An electric engine or jet of compressed air would be necessary only to overcome air friction, which would be the only drag that the car faces.

Even without room temperature superconductors, several nations have produced magnetic levitating trains (maglev) that hover above a set of rails containing magnets. Since the north poles of magnets repel other north poles, the magnets are arranged so that the bottom of the train contains magnets that allow them to float just above the tracks.

Germany, Japan, and China are leaders in this technology. Maglev trains have even set some world records. The first commercial maglev train

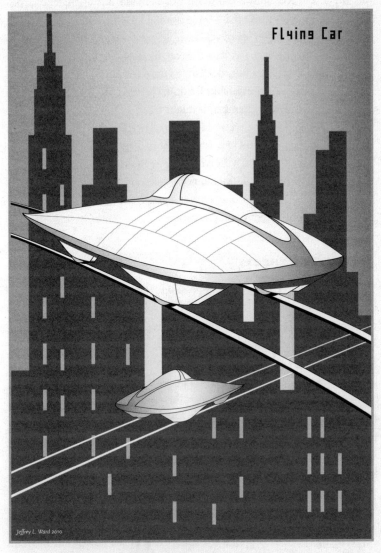

Flying Car

Room-temperature superconductors may one day give us flying cars and trains. These may float on rails or over superconducting pavement, without friction.

was the low-speed shuttle train that ran between Birmingham International Airport and Birmingham International Railway Station in 1984. The highest recorded maglev speed was 361 miles per hour, recorded in Japan on the MLX01 train in 2003. (Jet airplanes can fly faster, partly because there is less air resistance at high altitudes. Since a maglev train floats in air, most of its energy loss is in the form of air friction. However, if a maglev train were operating in a vacuum chamber, it might travel as fast as 4,000 miles per hour.) Unfortunately, the economics of maglev trains has prevented them from proliferating around the world. Room temperature superconductors might change all that. This could also revitalize the rail system in the United States, reducing the emission of greenhouse gases from airplanes. It is estimated that 2 percent of greenhouse gases come from jet engines, so maglev trains would reduce that amount.

ENERGY FROM THE SKY

By the end of the century, another possibility opens up for energy production: energy from space. This is called space solar power (SSP) and involves sending hundreds of space satellites into orbit around the earth, absorbing radiation from the sun, and then beaming this energy down to earth in the form of microwave radiation. The satellites would be based 22,000 miles above the earth, where they become geostationary, revolving around the earth as fast as the earth spins. Because there is eight times more sunlight in space than on the surface of the earth, this presents a real possibility.

At present, the main stumbling block to SSP is cost, mainly that of launching these space collectors. There is nothing in the laws of physics to prevent collecting energy directly from the sun, but it is a huge engineering and economic problem. But by end of the century, new ways of reducing the cost of space travel may put these space satellites within reach, as we will see in Chapter 6.

The first serious proposal for space-based solar power was made in 1968, when Peter Glaser, president of the International Solar Energy Society, proposed sending up satellites the size of a modern city to beam power down to the earth. In 1979, NASA scientists took a hard look at his proposal and estimated that the cost would be several hundred billion dollars, which killed the project.

But because of constant improvements in space technology, NASA continued to fund small-scale studies of SSP from 1995 to 2003. Its proponents maintain that it is only a matter of time before the technology and economics of SSP make it a reality. "SSP offers a truly sustainable, global-scale and emission-free electricity source," says Martin Hoffert, a physicist formerly at New York University.

There are formidable problems facing such an ambitious project, real and imaginary. Some people fear this project because the energy beamed down from space might accidentally hit a populated area, creating massive casualties. However, this fear is exaggerated. If one calculates the actual radiation hitting the earth from space, it is too small to cause any health hazard. So visions of a rogue space satellite sending death rays down to earth to fry entire cities is the stuff of a Hollywood nightmare.

Science fiction writer Ben Bova, writing in the *Washington Post* in 2009, laid out the daunting economics of a solar power satellite. He estimated that each satellite would generate 5 to 10 gigawatts of power, much more than a conventional coal-fired plant, and cost about eight to ten cents per kilowatt-hour, making it competitive. Each satellite would be huge, about a mile across, and cost about a billion dollars, roughly the cost of a nuclear plant.

To jump-start this technology, he asked the current administration to create a demonstration project, launching a satellite that could generate 10 to 100 megawatts. Hypothetically, it could be launched at the end of President Obama's second term in office if plans are started now.

Echoing these comments was a major initiative announced by the Japanese government. In 2009, the Japanese Trade Ministry announced a plan to investigate the feasibility of a space power satellite system. Mitsubishi Electric and other Japanese companies will join a $10 billion program to perhaps launch a solar power station into space that will generate a billion watts of power. It will be huge, about 1.5 square miles in area, covered with solar cells.

"It sounds like a science fiction cartoon, but solar power generation in space may be a significant alternative energy source in the century ahead as fossil fuel disappears," said Kensuke Kanekiyo of the Institute of Energy Economics, a government research organization.

Given the magnitude of this ambitious project, the Japanese government is cautious. A research group will first spend the next four years studying the scientific and economic feasibility of the project. If this group gives the green light, then the Japanese Trade Ministry and the Japanese Aerospace Exploration Agency plan to launch a small satellite in 2015 to test beaming down energy from outer space.

The major hurdle will probably not be scientific but economic. Hiroshi Yoshida of Excalibur KK, a space consulting company in Tokyo, warned, "These expenses need to be lowered to a hundredth of current estimates." One problem is that these satellites have to be 22,000 miles in space, much farther than satellites in near-earth orbits of 300 miles, so losses in transmission could be huge.

But the main problem is the cost of booster rockets. This is the same bottleneck that has stymied plans to return to the moon and explore Mars.

Unless the cost of rocket launches goes down significantly, this plan will die a quiet death.

Optimistically, the Japanese plan could go operational by midcentury. However, given the problems with booster rockets, more likely the plan will have to wait to the end of the century, when new generations of rocket drive down the cost. If the main problem with solar satellites is cost, then the next question is: Can we reduce the cost of space travel so that one day we might reach the stars?

> We have lingered long enough on the shores of the cosmic ocean. We are ready at last to set sail for the stars.
>
> —CARL SAGAN

6 FUTURE OF SPACE TRAVEL *To the Stars*

In powerful chariots, the gods of mythology roamed across the heavenly fields of Mount Olympus. On powerful Viking ships, the Norse gods sailed across the cosmic seas to Asgard.

Similarly, by 2100, humanity will be on the brink of a new era of space exploration: reaching for the stars. The stars at night, which seem so tantalizingly close yet so far, will be in sharp focus for rocket scientists by the end of the century.

But the road to building starships will be a rocky one. Humanity is like someone whose outstretched arms are reaching for the stars but whose feet are mired in the mud. On one hand, this century will see a new era for robotic space exploration as we send satellites to locate earthlike twins in space, explore the moons of Jupiter, and even take baby pictures of the big bang itself. However, the manned exploration of outer space, which has enthralled many generations of dreamers and visionaries, will be a source of some disappointment.

NEAR TERM (PRESENT TO 2030)

EXTRASOLAR PLANETS

One of the most stunning achievements of the space program has been the robotic exploration of outer space, which has vastly expanded the horizon of humanity.

Foremost among these robotic missions will be the search for earthlike planets in space that can harbor life, which is the holy grail of space science. So far, ground-based telescopes have identified about 500 planets orbiting in distant star systems, and new planets are being discovered at the rate of one planet every one to two weeks. The big disappointment, however, is that our instruments can identify only gigantic, Jupiter-sized planets that cannot sustain life as we know it.

To find planets, astronomers look for tiny wobbles in the path of a star. These alien solar systems can be likened to a spinning dumbbell, where the two balls revolve around each other; one end represents the star, clearly visible by telescope, while the other represents a Jupiter-sized planet, which is about a billion times dimmer. As the sun and Jupiter-sized planet spin around the center of the dumbbell, telescopes can clearly see the star wobbling. This method has successfully identified hundreds of gas giants in space, but it is too crude to detect the presence of tiny, earthlike planets.

The smallest planet found by these ground-based telescopes was identified in 2010 and is 3 to 4 times as massive as earth. Remarkably, this "superearth" is the first one to be in the habital zone of its sun—i.e., at the right distance to have liquid water.

All this changed with the launch of the Kepler Mission telescope in 2009 and the COROT satellite in 2006. These space probes look for tiny fluctuations in starlight, caused when a small planet moves in front of its star, blocking its light by a minuscule amount. By carefully scanning thousands of stars to look for these tiny fluctuations, the space probes will be able to detect perhaps hundreds of earthlike planets. Once identified, these planets can be quickly analyzed to see if they contain liquid water, perhaps the most precious commodity in space. Liquid water is the universal solvent, the mixing bowl where the first DNA probably got off the ground. If liquid-water oceans are found on these planets, it could alter our understanding of life in the universe.

Journalists in search of a scandal say, "Follow the money," but astronomers searching for life in space say, "Follow the water."

The Kepler satellite, in turn, will be replaced by other, more sensitive satellites, such as the Terrestrial Planet Finder. Although the launch date for the Terrestrial Planet Finder has been postponed several times, it remains the best candidate to further the goals of Kepler.

The Terrestrial Planet Finder will use much better optics to find earth-like twins in space. First, it will have a mirror four times larger and one hundred times more sensitive than that of the Hubble Space Telescope. Second, it will have infrared sensors that can nullify the intense radiation from a star by a factor of a million times, thereby revealing the presence of the dim planet that may be orbiting it. (It does this by taking two waves of radiation from the star and then carefully combining them so that they cancel each other out, thereby removing the unwanted presence of the star.)

So in the near future, we should have an encyclopedia of several thousand planets, of which perhaps a few hundred will be very similar to the earth in size and composition. This, in turn, will generate more interest in one day sending a probe to these distant planets. There will be an intense effort to see if these earthlike twins have liquid-water oceans and if there are any radio emissions from intelligent life-forms.

EUROPA—OUTSIDE THE GOLDILOCKS ZONE

There is also another tempting target for our probes within our solar system: Europa. For decades, it was believed that life in the solar system can exist only in the "Goldilocks zone" around the sun, where planets are not too hot or too cold to sustain life. The earth is blessed with liquid water because it orbits at the right distance from the sun. Liquid water will boil on a planet like Mercury, which is too close to the sun, and will freeze on a planet like Jupiter, which is too far. Since liquid water is probably the fluid in which DNA and proteins were first formed, it was long believed that life in the solar system can exist only on earth, or perhaps Mars.

But astronomers were wrong. After the *Voyager* spacecraft sailed past the moons of Jupiter, it became apparent that there was another place for life to flourish: under the ice cover of the moons of Jupiter. Europa, one of the moons of Jupiter discovered by Galileo in 1610, soon caught the attention of astronomers. Although its surface is permanently covered with ice, beneath that ice there is a liquid ocean. Because the ocean is much deeper on Europa than on earth, the total volume of the Europan ocean is estimated to be twice the volume of earth's oceans.

This was a bit of a shock, realizing that there is an abundant energy source in the solar system other than the sun. Underneath the ice, the surface of Europa is continually heated by tidal forces. As Europa tumbles in its orbit around Jupiter, that massive planet's gravity squeezes the moon in different directions, creating friction deep within its core. This friction creates heat, which in turn melts the ice and creates a stable ocean of liquid water.

This discovery means that perhaps the moons of distant gas giants are more interesting than the planets themselves. (This is probably one reason James Cameron chose a moon of a Jupiter-size planet for the site of his 2009 movie, *Avatar*.) Life, which was once thought to be quite rare, might actually flourish in the blackness of space on the moons of distant gas giants. Suddenly, the number of places where life might flourish has exploded by many times.

As a consequence of this remarkable discovery, the Europa Jupiter System Mission (EJSM) is tentatively scheduled for launch in 2020. It is designed to orbit Europa and possibly land on it. Beyond that, scientists have dreamed of exploring Europa by sending even more sophisticated machinery. Scientists have considered a variety of methods to search for life under the ice. One possibility is the Europa Ice Clipper Mission, which would drop spheres on the icy surface. The plume and debris cloud emerging from the impact site would then be carefully analyzed by a spacecraft flying through it. An even more ambitious program is to put a remote-control hydrobot submarine beneath the ice.

Interest in Europa has also been stoked by new developments under the ocean here on earth. Until the 1970s, most scientists believed that the sun was the only energy source that could make life possible. But in 1977, the *Alvin* submarine found evidence of new life-forms flourishing where no one suspected before. Probing the Galapagos Rift, it found giant tube worms, mussels, crustaceans, clams, and other life-forms using the heat energy from volcano vents to survive. Where there is energy, there might be life; and these undersea volcano vents have provided a new source of energy in the inky blackness of the sea floor. In fact, some scientists have suggested that the first DNA was formed not in some tide pool on the earth's coast but deep undersea near a volcano vent. Some of the most primitive forms of DNA (and perhaps the most ancient) have been found on the bottom of the ocean. If so, then perhaps volcano vents on Europa can provide the energy to get something like DNA off the ground.

One can only speculate about the possible life-forms that might form under Europa's ice. If they exist at all, they probably will be swimming creatures that use sonar, rather than light, for navigational purposes, so their view of the universe will be limited to living under the "sky" of ice.

LISA—BEFORE THE BIG BANG

Yet another space satellite that could create an upheaval in scientific knowledge is the Laser Interferometer Space Antenna (LISA) and its successors. These probes may be able to do the impossible: reveal what happened before the big bang.

Currently, we have been able to measure the rate at which the distant galaxies are moving away from us. (This is due to the Doppler shift, where light is distorted if the star moves toward or away from you.) This gives us the expansion rate of the universe. Then we "run the videotape backward," and calculate when the original explosion took place. This is very similar to the way you can analyze the fiery debris emanating from an explosion to determine when the explosion took place. That is how we determined that the big bang took place 13.7 billion years ago. What is frustrating, however, is that the current space satellite, the WMAP (Wilkinson Microwave Anisotropy Probe), can peer back only to less than 400,000 years after the original explosion. Therefore, our satellites can tell us only that there was a bang, but cannot tell us why it banged, what banged, and what caused the bang.

That is why LISA is creating such excitement. LISA will measure an entirely new type of radiation: gravity waves from the instant of the big bang itself.

Every time a new form of radiation was harnessed, it changed our worldview. When optical telescopes were first used by Galileo to map the planets and stars, they opened up the science of astronomy. When radio telescopes were perfected soon after World War II, they revealed a universe of exploding stars and black holes. And now the third generation of telescopes, which can detect gravitational waves, may open up an even more breathtaking vista, the world of colliding black holes, higher dimensions, and even a multiverse.

Tentatively, the launch date is being set for between 2018 and 2020. LISA consists of three satellites that will form a gigantic triangle 3 million miles across, connected by three laser beams. It will be the largest space instrument ever sent into orbit. Any gravity wave from the big bang still reverberating around the universe will jiggle the satellites a bit. This disturbance will change the laser beams, and then sensors will record the frequency and characteristics of the disturbance. In this way, scientists should be able to get within a trillionth of a second after the original big bang. (According to Einstein, space-time is like a fabric that can be curved and stretched. If there is a big disturbance, like colliding black holes or the big bang, then ripples can form and travel on this fabric. These ripples, or gravity waves, are too small to detect using ordinary instruments, but LISA is sensitive and large enough to detect vibrations caused by these gravity waves.)

Not only will LISA be able to detect radiation from colliding black holes, it might also be able to peer into the pre–big bang era, which was once thought to be impossible.

At present, there are several theories of the pre–big bang era coming from string theory, which is my specialty. In one scenario, our universe is a huge bubble of some sort that is continually expanding. We live on the skin of this gigantic bubble (we are stuck on the bubble like flies on flypaper). But our bubble universe coexists in an ocean of other bubble universes, making up the multiverse of universes, like a bubble bath. Occasionally, these bubbles might collide (giving us what is called the big splat theory) or they may fission into smaller bubbles and then expand (giving us what is called eternal inflation). Each of these pre–big bang theories predicts how the universe should release gravity radiation moments after the initial explosion. LISA can then measure the gravity radiation emitted after the big bang and compare it with the various predictions of string theory. In this way, LISA might be able to rule out or in some of these theories.

But even if LISA is not sensitive enough to perform this delicate task, perhaps the next generation of detectors beyond LISA (such as the Big Bang Observer) may be up to the task.

If successful, these space probes may answer the question that has defied explanation for centuries: Where did the universe originally come from? So in the near term, unveiling the origin of the big bang may be a distinct possibility.

MANNED MISSIONS TO SPACE

While robotic missions will continue to open new vistas for space exploration, the manned missions will face much greater hurdles. This is because, compared to manned missions, robotic missions are cheap and versatile; can explore dangerous environments; don't require costly life support; and most important, don't have to come back.

Back in 1969, it seemed as if our astronauts were poised to explore the solar system. Neil Armstrong and Buzz Aldrin had just walked on the moon, and already people were dreaming about going to Mars and beyond. It seemed as if we were on the threshold of the stars. A new age was dawning for humanity.

Then the dream collapsed.

As science fiction writer Isaac Asimov has written, we scored the touchdown, took our football, and then went home. Today, the old Saturn booster rockets are idling in museums or rotting in junkyards. An entire generation of top rocket scientists was allowed to dissipate. The momentum of the space race slowly dissipated. Today, you can find reference to the famous moon walk only in dusty history books.

What happened? Many things, including the Vietnam War, the Watergate scandal, etc. But, when everything is boiled down, it reduces to just one word: cost.

We sometimes forget that space travel is expensive, very expensive. It costs $10,000 to put a pound of anything just into near-earth orbit. Imagine John Glenn made of solid gold, and you can grasp the cost of space travel. To reach the moon would require about $100,000 per pound. And to reach Mars would require about $1,000,000 per pound (roughly your weight in diamonds).

All this, however, was covered up by the excitement and drama of competing with the Russians. Spectacular space stunts by brave astronauts hid the true cost of space travel from view, since nations were willing to pay dearly if their national honor was at stake. But even superpowers cannot sustain such costs over many decades.

Sadly, it has been over 300 years since Sir Isaac Newton first wrote down the laws of motion, and we are still dogged by a simple calculation. To hurl an object into near-earth orbit, you have to send it 18,000 miles per hour. And to send it into deep space, beyond the gravity field of the earth, you have to propel it 25,000 miles per hour. (And to reach this magic number of 25,000 miles per hour, we have to use Newton's third law of motion: for every action, there is an equal and opposite reaction. This means that the rocket can go rapidly forward because it spews out hot gases in the opposite direction, in the same way that a balloon flies around a room when you inflate it and then let it go.) So it is a simple step from Newton's laws to calculating the cost of space travel. There is no law of engineering or physics that prevents us from exploring the solar system; it's a matter of cost.

Worse, the rocket must carry its own fuel, which adds to its weight. Airplanes partially get around this problem because they can scoop oxygen from the air outside and then burn it in their engines. But since there is no air in space, the rocket must carry its own tanks of oxygen and hydrogen.

Not only is this the reason space travel is so expensive, it is also the reason we don't have jet packs and flying cars. Science fiction writers (not real scientists) glamorized the day when we would all put on jet packs and fly to work, or go on a Sunday day trip blasting off in our family flying car. Many people became disillusioned by futurists because these predictions never came to pass. (That is why we see a rash of articles and books with cynical titles like "*Where's My Jetpack?*") But a quick calculation shows the reason. Jet packs already exist; in fact, the Nazis used them briefly during World War II. But hydrogen peroxide, the common fuel used in jet packs, quickly runs out, so a typical flight in a jet pack lasts only a few minutes. Also, flying cars that use helicopter blades burn up an enormous amount of fuel, making them far too costly for the average suburban commuter.

CANCELING THE MOON PROGRAM

Because of the enormous cost of space travel, currently the future of the manned exploration of space is in flux. Former president George W. Bush presented a clear but ambitious plan for the space program. First, the space shuttle would be retired in 2010 and replaced in 2015 by a new rocket system called Constellation. Second, astronauts would return to the moon by 2020, eventually setting up a permanent manned base there. Third, this would pave the way for an eventual manned mission to Mars.

However, the economics of space travel have changed significantly since then, especially because the great recession has drained funds for future space missions. The Augustine Commission report, given to President Barack Obama in 2009, concluded that the earlier plan was unsustainable given current funding levels. In 2010, President Obama endorsed the findings of the Augustine report, canceling the space shuttle and its replacement that was to set the groundwork for returning to the moon. In the near term, without the rockets to send our astronauts into space, NASA will be forced to rely on the Russians. In the meantime, this provides an opportunity for private companies to create the rockets necessary to continue the manned space program. In a sharp departure from the past, NASA will no longer be building the rockets for the manned space program. Proponents of the plan say it will usher in a new age of space travel, when private enterprise takes over. Critics say the plan will reduce NASA to "an agency to nowhere."

LANDING ON AN ASTEROID

The Augustine report laid out what it called the flexible path, containing several modest objectives that did not require so much rocket fuel; for example, traveling to a nearby asteroid that happened to be floating by or traveling to the moons of Mars. Such an asteroid, it was pointed out, may not even be on our sky charts yet; it might be a wandering asteroid that might be discovered in the near future.

The problem, the Augustine report said, is that the rocket fuel for the landing and return mission from the moon, or especially from Mars, would be prohibitively expensive. But since asteroids and the moons of Mars have very low gravitational fields, these missions would not require so much rocket fuel. The Augustine report also mentioned the possibility of visiting the Lagrange points, which are the places in outer space where the gravitational pull of the earth and moon cancel each other out. (These points might serve as a cosmic dump, where ancient pieces of debris from the early solar system have collected, so by visiting them astronauts may find interesting rocks dating back to the formation of the earth-moon system.)

Landing on an asteroid would certainly be a low-cost mission, since asteroids have very weak gravitational fields. (This is also the reason asteroids are irregularly shaped, rather than round. In the universe, large objects—such as stars, planets, and moons—are all round because gravity pulls evenly. Any irregularity in the shape of a planet gradually disappears as gravity compresses the crust. But the gravity field of an asteroid is so weak that it cannot compress the asteroid into a sphere.)

One possibility is the asteroid Apophis, which will make an uncomfortably close pass in 2029. Apophis is about 1,000 feet across, the size of a large football stadium, and will come so close to the earth that it will actually pass beneath some of our satellites. Depending on how the orbit of the asteroid is distorted by this close pass, it may swing back to the earth in 2036, where there is a tiny chance (1 out of 100,000) that it might hit the earth. If this were to happen, it would hit with the force of 100,000 Hiroshima bombs, sufficient to destroy an area as large as France with firestorms, shock waves, and fiery debris. (By comparison, a much smaller object, probably the size of an apartment building, slammed into Tunguska, Siberia, in 1908, with the force of about 1,000 Hiroshima bombs, wiping out 1,000 square miles of forest and creating a shock wave felt thousands of miles away. It also created a strange glow seen over Asia and Europe, so that people in London could read the newspapers at night.)

Visiting Apophis would not strain the NASA budget, since the asteroid is coming near earth anyway, but landing on the asteroid might pose a problem. Since it has a weak gravity field, one would actually dock with the asteroid, rather than land on it in the traditional sense. Also, the asteroid is probably spinning irregularly, so precise measurements have to be made before landing. It would be interesting to test to see how solid the asteroid is. Some believe that an asteroid may be a collection of rock loosely held together by a weak gravity field. Others believe that it may be solid. Determining the consistency of an asteroid may be important one day, if we have to use nuclear weapons to blow one up. An asteroid, instead of being pulverized into a fine powder, might instead break up into several large pieces. If so, then the danger from these pieces might be greater than the original threat. A better idea may be to nudge the asteroid out of the way before it comes close to earth.

LANDING ON A MOON OF MARS

Although the Augustine report did not support a manned mission to Mars, one intriguing possibility is to send astronauts to visit the moons of Mars, Phobos and Deimos. These moons are much smaller than earth's moon and hence have a very low gravitational field. There are several advantages to landing on the moons of Mars, in addition to saving on cost.

1. First, these moons could be used as space stations. They would provide a cheap way of analyzing the planet from space without visiting it.

2. Second, they could eventually provide an easy way to access Mars. Phobos is less than 6,000 miles from the center of Mars, so a quick journey to the Red Planet can be made within a matter of hours.

3. These moons would probably have caves that could be used for a permanent manned base to protect against meteors and radiation. Phobos, in particular, has the huge Stickney crater on its side, indicating that the moon was probably hit by a huge meteor and almost blown apart. However, gravity slowly brought back the pieces and reassembled the moon. There are probably plenty of caves and gaps left over from this ancient collision.

BACK TO THE MOON

The Augustine report also mentioned a Moon First program, where we would go back to the moon, but only if more funding were available—at least $30 billion over ten years. Since that is unlikely, the moon program, in effect, is canceled, at least for the coming years.

The canceled moon mission was called the Constellation Program, which consisted of several major components. First was the booster rocket, the Ares, the first major U.S. booster rocket since the old Saturn rocket was mothballed back in the 1970s. On top of the Ares sat the Orion module, which could carry six astronauts to the space station or four astronauts to the moon. Then there was the Altair lander, which was supposed to actually land on the moon.

The old space shuttle, where the shuttle rocket was placed on the side of the booster rocket, had a number of design flaws, including the tendency of the rocket to shed pieces of foam. This had disastrous consequences for the Space Shuttle *Columbia,* which broke up on reentry in 2003, killing seven brave astronauts, because a piece of foam from the booster rocket hit the shuttle and made a hole in its wing during takeoff. Upon reentry, hot gases penetrated the hull of the *Columbia,* killing everyone inside and causing the ship to break up. In the Constellation, with the crew module placed directly on top of the booster rocket, this would no longer be a problem.

The Constellation program had been called "an Apollo program on steroids" by the press, since it looked very much like the moon rocket program of the 1970s. The Ares I booster was to be 325 feet tall, comparable to the 363-foot Saturn V rocket. It was supposed to carry the Orion module into space, replacing the old space shuttle. But for very heavy lifting, NASA was to use the Ares V rocket, which was 381 feet tall and capable of taking 207 tons of payload into space. The Ares V rocket would have been the backbone of any mission to the moon or Mars. (Although the Ares has been canceled, there is talk of perhaps salvaging some of these components for future missions.)

PERMANENT MOON BASE

Although the Constellation Program was canceled by President Obama, he left open several options. The Orion module, which was to have taken our astronauts back to the moon, is now being considered as an escape pod for the International Space Station. At some point in the future, when the economy recovers, another administration may want to set its sights on the moon again, including a moon base.

The task of establishing a permanent presence on the moon faces many obstacles. The first is micrometeorites. Because the moon is airless, rocks from space frequently hit it. We can see this by viewing its surface, pockmarked by meteorite collisions, some dating back billions of years.

I got a personal look at this danger when I was a graduate student at the University of California at Berkeley. Moon rocks brought back from space in the early 1970s were creating a sensation in the scientific community. I was invited into a laboratory that was analyzing moon rock under a microscope. The rock I saw looked ordinary, since moon rock very closely resembles earth rock, but under the microscope I got quite a shock. I saw tiny meteor craters in the rock, and inside them I saw even tinier craters. Craters inside craters inside craters, something I had never seen before. I immediately realized that without an atmosphere, even the tiniest microscopic piece of dirt, hitting you at 40,000 miles per hour, could easily kill you or at least penetrate your space suit. (Scientists understand the enormous damage created by these micrometeorites because they can simulate these impacts, and they have created huge gun barrels in their labs that can fire metal pellets to study these meteor impacts.)

One possible solution is to build an underground lunar base. Because of the moon's ancient volcanic activity, there is a chance our astronauts can find a lava tube that extends deep into the moon's interior. (Lava tubes are created by ancient lava flows that have carved out cavelike structures and tunnels underground.) In 2009, astronomers found a lava tube about the size of a skyscraper that might serve as a permanent base on the moon.

This natural cave could provide cheap protection for our astronauts against radiation from cosmic rays and solar flares. Even taking a transcontinental flight from New York to Los Angeles exposes us to a millirem of radiation per hour (equivalent to getting a dental X-ray). For our astronauts on the moon, the radiation might be so intense that they might need to live in underground bases. Without an atmosphere, a deadly rain of solar flares and cosmic rays would pose an immediate risk to astronauts, causing premature aging and even cancer.

Weightlessness is also a problem, especially for long missions in space. I had a chance to visit the NASA training center in Cleveland, Ohio, where extensive tests are done on our astronauts. In one test I observed, the subject was suspended in a harness so that his body was parallel to the ground. Then he began to run on a treadmill, whose tracks were vertical. By running on this treadmill, NASA scientists could simulate weightlessness while testing the endurance of the subject.

When I spoke to the NASA doctors, I learned that weightlessness was more damaging than I had previously thought. One doctor explained to me that after several decades of subjecting American and Russian astronauts to prolonged weightlessness, scientists now realize that the body undergoes significant changes: degradation occurs in the muscles, bones, and cardiovascular system. Our bodies evolved over millions of years while living in the earth's gravitational field. When placed in a weaker gravitational field for long periods of time, all our biological processes are thrown into disarray.

Russian astronauts who have spent about a year in space are so weak when they come back to earth that they can barely crawl. Even if they exercise daily in space, their muscles atrophy, their bones lose calcium, and their cardiovascular systems begin to weaken. Some of the astronauts take months to recover from this damage, some of which may be permanent. A trip to Mars, which might take two years, may drain the strength of our astronauts so they cannot perform their mission when they arrive. (One solution to this problem is to spin the spacecraft, which creates artificial gravity inside the ship. This is the same reason that you can spin a pail of water over your head without the water spilling out. But this is prohibitively expensive because of the heavy machinery necessary to spin the craft. Every pound of extra weight adds $10,000 to the cost of the mission.)

WATER ON THE MOON

One game changer has been the discovery of ancient ice on the moon, probably left over from ancient comet impacts. In 2009, NASA's lunar crater observation and sensing satellite (LCROSS) probe and its Centaur booster rocket slammed into the moon's south polar region. They hit the moon at 5,600 miles per hour, creating a plume almost a mile high, and a crater about 60 feet across. Although TV audiences were disappointed that the LCROSS impact did not create a spectacular explosion as predicted, it yielded a wealth of scientific data. About 24 gallons of water were found in that plume. Then, in 2010, scientists made the shocking announcement that 5 percent of the debris contained water, so the moon was actually wetter than parts of the Sahara desert.

This could be significant, because it might mean that future astronauts can harvest underground ice deposits for rocket fuel (by extracting the hydrogen in the water), for breathing (by extracting the oxygen), for shielding (since water can absorb radiation), and for drinking once it is purified. So this discovery could shave hundreds of millions of dollars off any mission to the moon.

This discovery may mean that it will be possible for our astronauts to live off the land, harvesting ice and minerals on the moon to create and supply a permanent base.

MIDCENTURY (2030 TO 2070)

MISSION TO MARS

President Obama, when he journeyed to Florida in 2010 to announce the cancellation of the moon program, held out the prospects of a mission to Mars instead. He supported funding for a yet-unspecified heavy booster rocket that may one day send astronauts into deep space beyond the moon. He mused that he might see the day, perhaps in the mid-2030s, when our astronauts would walk on Mars. Some astronauts, like Buzz Aldrin, have been enthusiastic supporters of the Obama plan, because it would skip the moon. Aldrin once told me that the United States has already been to the moon, and hence the real adventure lies in going to Mars.

Of all the planets in the solar system, only Mars seems to resemble earth enough to harbor some form of life. (Mercury, which is scorched by the sun, is probably too hostile to have life as we know it. And the gas giants—Jupiter, Saturn, Uranus, and Neptune—are too cold to support life. Venus is a twin of the earth, but a runaway greenhouse effect has created a hellhole: temperatures soar to 900°F, its mostly carbon dioxide atmosphere is 100 times denser than ours, and it rains sulfuric acid. Walking on the Venusian surface, you would suffocate, be crushed to death, and your remains would be incinerated by the heat and dissolved by the sulfuric acid.)

Mars, on the other hand, was once a wet planet, like earth, with oceans and riverbeds that have long since vanished. Today, it is a frozen desert, devoid of life. Perhaps microbial life once flourished there billions of years ago or may still live underground in hot springs.

Once our nation has made a firm commitment to go to Mars, it may take another twenty to thirty years to actually complete the mission. But getting to Mars will be much more difficult than reaching the moon. In contrast to the moon, Mars represents a quantum leap in difficulty. It takes only three days to reach the moon. It takes six months to a year to reach Mars.

In July 2009, NASA scientists gave a rare look at what a realistic Mars mission might look like. Astronauts would take approximately six months or more to reach Mars, then spend eighteen months on the planet, then take another six months for the return voyage.

Altogether, about 1.5 million pounds of equipment would need to be sent to Mars, more than the amount needed for the $100 billion space station. To save on food and water, the astronauts would have to purify their own waste and then use it to fertilize plants during the trip and while on Mars. With no air, soil, or water, everything must be brought from earth. It will be impossible to live off the land, since there is no oxygen, liquid water, animals, or plants on Mars. The atmosphere is almost pure carbon dioxide, with an atmospheric pressure only 1 percent that of earth. Any rip in a space suit would create rapid depressurization and death.

The mission would be so complex that it would have to be broken down into several steps. Since carrying rocket fuel for the return mission back to earth would be costly, a separate rocket might be sent to Mars ahead of time carrying rocket fuel to be used for refueling the spacecraft. (Or, if enough oxygen and hydrogen could be extracted from the ice on Mars, this might be used for rocket fuel as well.)

Once on Mars, it might take weeks for the astronauts to get accustomed to living on another planet. The day/night cycle is about the same as on earth (a day on Mars is 24.6 hours). But a year is almost twice as long. The temperature on Mars never goes above the melting point of ice. The dust storms on Mars are ferocious. The sand of Mars has the consistency of talcum powder, and dust storms that engulf the entire planet are common.

TERRAFORM MARS?

Assuming that astronauts visit Mars by midcentury and establish a primitive Martian outpost, there is the possibility that astronauts might consider terraforming Mars, that is, transforming the planet to make it more hospitable for life. This would begin late in the twenty-first century, at the earliest, or more likely early in the twenty-second.

Scientists have analyzed several ways in which Mars might be terraformed. Perhaps the simplest way would be to inject methane gas or other greenhouse gases into the atmosphere. Since methane gas is an even more potent greenhouse gas than carbon dioxide, the methane gas might be able to trap sunlight, raising the surface temperature of Mars to above the melting point of ice. In addition to methane, other greenhouse gases have been analyzed for possible terraforming experiments, such as ammonia and chlorofluorocarbons.

Once the temperature starts to rise, the underground permafrost may begin to thaw out, for the first time in billions of years. As the permafrost melts, riverbeds would begin to fill up with water. Eventually, lakes and even oceans might form again on Mars as the atmosphere thickens. This would release more carbon dioxide, setting off a positive feedback loop.

In 2009 it was discovered that methane gas naturally escapes from the Martian surface. The source of this gas is still a mystery. On earth, most of the methane gas is due to the decay of organic materials. But on Mars, the methane gas may be a by-product of geologic processes. If one can locate the source of this methane gas, then it might be possible to increase its output and hence alter the atmosphere.

Another possibility is to deflect a comet into the Martian atmosphere. If one can intercept a comet far enough away, then even a small nudge by a rocket engine, an impact with a probe, or even the tug of the gravity of a spaceship might be enough to deflect it. Comets are made mainly of water ice and periodically race through our solar system. (Halley's comet, for example, consists of a core—resembling a peanut—that is roughly twenty miles across, made mainly of ice and rock.) As the comet gradually gets closer to the surface of Mars, it would encounter friction from the atmosphere, causing the comet to slowly disintegrate, releasing water into the atmosphere in the form of steam.

If comets are not available, it could also be possible to deflect one of the ice moons of Jupiter or perhaps an asteroid that contains ice, such as Ceres, which is believed to be 20 percent water. (These moons and asteroids would be harder to deflect, since they are usually in stable orbits.) Instead of having the comet, moon, or asteroid slowly decay in its orbit around Mars, releasing water vapor, another choice would be to maneuver them into a controlled impact on the Martian ice caps. The polar regions of Mars are made of frozen carbon dioxide, which disappears during the summer months, and ice, which makes up the permanent part of the ice caps. If the comet, moon, or asteroid hits the ice caps, they can release a tremendous amount of heat and vaporize the dry ice. Since carbon dioxide is a greenhouse gas, this would thicken the atmosphere and help to accelerate global warming on Mars. It might also create a positive feedback loop. The more carbon dioxide is released from the ice caps, the warmer the planet becomes, which in turn releases even more carbon dioxide.

Another suggestion is to detonate nuclear bombs directly on the ice caps. The drawback is that the resulting liquid water might contain radioactive fallout. Or we could try to create a fusion reactor that can melt the polar ice caps. Fusion plants use water as a basic fuel, and there is plenty of frozen water on Mars.

Once the temperature of Mars rises to the melting point of ice, pools of water may form, and certain forms of algae that thrive on earth in the Antarctic may be introduced to Mars. They might actually thrive in the atmosphere of Mars, which is 95 percent carbon dioxide. They could also be genetically modified to maximize their growth on Mars. These algae pools could accelerate terraforming in several ways. First, they could convert carbon dioxide into oxygen. Second, they would darken the surface color of Mars, so that it absorbs more heat from the sun. Third, since they grow by themselves without any prompting from the outside, it would be a relatively cheap way to change the environment of the planet. Fourth, the algae can be harvested for food. Eventually these algae lakes would create soil and nutrients that may be suitable for plants, which in turn would accelerate the production of oxygen.

Scientists have also looked into the possibility of building solar satellites surrounding the planet, reflecting sunlight onto Mars. Solar satellites by themselves might be able to heat the Martian surface above freezing. Once this happens and the permafrost begins to melt, the planet would naturally continue to warm on its own.

ECONOMIC BENEFIT?

One should have no illusions that we will benefit immediately from an economic bonanza by colonizing the moon and Mars. When Columbus sailed to the New World in 1492, he opened the door to a historic economic windfall. Soon, the conquistadors were sending back huge quantities of gold that they plundered from Native Americans, and settlers were sending valuable raw materials and crops back to the Old World. The cost of sending expeditions to the New World was more than offset by the fabulous fortunes that could be made.

But colonies on the moon or Mars are quite different. There is no air, liquid water, or fertile soil, so everything would have to be brought by rocket ship, which is prohibitively expensive.

Furthermore, there is little military value to colonizing the moon, at least for the near term. This is because it takes three days on average to reach the moon from the earth or vice versa, but a nuclear war can be fought in just ninety minutes by intercontinental ballistic missiles. A space cavalry on the moon would not reach the battle on earth in time to make a difference. Hence, the Pentagon has not funded any crash program to weaponize the moon.

This means that if we do initiate large-scale mining operations on other worlds, it will be for the benefit of space colonies, not for the earth. Colonists will extract the metals and minerals for their own use, since it would cost too much to transport them to earth. Mining operations in the asteroid belt would become economic only when we have self-sustaining colonies that can use these raw materials themselves, which won't happen until late in this century or, more likely, beyond.

SPACE TOURISM

But when might the average civilian go into space? Some visionaries, like the late Gerard O'Neill of Princeton University, dreamed of a space colony as a gigantic wheel, including living units, water-purification plants, air-recycling units, etc., established to solve overpopulation on earth. But in the twenty-first century, the idea that space colonies would relieve the population problem is fanciful at best. For the majority of the human race, earth will be our only home for at least a century or more.

However, there is one way in which the average person may realistically go into space: as a tourist. Some entrepreneurs, who criticize the enormous waste and bureaucracy of NASA, think they can drive down the cost of space travel using market forces. Already, Burt Rutan and his investors won the $10 million Ansari X Prize on October 4, 2004, by having launched SpaceShipOne twice within two weeks to just over 62 miles above the earth. SpaceShipOne is the first rocket-powered spacecraft to have successfully completed a privately funded venture into space. Development costs were about $25 million. Microsoft billionaire Paul Allen helped to underwrite the project.

Now, with SpaceShipTwo, Rutan expects to begin tests to make commercial spaceflight a reality. Billionaire Richard Branson of Virgin Atlantic has created Virgin Galactic, with a spaceport in New Mexico and a long list of people who will spend $200,000 to realize their dream of flying into space. Virgin Galactic, which will be the first major company to offer commercial flights into space, has already ordered five SpaceShipTwo rockets. If successful, this might drive down the cost of space travel by a factor of ten.

SpaceShipTwo uses several methods to cut costs. Instead of huge booster rockets to carry the payload into space, Rutan places his spaceship atop an airplane, so that it can piggyback on a standard air-breathing plane. This way, you simply consume the oxygen in the atmosphere to reach high altitudes. Then, at about 10 miles above the earth, the spaceship separates from the airplane and turns on its rocket engines. Although the spaceship cannot orbit the earth, it has enough fuel to reach almost 70 miles above the earth, above most of the atmosphere, so passengers can see the sky turn purple and then black. Its engines are powerful enough to hit Mach 3, or three times the speed of sound (roughly 2,200 miles per hour). This is certainly not fast enough to put a rocket into orbit (you need to hit 18,000 miles per hour for that), but it is enough to take you to the edge of the atmosphere and the threshold of outer space. In the near future, perhaps a trip to space may cost no more than a safari in Africa.

(However, to go completely around the earth, you would need to pay considerably more to take a trip aboard the space station. I once asked Microsoft billionaire Charles Simonyi how much it cost him to get a ticket to the Space Station. Media reports estimated that it cost $20 million. He said he was reluctant to give the precise cost, but he told me that the media reports were not far off. He had such a good time that he actually went into space twice. So space travel, even into the near future, will still be the province of the well-off.)

Space tourism, however, got a shot in the arm in September 2010, when the Boeing Corporation announced that it, too, was entering the business, with commercial flights for tourists planned as early as 2015. This would bolster President Obama's decision to turn over the manned spaceflight program to private industry. Boeing's plan calls for launches to the International Space Station from Cape Canaveral, Florida, each involving four crew members, which would leave free up to three seats for space tourists. Boeing, however, was blunt about the financing for private ventures into space: the taxpayer would have to pay most of the bill. "This is an uncertain market," says John Elbon, program manager for Boeing's commerical crew effort. "If we had to do this with Boeing investment only and the risk factors were in there, we wouldn't be able to close the business case."

WILD CARDS

The punishing cost of space travel has hindered both commercial and scientific progress, so we need a revolutionary new design. By midcentury, scientists and engineers will be perfecting new booster-rocket technologies to drive down the cost of space travel.

Physicist Freeman Dyson has narrowed down some experimental technologies that may one day open up the heavens for the average person. These proposals are all high risk, but they might drastically reduce the cost. The first is the laser propulsion engine; this fires a high-power laser beam at the bottom of a rocket, causing a mini-explosion whose shock wave pushes the rocket upward. A steady stream of rapid-fire laser blasts vaporizes water, which propels the rocket into space. The great advantage of the laser propulsion system is that the energy comes from a ground-based system. The laser rocket contains no fuel whatsoever. (Chemical rockets, by contrast, waste much of their energy lifting the weight of their fuel into space.)

The technology for the laser propulsion system has already been demonstrated, and the first successful test of a model was carried out in 1997. Leik Myrabo of Rensselaer Polytechnic Institute in New York has created workable prototypes of this rocket, which he calls the lightcraft technology demonstrator. One early design was six inches in diameter and weighed two ounces. A 10-kilowatt laser generated a series of laser bursts on the bottom of the rocket, creating a machine-gun sound as the air bursts pushed the rocket at an acceleration of 2 g's (twice the earth's gravitational acceleration, or 64 feet per second squared). He has been able to build lightcraft rockets that have risen more than 100 feet into the air (equivalent to the early liquid-fueled rockets of Robert Goddard in the 1930s).

Dyson dreams of the day when laser propulsion systems can place heavy payloads into earth orbit for just $5 per pound, which would truly revolutionize space travel. He envisions a giant, 1,000-megawatt laser that can boost a two-ton rocket into orbit. (That is the power output of a standard nuclear power plant.) The rocket consists of the payload and a tank of water on the bottom, which slowly leaks water through tiny pores. The payload and the water tank each weigh one ton. As the laser beam strikes the bottom of the rocket, the water instantly vaporizes, creating a series of shock waves that push the rocket toward space. The rocket attains an acceleration of 3 g's and it leaves the earth's gravitational pull within six minutes.

Because the rocket carries no fuel, there is no danger of a catastrophic booster-rocket explosion. Chemical rockets, even fifty years into the space age, still have a failure rate of about 1 percent. And these failures are spectacular, with the volatile oxygen and hydrogen fuel creating huge fireballs and raining down debris all over the launch site. This system, by contrast, is simple, safe, and can be used repeatedly with a very small downtime, using only water and a laser.

Furthermore, the system would eventually pay for itself. If it can launch half a million spacecraft per year, the fees from these launches could easily pay for the operating costs as well as its development costs. Dyson, however, realizes that this dream is many decades into the future. The basic research on these huge lasers requires funding far beyond that of a university. Unless the research is underwritten by a large corporation or by the government, the laser propulsion system will never be built.

Here is where the X Prize may help. I once spoke with Peter Diamandis, who created the X Prize back in 1996, and he was well aware of the limitations of chemical rockets. Even SpaceShipTwo, he admitted to me, faced the problem that chemical rockets are an expensive way to escape the earth's gravity. As a consequence, a future X Prize will be given to someone who can create a rocket propelled by a beam of energy. (But instead of using a laser beam, it would use a similar source of electromagnetic energy, a microwave beam.) The publicity of the X Prize and the lure of a multimillion-dollar prize might be enough to spark interest among entrepreneurs and inventors to create nonchemical rockets, such as the microwave rocket.

There are other experimental rocket designs, but they involve different risks. One possibility is the gas gun, which fires projectiles out of a huge gun, somewhat similar to the rocket in Jules Verne's novel *From the Earth to the Moon*. Verne's rocket, however, would never fly, because gunpowder cannot shoot a projectile to 25,000 miles per hour, the velocity necessary to escape the earth's gravity. The gas gun, by contrast, uses high-pressure gas in a long tube to blast projectiles at high velocities. The late Abraham Hertzberg at the University of Washington in Seattle built a gun prototype that is four inches in diameter and thirty feet long. The gas inside the gun is a mixture of methane and air pressurized to twenty-five times atmospheric pressure. When the gas is ignited, the payload rides along the explosion at a remarkable 30,000 g's, an acceleration so great that it can flatten most metallic objects.

Hertzberg has proven that the gas gun can work. But to launch a payload into outer space, the tube must be much longer, about 750 feet, and must use different gases along the trajectory. Up to five different stages with different gases must be used to propel the payload to escape velocity.

The gas gun's launch costs may be even lower than those of the laser propulsion system. However, it is much too dangerous to launch humans in this way; only solid payloads that can withstand the intense acceleration will be launched.

A third experimental design is the slingatron, which, like a ball on a string, whirls payloads in a circle and then slings them into the air.

A prototype was built by Derek Tidman, who constructed a tabletop model that could hurl an object to 300 feet per second in a few seconds. The slingatron consists of a doughnut-shaped tube three feet in diameter. The tubing itself is one inch in diameter and contains a small steel ball. As the ball rolls around the tube, small motors push the ball so it moves increasingly fast.

A real slingatron that can hurl a payload into outer space must be significantly larger—hundreds or thousands of feet in diameter, capable of pumping energy into the ball until it reaches a speed of 7 miles per second. The ball would leave the slingatron with an acceleration of 1,000 g's, still enough to flatten most objects. There are many technical questions that have to be solved, the most important being the friction between the ball and the tube, which must be minimal.

All three of these designs will take decades to perfect, but only if funds from government or private industry are provided. Otherwise, these prototypes will always remain on the drawing board.

FAR FUTURE (2070 TO 2100)

SPACE ELEVATOR

By the end of this century, nanotechnology might even make possible the fabled space elevator. Like Jack and the beanstalk, we might be able to climb into the clouds and beyond. We would enter an elevator, push the up button, and then ascend along a carbon nanotube fiber that is thousands of miles long. This could turn the economics of space travel upside down.

Back in 1895, Russian physicist Konstantin Tsiolkovsky was inspired by the building of the Eiffel Tower, then the tallest structure of its kind in the world. He asked himself a simple question: Why can't you build an Eiffel Tower to outer space? If it was tall enough, he calculated, then it would never fall down, held up by the laws of physics. He called it a "celestial castle" in the sky.

Think of a ball on a string. By whipping the ball around, centrifugal force is enough to keep the ball from falling. Likewise, if a cable is sufficiently long, then centrifugal force will prevent it from falling back to earth. The spin of the earth would be sufficient to keep the cable in the sky. Once this cable is stretched into the heavens, any elevator cab that rides along this cable could take a ride into space.

On paper, this trick seems to work. But unfortunately, when using Newton's laws of motion to calculate the tension on the cable, you find that it is greater than the tensile strength of steel: the cable will snap, making a space elevator impossible.

Over the decades, the idea of a space elevator was periodically revived, only to be rejected for this reason. In 1957, Russian scientist Yuri Artsutanov proposed an improvement, suggesting that the space elevator be built top-down instead of bottom-up, that is, a spaceship would first be sent into orbit, and then a cable would descend to and be anchored in the earth. Also, science fiction writers popularized the idea of space elevators in Arthur C. Clarke's 1979 novel *The Fountains of Paradise* and Robert Heinlein's 1982 novel *Frida*.

Carbon nanotubes have helped revive this idea. These nanotubes, as we have seen, have some of the greatest tensile strengths of any material. They are stronger than steel, with enough strength to withstand the tension found in a space elevator.

The problem, however, is creating a pure carbon nanotube cable that is 50,000 miles long. This is a huge hurdle, since so far scientists have been able to create only a few centimeters of pure carbon nanotubes. It is possible to weave together billions of strands of carbon nanotubes to create sheets and cables, but these carbon nanotube fibers are not pure; they are

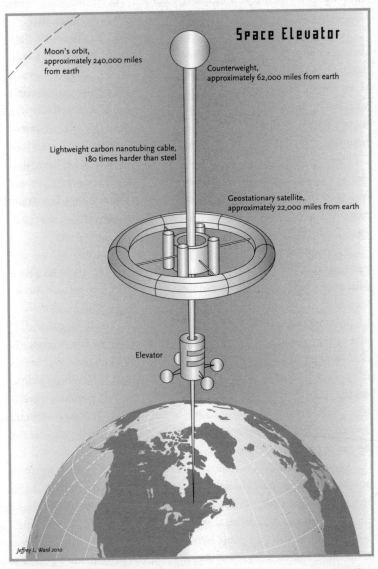

Space Elevator

Moon's orbit, approximately 240,000 miles from earth

Counterweight, approximately 62,000 miles from earth

Lightweight carbon nanotubing cable, 180 times harder than steel

Geostationary satellite, approximately 22,000 miles from earth

Elevator

Jeffrey L. Ward 2010

A space elevator to the heavens may one day vastly reduce the cost of space travel. The key to the space elevator may be nanotechnology.

fibers that have been pressed and woven together. The challenge is to create a carbon nanotube in which every atom of carbon is correctly in place.

In 2009, scientists at Rice University announced a breakthrough. Their fibers are not pure but composite (that is, they are not suitable for a space elevator), but their method is versatile enough to create carbon nanotubes of any length. They discovered, by trial and error, that these carbon nanotubes can be dissolved in a solution of chlorosulphonic acid, and then shot out of a nozzle, similar to a shower head. This method can produce carbon nanotube fibers that are 50 micrometers thick and hundreds of meters long.

One commercial application would be for electrical power lines, since carbon nanotubes conduct electricity better than copper, are lighter, and fail less often. Rice engineering professor Matteo Pasquali says, "For transmission lines you need to make tons, and there are no methods now to do that. We are one miracle away."

Although these cables are not pure enough to qualify for use in a space elevator, this research points to the day when one might be able to grow pure strands of carbon nanotubes, strong enough to take us into the heavens.

Assuming that in the future one will be able to create long strands of pure carbon nanotubes, there are still practical problems. For example, the cable will extend far beyond the orbit of most satellites, meaning that the orbits of satellites, after many passes around the earth, will eventually intersect the space elevator and cause a crash. Since satellites routinely travel at 18,000 miles per hour, an impact could be catastrophic. This means that the elevator has to be equipped with special rockets to move the cable out of the way of passing satellites.

Another problem is turbulent weather, such as hurricanes, lightning storms, and high winds. The space elevator must be anchored to the earth, perhaps on an aircraft carrier or oil platform sitting in the Pacific, but it must be flexible to avoid being damaged by the powerful forces of nature.

There must also be a panic button and escape pod in case of a break in the cable. If something snaps the cable, the elevator cab must be able to glide or parachute back to the earth's surface in order to save the passengers.

To jump-start research in space elevators, NASA has encouraged several contests. A total of $2 million in prizes is awarded through NASA's Space Elevator Games. According to the rules set down by NASA, to win the Beam Power Challenge, you must create a device weighing no more than 50 kilograms that can climb up a tether at the speed of 2 meters per second for a distance of 1 kilometer. What makes this challenge so difficult is that the device cannot have fuel, batteries, or an electrical cord. The energy must be beamed to the device from the outside.

I had a chance to see firsthand the enthusiasm and energy of engineers working on the space elevator and dreaming of claiming the prize. I flew to Seattle to meet young, enterprising engineers in a group called Laser-Motive. They had heard the siren call of NASA's contest and then began to create prototypes that may one day activate the space elevator.

I entered a large warehouse that they had rented to test out their ideas. On one side of the warehouse, I saw a powerful laser, capable of firing an intense beam of energy. On the other side of the warehouse, I saw their space elevator. It was a box about three feet wide, with a large mirror. The laser beam would hit the mirror and be deflected onto a series of solar cells that would convert the laser energy into electricity. This would trigger a motor, and the elevator car would gradually climb a short cable. In this way, you would not need electrical cables dangling from the space elevator to provide its energy. You would just fire a laser at the elevator from the earth, and the elevator would climb the cable by itself.

The laser was so powerful, we all had to wear special goggles to protect our eyes. It took numerous trial runs, but they finally were able fire the laser and send the device climbing the cable. At least in theory, one aspect of the space elevator had been solved.

Initially, the task was so difficult that no one won the prize. However, in 2009 LaserMotive claimed the prize. The contest took place at Edwards Air Force Base in the Mojave Desert in California. A helicopter flew over the desert, holding up a long cable. The LaserMotive team was able to make their elevator climb the cable four times in two days, with the best time being 3 minutes and 48 seconds. So all the hard work I had seen finally paid off for these young engineers.

STARSHIPS

By the end of the century, even despite recent setbacks in funding for manned space missions, scientists will likely have set up outposts on Mars and perhaps in the asteroid belt. Next, they will set their sights on an actual star. Although an interstellar probe is hopelessly beyond reach today, within 100 years it might become a reality.

The first challenge is to find a new propulsion system. For a conventional chemical rocket, it would take about 70,000 years to reach the nearest star. For example, the two *Voyager* spacecrafts, launched in 1977, have set a world record for an object sent into deep space. They are currently about 10 billion miles into space but only a tiny fraction of the way to the stars.

Several designs and propulsions systems have been proposed for an interstellar craft:

- solar sail
- nuclear rocket
- ramjet fusion
- nanoships

I had a chance to meet one of the visionaries of the solar sail when I visited the NASA Plum Brook Station in Cleveland, Ohio. There, engineers have built the world's largest vacuum chamber for testing space satellites. The chamber is truly cavernous: it is 100 feet across and 122 feet tall, large enough to contain several multistory apartment buildings and big enough to test satellite and rocket parts in the vacuum of space. Walking into the chamber, I felt overwhelmed by the enormity of the project. But I also felt privileged to be walking in the very same chamber where many of the United States' landmark satellites, probes, and rockets have been tested.

There, I met one of the leading proponents of the solar sail, NASA scientist Les Johnson. He told me that ever since he was a kid reading science fiction, he dreamed of building rockets that could reach the stars. Johnson has even written the basic textbook on solar sails. Although he thinks it might be accomplished within a few decades, he is resigned to the fact that an actual starship may not be built until long after he has passed away. Like the masons who built the great cathedrals of the Middle Ages, Johnson realizes that it may take several human life spans to build a ship that can reach the stars.

The solar sail takes advantage of the fact that, although light has no mass, it has momentum, and hence can exert pressure. Although light pressure from the sun is extremely tiny, too small to be felt by our hands, it is enough to drive a starship if the sail is big enough and we wait long enough. (Sunlight is eight times more intense in space than on the earth.)

Johnson told me his goal is to create a gigantic solar sail, made of very thin but resilient plastic. The sail would be several miles across and built in outer space. Once assembled, it would slowly revolve around the sun, gaining more and more momentum as it moves. After several years orbiting the sun, the sail would spiral out of the solar system and on to the stars. Such a solar sail, he told me, could send a probe to 0.1 percent the speed of light and perhaps reach the nearest star in four hundred years.

In order to cut down the time necessary to reach the stars, Johnson has looked into ways to add an extra boost to the solar sail. One possibility is to put a huge battery of lasers on the moon. The laser beams would hit the sail and give it added momentum as it sailed to the stars.

One problem with a solar sail–driven spaceship is that it is difficult to stop and reverse, since light moves outward from the sun. One possibility is to reverse the direction of the sail and use the destination star's light pressure to slow down the spacecraft. Another possibility is to sail around the distant star, using the star's gravity to create a slingshot effect for the return voyage. And yet another possibility is to land on a moon, build laser batteries, and then sail back on the star's light and the laser beams from that moon.

Although Johnson has stellar dreams, he realizes that the reality is much more modest. In 1993, the Russians deployed a sixty-foot Mylar reflector in space from the Mir space station, but it was only to demonstrate deployment. A second attempt failed. In 2004, the Japanese successfully launched two solar sail prototypes, but again it was to test deployment, not propulsion. In 2005, there was an ambitious attempt by the Planetary Society, Cosmos Studios, and the Russian Academy of Sciences to deploy a genuine solar sail called Cosmos 1. It was launched from a Russian submarine. However, the Volna rocket misfired and failed to reach orbit. And in 2008, a team from NASA tried to launch a solar sail called NanoSail-D, but it was lost when the Falcon 1 rocket failed.

But finally, in May 2010, the Japan Aerospace Exploration Agency successfully launched the IKAROS, the first spacecraft to use solar-sail technology in interplanetary space. It has a square-shaped sail, 20 meters (60 feet) on the diagonal, and uses solar-sail propulsion to travel on its way to Venus. The Japanese eventually hope to send another ship to Jupiter using solar-sail propulsion.

NUCLEAR ROCKET

Scientists have also considered using nuclear energy to drive a starship. Beginning in 1953, the Atomic Energy Commission began to look seriously at rockets carrying atomic reactors, beginning with Project Rover. In the 1950s and 1960s, experiments with nuclear rockets ended mainly in failure. They tended to be unstable and too complex to handle properly. Also, an ordinary fission reactor, one can easily show, simply does not produce enough energy to drive a starship. A typical nuclear power plant produces about a billion watts of power, not enough to reach the stars.

But in the 1950s, scientists proposed using atomic and hydrogen bombs, not reactors, to drive a starship. The Orion Project, for example, proposed a rocket propelled by a succession of nuclear blast waves from a stream of atomic bombs. A starship would drop a series of atomic bombs out its back, creating a series of powerful blasts of X-rays. This shock wave would then push the starship forward.

In 1959, physicists at General Atomics estimated that an advanced version of Orion would weigh 8 million tons, with a diameter of 400 meters, and be powered by 1,000 hydrogen bombs.

One enthusiastic proponent of the Orion project was physicist Freeman Dyson. "For me, Orion meant opening up the whole solar system to life. It could have changed history," he says. It would also have been a convenient way to get rid of atomic bombs. "With one trip, we'd have got rid of 2,000 bombs," he says.

What killed Project Orion, however, was the Nuclear Test Ban Treaty of 1963, which prohibited aboveground testing of nuclear weapons. Without tests, physicists could not refine the design of the Orion, and the idea died.

RAMJET FUSION

Yet another proposal for a nuclear rocket was made by Robert W. Bussard in 1960; he envisioned a fusion engine similar to an ordinary jet engine. A ramjet engine scoops air in the front and then mixes it with fuel internally. By igniting the mixture of air and fuel, a chemical explosion occurs that creates thrust. He envisioned applying the same basic principle to a fusion engine. Instead of scooping air, the ramjet fusion engine would scoop hydrogen gas, which is found everywhere in interstellar space. The hydrogen gas would be squeezed and heated by electric and magnetic fields until the hydrogen fused into helium, releasing enormous amounts of energy in the process. This would create an explosion, which then creates thrust. Since there is an inexhaustible supply of hydrogen in deep space, the ramjet fusion engine can conceivably run forever.

Designs for the ramjet fusion rocket look like an ice cream cone. The scoop traps hydrogen gas, which is sent into the engine, where it is heated and fused with other hydrogen atoms. Bussard calculated that if a 1,000-ton ramjet engine can maintain the acceleration of 32 feet per second squared (or the gravity felt on the earth), then it will approach 77 percent of the speed of light in just one year. Since the ramjet engine can run forever, it could theoretically leave our galaxy and reach the Andromeda galaxy, 2,000,000 light-years from earth, in just 23 years as measured by the astronauts in the rocket ship. (As stated by Einstein's theory of relativity, time slows down in a speeding rocket, so millions of years may have passed on earth, but the astronauts will have aged only 23 years.)

There are several problems facing the ramjet engine. First, since mainly protons exist in interstellar space, the fusion engine must burn pure hydrogen fuel, which does not produce that much energy. (There are many ways in which to fuse hydrogen. The method preferred on earth is to fuse deuterium and tritium, which has a large yield of energy. But in outer space, hydrogen is found as a single proton, and hence ramjet engines can only fuse protons with protons, which does not yield as much energy as deuterium-tritium fusion.) However, Bussard showed that if one modifies the fuel mixture by adding some carbon, the carbon acts as a catalyst to create enormous amounts of power, sufficient to drive a starship.

Second, the scoop would have to be huge—on the order of 160 kilometers—in order to collect enough hydrogen, so it would have to be assembled in space.

There is another problem that is still unresolved. In 1985, engineers Robert Zubrin and Dana Andrews showed that the drag felt by the ram-

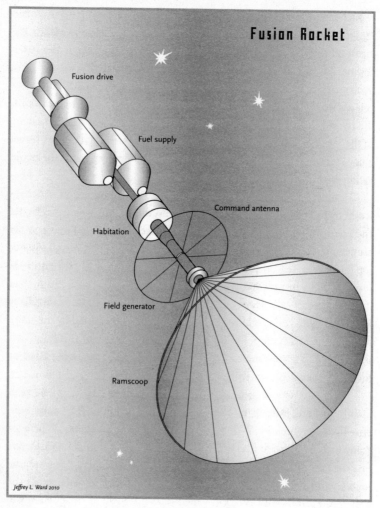

Fusion Rocket

Fusion drive

Fuel supply

Command antenna

Habitation

Field generator

Ramscoop

Jeffrey L. Ward 2010

A ramjet fusion engine, because it scoops hydrogen from interstellar space, can theoretically run forever.

jet engine would be large enough to prevent it from accelerating to near light speed. The drag is caused by the resistance that the starship encounters when it moves in a field of hydrogen atoms. However, their calculation rests heavily on certain assumptions that may not apply to ramjet designs of the future.

At present, until we have a better grasp of the fusion process (and also drag effects from ions in space), the jury is still out on ramjet fusion engines. But if these engineering problems can be solved, then the ramjet fusion rocket will definitely be on the short list.

ANTIMATTER ROCKETS

Another distinct possibility is to use the greatest energy source in the universe, antimatter, to power your spaceship. Antimatter is the opposite of matter, with the opposite charge; for example, an electron has negative charge, but an antimatter electron (the positron) has positive charge. It will also annihilate upon contact with ordinary matter. In fact, a teaspoon of antimatter has enough energy to destroy the entire New York metropolitan area.

Antimatter is so powerful that Dan Brown had the villains in his novel *Angels and Demons* build a bomb to blow up the Vatican using antimatter stolen from CERN, outside Geneva, Switzerland. Unlike a hydrogen bomb, which is only 1 percent efficient, an antimatter bomb would be 100 percent efficient, converting matter into energy via Einstein's equation $E = mc^2$.

In principle, antimatter makes the ideal rocket fuel for a starship. Gerald Smith of Pennsylvania State University estimates that 4 milligrams of antimatter will take us to Mars, and perhaps a hundred grams will take us to the nearby stars. Pound for pound, it releases a billion times more energy than rocket fuel. An antimatter engine would look rather simple. You just drop a steady stream of antimatter particles down the rocket chamber, where it combines with ordinary matter and causes a titanic explosion. The explosive gas is then shot out one end of the chamber, creating thrust.

We are still far from that dream. So far, physicists have been able to create antielectrons and antiprotons, as well as antihydrogen atoms, with antielectrons circulating around antiprotons. This was done at CERN and also at the Fermi National Accelerator Laboratory (Fermilab), outside Chicago, in its Tevatron, the second-largest atom smasher, or particle accelerator, in the world (second only to the Large Hadron Collider at CERN). Physicists at both labs slammed a beam of high-energy particles at a target, creating a shower of debris that contained antiprotons. Powerful magnets were used to separate the antimatter from ordinary matter. These antiprotons were then slowed down and antielectrons were allowed to mix with them, creating antihydrogen atoms.

One man who has thought long and hard about the practicalities of antimatter is Dave McGinnis, a physicist at Fermilab. While standing next to the Tevatron, he explained to me the daunting economics of antimatter. The only known way to produce steady quantities of antimatter, he emphasized to me, is to use an atom smasher like the Tevatron; these machines are extremely expensive and produce only minuscule amounts of antimatter. For example, in 2004, the atom smasher at CERN produced several trillionths of a gram of antimatter at a cost of $20 million. At that rate, it would bankrupt the entire economy of earth to produce enough antimatter to power a starship. Antimatter engines, he stressed to me, are not a far-fetched concept. They are certainly within the laws of physics. But the cost of building one would be prohibitive for the near future.

One reason antimatter is so prohibitively expensive is because the atom smashers necessary to produce it are notoriously expensive. However, these atom smashers are all-purpose machines, designed mainly to produce exotic subatomic particles, not the more common antimatter particles. They are research tools, not commercial machines. It is conceivable that costs could be brought down considerably if one designs a new type of atom smasher specifically to produce copious amounts of antimatter. Then, by mass-producing these machines, it might be possible to create sizable quantities of antimatter. Harold Gerrish of NASA believes that the cost of antimatter might eventually go down to $5,000 per microgram.

Another possibility lies in finding an antimatter meteorite in outer space. If such an object were found, it could supply enough energy to power a starship. In fact, the European satellite PAMELA (Payload for Antimatter Matter Exploration and Light-Nuclei Astrophysics) was launched in 2006 specifically to look for naturally occurring antimatter in outer space.

If large quantities of antimatter are found in space, one can envision using large electromagnetic nets to collect it.

So although antimatter interstellar rockets are certainly within the laws of physics, it may take until the end of the century to drive down the cost. But if this can be done, then antimatter rockets would be on everyone's short list of starships.

NANOSHIPS

When we are dazzled by the special effects in *Star Wars* or *Star Trek*, we immediately envision a huge, futuristic starship bristling with all the latest high-tech gadgets. Yet another possibility lies in using nanotechnology to create tiny starships, perhaps no larger than a thimble, a needle, or even smaller. We have this prejudice that a starship must be huge, like the *Enterprise*, and capable of supporting a crew of astronauts. But the essential functions of a starship may be miniaturized by nanotechnology so that perhaps millions of tiny nanoships might be launched to the nearby stars, only a fraction of which actually make it. Once they arrive on a nearby moon, they might create a factory to make unlimited copies of themselves.

Vint Cerf, one of the original creators of the Internet, envisions tiny nanoships that can explore not just the solar system but eventually the stars themselves. He says, "The exploration of the solar system will be made more effective through the construction of small but powerful nano-scale devices that will be easy to transport and deliver to the surface, below the surface, and into the atmospheres of our neighboring planets and satellites. . . . One can even extrapolate these possibilities to interstellar exploration."

In nature, mammals produce just a few offspring and make sure that all survive. Insects produce large quantities of offspring, only a tiny fraction of which survive. Both strategies can keep the species alive for millions of years. Likewise, instead of sending a single, expensive starship to the stars, one can send millions of tiny starships, each costing a penny and requiring very little rocket fuel.

This concept is patterned after a very successful strategy found in nature: swarming. Birds, bees, and other flying animals fly in flocks or swarms. Not only is there safety in numbers but the swarm also acts as an early warning system. If a dangerous disturbance happens in one part of the swarm, such as an attack by a predator, the message is quickly relayed to the rest of the swarm. They are also quite efficient in energy. When birds fly in a characteristic V pattern, the wake and turbulence created by this formation reduce the energy necessary for each bird to fly.

Scientists characterize a swarm as a "superorganism," one that appears to have an intelligence of its own, independent of the abilities of any single individual. Ants, for example, have a very simple nervous system and a tiny brain, but together they can create complex anthills. Scientists hope to incorporate some of these lessons from nature by designing swarm-bots that might one day journey to other planets and stars.

This is similar to the hypothetical concept of smart dust being pursued by the Pentagon: billions of particles sent into the air, each one with tiny sensors to do reconnaissance. Each sensor is not very intelligent, but collectively they can relay back mountains of information. The Pentagon's DARPA has funded this research for possible military applications, such as monitoring enemy positions on the battlefield. In 2007 and 2009, the Air Force released position papers detailing plans for the coming decades, outlining everything from advanced versions of the Predator (which today cost $4.5 million apiece) to swarms of tiny sensors smaller than a moth costing pennies.

Scientists are also interested in this concept. They might want to spray smart dust to instantly monitor thousands of locations during hurricanes, thunderstorms, volcanic eruptions, earthquakes, floods, forest fires, and other natural phenomena. In the movie *Twister*, for example, we see a band of hardy storm chasers risking life and limb to place sensors around a tornado. This is not very efficient. Instead of having a handful of scientists placing a few sensors during a volcanic eruption or tornado to measure temperature, humidity, and wind velocity, smart dust can give you data from thousands of different positions at once over hundreds of miles. When fed into a computer, this data can instantly give you real-time information about the evolution of a hurricane or volcano in three dimensions. Commercial ventures have already been set up to market these tiny sensors, some no larger than the head of a pin.

Another advantage of nanoships is that they require very little fuel to send them into space. Instead of using huge booster rockets that can reach only 25,000 miles per hour, it is relatively easy to send tiny objects into space at incredible velocities. In fact, it is easy to send subatomic particles near the speed of light using ordinary electric fields. These nanoparticles carry a small electric charge and can be easily accelerated by electric fields.

Instead of using enormous resources to send a probe to another moon or planet, a single probe might have the ability to self-replicate, and thus create an entire factory or even moon base. These self-replicating probes can then blast off to explore other worlds. (The problem is to create the first self-replicating nanoprobe, which is still in the distant future.)

In 1980, NASA took the idea of self-replicating robot probes seriously enough to convene a special study, called Advanced Automation for Space Missions, which was conducted at the University of Santa Clara and looked into several possibilities. One explored by NASA scientists was to send small, self-replicating robots to the moon. There, the robot would use the soil and create unlimited copies of itself.

Most of the report was devoted to the details of constructing a chemical factory to process moon rocks (called regoliths). The robot, for example, might land on the moon, disassemble itself, then rearrange its parts to create a new factory, much like a toy transformer robot. For example, the robot might create large parabolic mirrors to focus sunlight and begin melting the regoliths. It would then use hydrofluoric acid leeching to begin processing the regoliths to extract usable minerals and metals. The metals could then be fabricated into the moon base. Eventually, the robot would construct a small moon factory to reproduce itself.

Building on this report, in 2002, NASA's Institute for Advanced Concepts began funding a series of projects based on these self-replicating robots. One scientist who has taken seriously the proposal of a starship on a chip is Mason Peck of Cornell University.

I had a chance to visit Peck in his laboratory, where you could see his workbench filled with components that may eventually be sent into orbit. Next to his workbench was a small, clean room, with walls draped in plastic, where delicate satellite components are assembled.

His vision of space exploration is quite different from the one given to us by Hollywood movies. He envisions a microchip, one centimeter in size and weighing one gram, that could be accelerated to 1 percent to 10 percent of the speed of light. He takes advantage of the slingshot effect that NASA uses to hurl spacecraft to enormous velocities. This gravity-assist maneuver involves sending a spacecraft around a planet, like a rock from a slingshot, thereby using the planet's gravity to increase the spacecraft's speed.

But instead of gravity, Peck wants to use magnetic forces. His idea is to send a microchip spaceship whipping around Jupiter's magnetic field, which is 20,000 times greater than the earth's field. He plans to accelerate his nanostarship with the magnetic force that is used to hurl subatomic particles to trillions of electron volts in our atom smashers.

He showed me a sample chip that he thought one day might be hurled around Jupiter. It was a tiny square, smaller than your fingertip, crammed with scientific circuitry. His starship would be simple. On one side of the chip, there is a solar cell to provide energy for communication. On the other side, there is a radio transmitter, camera, and other sensors. The device has no engine, since it is propelled using only Jupiter's magnetic field. (NASA's Institute for Advanced Concepts, which funded this and other innovative proposals for the space program since 1998, was unfortunately closed in 2007 due to budget cuts.)

So Peck's vision of a starship is a sharp departure from the usual one found in science fiction, where huge starships lumber into space piloted by a crew of daring astronauts. For example, if a base were set up on a moon of Jupiter, then scores of these tiny chips could be fired into orbit around that giant planet. If a battery of laser canons were also built on this moon, then these chips could be accelerated by hitting them with laser light, increasing their velocity until they reached a fraction of the speed of light.

I then asked him a simple question: Can you reduce your chips to the size of molecules using nanotechnology? Then, instead of using Jupiter's magnetic fields to accelerate these chips, you could use atom smashers based on our own moon to fire molecular-sized probes at near the speed of light. He agreed that this would be a real possibility, but that he hadn't worked out the details yet.

So, we took out a sheet of paper and together began to crank out the equations for this possibility. (This is how we research scientists interact with one another, by going to the blackboard or taking out a sheet of paper to solve a problem by writing down the equations.) We wrote down the equations for the Lorentz force, which Peck uses to accelerate his chips around Jupiter, but then we reduced the chips to the size of molecules and placed them into a hypothetical accelerator similar to the Large Hadron Collider at CERN. We could quickly see that the equations allowed for such a nanostarship to accelerate to nearly the speed of light, using only a conventional atom smasher based on the moon. Because we were reducing the size of our starship from a chip to a molecule, we could reduce the size of our accelerator from the size of Jupiter to a conventional atom smasher. It seemed like this idea was a real possibility.

But after analyzing the equations, we both agreed that the only problem was the stability of these delicate nanostarships. Would the acceleration eventually rip these molecules apart? Like a ball whipping around on a string, these molecules would experience centrifugal forces as they were accelerated to near light speed. Also, these molecules would be electrically charged, so that even electrical forces might rip them apart. We both concluded that nanoships were a definite possibility, but it might take decades of more research to reduce Peck's chips to the size of a molecule and reinforce them so that they don't disintegrate when accelerated to near light speed.

So Mason Peck's dream is to send a swarm of chips to the nearest star, hoping that some of them actually make it across interstellar space. But what do they do when they arrive?

This is where the work of Pei Zhang of Carnegie Mellon University in Silicon Valley comes in. He has created a fleet of minihelicopters that may one day wind up on another planet. He proudly showed me his fleet of swarm-bots, which resemble toy helicopters. But looks are deceptive. I could see that at the center of each was a chip crammed with sophisticated circuitry. With one push of a button, he released four swam-bots into the air, where they flew in all directions and sent back information. Soon, I was surrounded by swarm-bots.

The purpose of these swarm-bots, he told me, is to provide crucial assistance during emergencies, like fires and explosions, by doing reconnaissance and surveillance. Eventually, these swarm-bots could be outfitted with TV cameras and sensors that can detect temperature, pressure, wind direction, etc., information that may prove critical during an emergency. Thousands of swarm-bots could be released over a battlefield, a fire, or even an extraterrestrial terrain. These swarm-bots also communicate with one another. If one of them hits an obstacle, it radios the information to the other swarm-bots.

So one vision of space travel might be that thousands of cheap, disposable chips devised by people like Mason Peck are fired at the nearest star at nearly the speed of light. Once a handful of them reach their destination, they sprout wings and blades and fly over the alien terrain, just like Pei Zhang's fleet of swarm-bots. They would then radio information back to earth. Once promising planets are found, a second generation of swarm-bots might be sent to create factories on these planets that then create more copies of these swarm-bots, which then fly to the next star. Then the process continues indefinitely.

EXODUS EARTH?

By 2100, it is likely that we will have sent astronauts to Mars and the asteroid belt, explored the moons of Jupiter, and begun the first steps to send a probe to the stars.

But what about humanity? Will we have space colonies to relieve the world population by finding a new home in outer space? Will the human race begin to leave the earth by 2100?

No. Given the cost, even by 2100 and beyond, the majority of the human race will not board a spaceship to visit the other planets. Although a handful of astronauts will have created tiny outposts among the planets, humanity itself will be stuck on earth.

Given the fact that earth will be the home of humanity for centuries to come, this raises another question: How will civilization itself evolve? How will science affect our lifestyle, our jobs, and our society? Science is the engine of prosperity, so how will it reshape civilization and wealth in the future?

> Technology and ideology are shaking the foundations of twenty-first-century capitalism. Technology is making skills and knowledge the only sources of sustainable strategic advantage.
>
> —LESTER THUROW

7 FUTURE OF WEALTH *Winners and Losers*

In mythology, the rise and fall of great empires depended on the strength and cunning of one's armies. The great generals of the Roman Empire worshipped at the temple of Mars, the god of war, before decisive military campaigns. The legendary exploits of Thor inspired the Vikings into heroic battles. The ancients built huge temples and monuments dedicated to the gods, commemorating victories in battle against their enemies.

But when we analyze the actual rise and decline of great civilizations, we find an entirely different story.

If you were an alien from Mars visiting earth in the year 1500 and viewed all the great civilizations, which would you think would eventually dominate the word? The answer would be easy: any civilization but the European one.

In the east, you would see the great Chinese civilization, which had lasted for millennia. The long list of inventions pioneered by the Chinese is without parallel: paper, the printing press, gunpowder, the compass, etc. Its scientists are the best on the planet. Its government is unified and the mainland is at peace.

In the south, you have the Ottoman Empire, which came within a hair-breadth of overrunning Europe. The great Muslim civilization invented algebra, produced advances in optics and physics, and named the stars. Art and science flourish. Its great armies face no credible opposition. Istanbul is one of the world's great centers for scientific learning.

Then you have the pitiful European countries, which are racked by religious fundamentalism, witch trials, and the Inquisition. Western Europe, in precipitous decline for a thousand years since the collapse of the Roman Empire, is so backward that it is a net importer of technology. It is a medieval black hole. Most of the knowledge of the Roman Empire has long since vanished, replaced by stifling religious dogma. Opposition or dissent is frequently met with torture or worse. Moreover, the city-states of Europe are constantly at war with one another.

So what happened?

Both the great Chinese and Ottoman empires are entering a 500-year-period of technological stagnation, while Europe is beginning an unprecedented embrace of science and technology.

Beginning in 1405, the Yongle emperor of China ordered a massive naval armada, the largest the world had ever seen, to explore the world. (The three puny naval ships of Columbus would have fit nicely on the deck of just one of these colossal vessels.) Seven massive expeditions were launched, each larger than the previous one. This fleet sailed around the coast of Southeast Asia and reached Africa, Madagascar, and perhaps even beyond that. The fleet brought back a rich bounty of goods, delicacies, and exotic animals from the far reaches of the earth. There are remarkable ancient woodcuts of African giraffes being paraded at a Ming Dynasty zoo.

But the rulers of China were also disappointed. Was that all there was? Where were the great armies that could rival the Chinese? Were exotic foods and strange animals all that the rest of the world could offer? Losing interest, the subsequent rulers of China let their great naval fleet decay and eventually burn. China gradually isolated itself from the outside world, stagnating as the world lunged forward.

A similar attitude settled in the Ottoman Empire. Having conquered most of the world they knew, the Ottomans turned inward, into religious fundamentalism and centuries of stagnation. Mahathir Mohamad, the former prime minister of Malaysia, has said, "The great Islamic civilization went into decline when Muslim scholars interpreted knowledge acquisition, as enjoined by the Qur'an, to mean only knowledge of religion, and that other knowledge was un-Islamic. As a result, Muslims gave up the study of science, mathematics, medicine, and other so-called worldly disciplines. Instead, they spent much time debating on Islamic teachings and interpretations, on Islamic jurisprudence and Islamic practices, which led to a breakup of the Ummah and the founding of numerous sects, cults, and schools."

In Europe, however, a great awakening was beginning. Trade brought in fresh, revolutionary ideas, accelerated by Gutenberg's printing press. The power of the Church began to weaken after a millennium of domination. The universities slowly turned their attention away from interpreting obscure passages of the Bible to applying the physics of Newton and the chemistry of Dalton and others. Historian Paul Kennedy of Yale adds one more factor to the meteoric rise of Europe: the constant state of war between nearly equal European powers, none of which could ever dominate the Continent. Monarchs, constantly at war with one another, funded science and engineering to further their territorial ambitions. Science was not just an academic exercise but a way to create new weapons and new avenues of wealth.

Soon, the rise of science and technology in Europe began to weaken the power of China and the Ottoman Empire. The Muslim civilization, which had prospered for centuries as a gateway for trade between the East and the West, faltered as European sailors forged trade routes to the New World and the East—especially around Africa, bypassing the Middle East. And China found itself being carved up by European gunboats that ironically exploited two pivotal Chinese inventions, gunpowder and the compass.

The answer to the question "What happened?" is clear. Science and technology happened. Science and technology are the engines of prosperity. Of course, one is free to ignore science and technology, but only at your peril. The world does not stand still because you are reading a religious text. If you do not master the latest in science and technology, then your competitors will.

MASTERY OF THE FOUR FORCES

But precisely how did Europe, the dark horse, suddenly sprint past China and the Muslim world after centuries of ignorance? There are both social and technological factors in this remarkable upset.

When analyzing world history after 1500, one realizes that Europe was ripe for the next great advance, with the decline of feudalism, the rise of a merchant class, and the vibrant winds of the Renaissance. Physicists, however, view this great transition through the lens of the four fundamental forces that rule the universe. These are the fundamental forces that can explain everything around us, from machines, rockets, and bombs to the stars and the universe itself. Changing social trends may have set the stage for this transition, but it was the mastery of these forces in Europe that finally propelled it to the forefront of world powers.

The first force is gravity, which holds us anchored to the ground, prevents the sun from exploding, and holds the solar system together. The second is the electromagnetic force, which lights up our cities, energizes our dynamos and engines, and powers our lasers and computers. The third and fourth forces are the weak and strong nuclear forces, which hold the nucleus of the atom together, light the stars in the heavens, and create the nuclear fire at the center of our sun. All four forces were unraveled in Europe.

Each time one of these forces was understood by physicists, human history changed, and Europe was ideally suited to exploit that new knowledge. When Isaac Newton witnessed an apple fall and gazed at the moon, he asked himself a question that forever changed human history: If an apple falls, then does the moon also fall? In a brilliant stroke of insight when he was twenty-three years old, he realized that the forces that grab an apple are the same that reach out to the planets and comets in the heavens. This allowed him to apply the new mathematics he had just invented, the calculus, to plot the trajectory of the planets and moons, and for the first time to decode the motions of the heavens. In 1687, he published his masterpiece, *Principia*, arguably the most important book of science ever written, ranking among the most influential books in all human history.

More important, Newton introduced a new way of thinking, a mechanics by which one could compute the motion of moving bodies via forces. No longer were we subject to the whims of spirits, demons, and ghosts;

instead objects moved because of well-defined forces that could be measured and harnessed. This led to Newtonian mechanics, by which scientists could accurately predict the behavior of machines; this in turn paved the way for the steam engine and the locomotive. The intricate dynamics of complex steam-powered machines could be broken down systematically, bolt by bolt, lever by lever, by Newton's laws. So Newton's description of gravity helped to pave the way for the Industrial Revolution in Europe.

Then in the 1800s, again in Europe, Michael Faraday, James Clerk Maxwell, and others harnessed the second great force, electromagnetism, which ushered in the next great revolution. When Thomas Edison built generators at the Pearl Street Station in Lower Manhattan and electrified the first street on earth, he opened the gateway to the electrification of the entire planet. Today, from outer space, we can view the earth at night, with entire continents set ablaze. Gazing at the earth from space, any alien would immediately realize that earthlings had mastered electromagnetism. We dearly appreciate our dependence on it any time there is a power blackout. In an instant, we are suddenly thrown over 100 years back into the past, without credit cards, computers, lights, elevators, TV, radio, the Internet, motors, etc.

Last, the nuclear forces, also mastered by European scientists, are changing everything around us. Not only can we unlock the secrets of the heavens, revealing the power source that fires the stars, but we can also unravel inner space, using this knowledge for medicine through MRI, CAT, and PET scans; radiation therapy; and nuclear medicine. Because the nuclear forces govern the immense power stored within the atom, the nuclear forces can ultimately determine the fate of humanity, whether we will prosper by harnessing the unlimited power of fusion or die in a nuclear inferno.

NEAR FUTURE (PRESENT TO 2030)

FOUR STAGES OF TECHNOLOGY

The combination of changing social conditions and the mastery of the four forces propelled Europe to the forefront of nations. But technologies are dynamic, changing all the time. They are born, evolve, and rise and fall. To see how specific technologies will change in the near future, it is useful to see how technologies obey certain laws of evolution.

Mass technologies usually evolve in four basic stages. This can be seen in the evolution of paper, running water, electricity, and computers. In stage I, the products of technology are so precious that they are closely guarded. Paper, when it was invented in the form of papyrus by the ancient Egyptians and then by the Chinese thousands of years ago, was so precious that one papyrus scroll was closely guarded by scores of priests. This humble technology helped to set into motion ancient civilization.

Paper entered stage II around 1450, when Gutenberg invented printing from movable type. This made possible the "personal book," so that one person could possess one book containing the knowledge of hundreds of scrolls. Before Gutenberg, there were only 30,000 books in all Europe. By 1500, there were 9 million books, stirring up intense intellectual ferment and stimulating the Renaissance.

But around 1930, paper hit stage III, when the cost fell to a penny a sheet. This made possible the personal library, where one person could possess hundreds of books. Paper became an ordinary commodity, sold by the ton. Paper is everywhere and nowhere, invisible and ubiquitous. Now we are in stage IV, where paper is a fashion statement. We decorate our world with paper of all colors, shapes, and sizes. The largest source of urban waste is paper. So paper evolved from being a closely guarded commodity to being waste.

The same applies to running water. In ancient times, in stage I, water was so precious that a single well had to be shared by an entire village. This lasted for thousands of years, until the early 1900s, when personal plumbing was gradually introduced and we entered stage II. After World War II, running water entered stage III and became cheap and available to an expanding middle class. Today, running water is in stage IV, a fashion statement, appearing in numerous shapes, sizes, and applications. We decorate our world with water, in the form of fountains and displays.

Electricity also went through the same stages. With the pioneering work of Thomas Edison and others, in stage I a factory shared a single lightbulb and electric motor. After World War I, we entered stage II with the personal lightbulb and personal motor. Today, electricity has disappeared; it is everywhere and nowhere. Even the word "electricity" has pretty much disappeared from the English language. At Christmas, we use hundreds of blinking lights to decorate our homes. We assume that electricity is hidden in the walls, ubiquitous. Electricity is a fashion statement, lighting up Broadway and decorating our world.

In stage IV, both electricity and running water have become utilities. They are so cheap, and we consume so much of them, that we meter the amount of electricity and water that runs into our home.

The computer follows the same pattern. Companies that understood this thrived and prospered. Companies that didn't were driven almost to bankruptcy. IBM dominated stage I with the mainframe computer in the 1950s. One mainframe computer was so precious that it was shared by 100 scientists and engineers. However, the management of IBM failed to appreciate Moore's law, so they almost went bankrupt when we entered stage II in the 1980s, with the coming of the personal computer.

But even personal computer manufacturers got complacent. They envisioned a world with stand-alone computers on every desk. They were caught off guard with the coming of stage III, Internet-linked computers by which one person could interact with millions of computers. Today, the only place you can find a stand-alone computer is in a museum.

So the future of the computer is to eventually enter stage IV, where it disappears and gets resurrected as a fashion statement. We will decorate our world with computers. The very word *computer* will gradually disappear from the English language. In the future, the largest component of urban waste will not be paper but chips. The future of the computer is to disappear and become a utility, sold like electricity and water. Computer chips will gradually disappear as computation is done "in the clouds."

So the evolution of computers is not a mystery; it is following the well-worn path of its predecessors, like electricity, paper, and running water.

But the computer and the Internet are still evolving. Economist John Steele Gordon was asked if this revolution is over. "Heavens, no. It will be a hundred years before it fully plays out, just like the steam engine. We are now at the point with the Internet that they were with the railroad in 1850. It's just the beginning."

Not all technologies, we should point out, enter stages III and IV. For example, consider the locomotive. Mechanized transportation entered stage I in the early 1800s with the coming of the steam-driven locomotive.

A hundred people would share a single locomotive. We entered stage II with the introduction of the "personal locomotive," otherwise known as the car, in the early 1900s. But the locomotive and the car (essentially a box on rails or wheels) have not changed much in the past decades. What has changed are refinements, such as more powerful and efficient engines as well as intelligence. So technologies that cannot enter stages III and IV will be embellished; for example, they will have chips placed in them so they become intelligent. Some technologies evolve all the way to stage IV, like electricity, computers, paper, and running water. Others stay stuck at an intermediate stage, but they continue to evolve by having incremental improvements such as chips and increased efficiency.

WHY BUBBLES AND CRASHES?

But today, in the wake of the great recession of 2008, some voices can be heard saying that all this progress was an illusion, that we have to return to the simpler days, that there is something fundamentally flawed with the system.

When taking the long view of history, it is easy to point to the unexpected, with colossal bubbles and crashes that seem to come out of nowhere. They seem random, a by-product of the fickleness of fate and human folly. Historians and economists have written voluminously about the crash of 2008, trying to make sense out of it by examining a variety of causes, such as human nature, greed, corruption, lack of regulation, weaknesses in oversight, etc.

However, I have a different way of looking at the great recession, looking through the lens of science. In the long term, science is the engine of prosperity. For example, *The Oxford Encyclopedia of Economic History* cites studies that "attribute 90 percent of income growth in England and the United States after 1780 to technological innovation, not mere capital accumulation."

Without science, we would be thrown back millennia into the dim past. But science is not uniform; it comes in waves. One seminal breakthrough (for example, the steam engine, the lightbulb, the transistor) often causes a cascade of secondary inventions that then create an avalanche of innovation and progress. Since they create vast amounts of wealth, these waves should be reflected in the economy.

The first great wave was steam power, which eventually led to the creation of the locomotive. Steam power fed the Industrial Revolution, which would turn society upside down. Fabulous wealth was created by steam power. But under capitalism, wealth is never stagnant. Wealth has to go somewhere. Capitalists are ceaselessly hunting for the next break, and will shift this wealth to invest in even more speculative schemes, sometimes with catastrophic results.

In the early 1800s, much of the excess wealth generated by steam power and the Industrial Revolution went into locomotive stocks on the London Stock Exchange. In fact, a bubble began to form, with scores of locomotive companies appearing on the London Exchange. Virginia Postrel, business writer for the *New York Times*, writes, "A century ago, railroad companies accounted for half the securities listed on the New York Stock Exchange." Since the locomotive was still in its infancy, this bubble was unsustainable, and it finally popped, creating the Crash of 1850, one of the great collapses in the history of capitalism. This was followed by a series of minicrashes that occurred nearly every decade, created by the excess wealth spawned by the Industrial Revolution.

There is irony here: the heyday of the railroad would be the 1880s and 1890s. So the Crash of 1850 was due to speculative fever and the wealth created by science, but the real job of railing the world would take many more decades to mature.

Thomas Friedman writes, "In the 19th century, America had a railroad boom, bubble and bust. . . . But even when that bubble burst, it left America with an infrastructure of railroads that made transcontinental travel and shipping dramatically easier and cheaper."

Instead of capitalists learning this lesson, this cycle began to repeat soon afterward. A second great wave of technology spread, led by the electric and automotive revolutions of Edison and Ford. The electrification of the factory and household, as well as the proliferation of the Model T, once again created fabulous wealth. As always, excess wealth had to go somewhere. In this case, it went into the U.S. Stock Exchange, in the form of a bubble in utility and automotive stocks. People ignored the lesson of the Crash of 1850, since that had happened eighty years earlier in the dim past. From 1900 to 1925, the number of automobile start-up companies hit 3,000, which the market simply could not support. Once again, this bubble was unsustainable. For this and other reasons, the bubble popped in 1929, creating the Great Depression.

But the irony here is that the paving and electrification of America and Europe would not take place until after the crash, during the 1950s and 1960s.

More recently, we had the third great wave of science, the coming of high tech, in the form of computers, lasers, space satellites, the Internet, and electronics. The fabulous wealth created by high tech had to go somewhere. In this case, it went into real estate, creating a huge bubble. With the value of real estate exploding through the roof, people began to borrow against the value of their homes, using them as piggy banks, which further accelerated the bubble. Unscrupulous bankers fueled this bubble by giving away home mortgages like water. Once again, people ignored the lesson of the crashes of 1850 and 1929, which happened 160 and 80 years in the past. Ultimately, this new bubble could not be sustained, and we had the crash of 2008 and the great recession.

Thomas Friedman writes, "The early 21st century saw a boom, bubble and now a bust around financial services. But I fear all it will leave behind are a bunch of empty Florida condos that never should have been built, used private jets that the wealthy can no longer afford and the dead derivative contracts that no one can understand."

But in spite of all the silliness that accompanied the recent crash, the irony here is that the wiring and networking of the world will take place after the crash of 2008. The heyday of the information revolution is yet to come.

This leads to the next question: What is the fourth wave? No one can be sure. It might be a combination of artificial intelligence, nanotechnology, telecommunications, and biotechnology. As with previous cycles, it may take another eighty years for these technologies to create a tidal wave of fabulous wealth. Around the year 2090, hopefully people will not ignore the lesson of the previous eighty years.

MIDCENTURY (2030 TO 2070)

WINNERS AND LOSERS: JOBS

But as technologies evolve, they create abrupt changes in the economy that sometimes lead to social dislocations. In any revolution, there are winners and losers. This will become more evident by midcentury. We no longer have blacksmiths and wagonmakers in every village. Moreover, we do not mourn the passing of many of these jobs. But the question is: What jobs will flourish by midcentury? How will the evolution of technology change the way we work?

We can partially determine the answer by asking a simple question: What are the limitations of robots? As we have seen, there are at least two basic stumbling blocks to artificial intelligence: pattern recognition and common sense. Therefore, the jobs that will survive in the future are, in the main, those that robots cannot perform—ones that require these two abilities.

Among blue-collar workers, the losers will be workers who perform purely repetitive tasks (like autoworkers on the factory line) because robots excel at this. Computers give the illusion that they possess intelligence, but that is only because they can add millions of times faster than we can. We forget that computers are just sophisticated adding machines, and repetitive work is what they do best. That is why some automobile assembly-line workers have been among the first to suffer from the computer revolution. This means that any factory work that can be reduced to a set of scripted, repetitive motions will eventually disappear.

Surprisingly, there is a large class of blue-collar work that *will* survive the computer revolution and even flourish. The winners will be those who perform nonrepetitive work that requires pattern recognition. Garbage collectors, police officers, construction workers, gardeners, and plumbers will all have jobs in the future. Garbage collectors, in order to pick up the trash at different homes and apartments, have to recognize the garbage bags, place them in the truck, and haul them out to the waste yard. But every piece of trash requires a different method of disposal. For construction workers, every task requires different tools, blueprints, and instructions. No two construction sites or two tasks are the same. Police officers have to analyze a variety of crimes in different situations. Moreover, they also have to understand the motives and methods of criminals, which is far beyond the ability of any computer. Similarly, every garden and sink is different, requiring different skills and tools of the plumber.

Among white-collar workers, the losers will be those involved in middleman work taking inventory and "bean counting." This means low-level agents, brokers, tellers, accountants, etc., will be increasingly thrown out of work as their jobs disappear. These jobs are called "the friction of capitalism." Already, one can buy a plane ticket by scanning the Web for the best prices, bypassing a travel agent.

Merrill Lynch, for example, famously stated that it would never adopt online stock trading. It would always do stock trading the old-fashioned way. John Steffens, Merrill's brokerage chief, said, "The do-it-yourself model of investing, centered on Internet trading, should be regarded as a serious threat to America's financial lives." So it was humiliating, therefore, when it was finally forced by market forces to adopt online trading in 1999. "Rarely in history has the leader in an industry felt compelled to do an about-face and, virtually overnight, adopt what is essentially a new business model," wrote Charles Gasparino of ZDNet news.

This also means that the corporate pyramid will be thinned out. Since the people at the very top can interact directly with the sales force and representatives in the field, there is less need for middlemen to carry out orders from the top. In fact, such job reductions occurred when the personal computer first entered the office.

So how will middlemen survive in the future? They will have to add value to their work and provide the one commodity that robots cannot deliver: common sense.

For example, in the future, you will be able to buy a house on the Internet via your watch or contact lens. But no one is going to buy a house this way, since this is one of the most important financial transactions you will perform in your life. For important purchases like a home, you want to talk to a human who can tell you where the good schools are, where the crime rate is low, how the sewer system works, etc. For this, you want to talk to a skilled agent who adds value.

Similarly, low-level stockbrokers are being thrown out of work by online trading, but stockbrokers who give reasoned, wise investment advice will always be in demand. Brokerage jobs will continue to dry up unless they offer value-added services, such as the wisdom of top market analysts and economists and the inside knowledge of experienced brokers. In an era when online trading mercilessly drives down the cost of stock trades, stockbrokers will survive only if they can also market their intangible qualities, such as experience, knowledge, and analysis.

So among white-collar workers, the winners will be those who can provide useful common sense. This means workers involved with creativity—artwork, acting, telling jokes, writing software, leadership, analysis, science, creativity—qualities that "make us human."

People in the arts will have jobs, since the Internet has an insatiable appetite for creative art. Computers are great at duplicating art and helping artists to embellish art, but they are miserable at originating new forms of it. Art that inspires, intrigues, evokes emotions, and thrills us is beyond the capability of a computer, because all these qualities involve common sense.

Novelists, scriptwriters, and playwrights will have jobs, since they have to convey realistic scenes, human conflicts, and human triumphs and defeats. For computers, modeling human nature, which involves understanding motives and intentions, is beyond their capability. Computers are not good at determining what makes us cry or laugh, since they cannot cry or laugh on their own, or understand what is funny or sad.

People involved in human relations, such as lawyers, will have jobs.

Although a robolawyer can answer rudimentary questions about the law, the law itself is constantly changing, depending on shifting social standards and mores. Ultimately, the interpretation of the law boils down to a value judgment, where computers are deficient. If the law were cut-and-dried with clear-cut interpretations, there would be no need for courts, judges, and juries. A robot cannot replace a jury, since juries often represent the mores of a specific group, which are constantly shifting with time. This was most apparent when Supreme Court justice Potter Stewart once had to define pornography. He failed to do so, but concluded, "I know it when I see it."

Furthermore, it will probably be illegal for robots to replace the justice system, since our laws have enshrined a fundamental principle: that juries be made up of our peers. Since robots cannot be our peers, it will be illegal for them to replace the justice system.

On the surface, laws may seem exacting and well-defined, with precise and rigorous wording and arcane-sounding titles and definitions. But this is only an appearance, since the interpretations of these definitions constantly shifts. The U.S. Constitution, for example, appears to be a well-defined document, yet the Supreme Court is constantly split down the middle on controversial questions. It is forever reinterpreting every word and phrase in the Constitution. The changing nature of human values can be easily seen simply by looking at history. For example, the U.S. Supreme Court in 1857 ruled that slaves could never become citizens of the United States. In some sense, it took a civil war and the death of thousands to overturn that decision.

Leadership will also be a prized commodity in the future. In part, leadership consists of sizing up all the available information, viewpoints, and options and then choosing the most appropriate one, consistent with certain goals. Leadership becomes especially complicated because it deals with inspiring and providing guidance to human workers, who have their own personal strengths and weaknesses. All these factors require a sophisticated understanding of human nature, market forces, etc., that is beyond the ability of any computer.

FUTURE OF ENTERTAINMENT

This also means that entire industries, such as entertainment, are undergoing a profound upheaval. For example, the music industry since time immemorial was based on individual musicians who went from town to town, making personal appearances. Entertainers were constantly on the road, setting up shop one day and then moving on to the next village. It was a hard life, with little financial reward. This age-old pattern changed abruptly when Thomas Edison invented the phonograph and forever changed the way we hear music. Suddenly, one singer could produce records sold by the millions and derive revenue on a previously unimaginable scale. Within a single generation, rock singers would become the nouveau riche of society. Rock stars, who might have been lowly waiters in a previous generation, became the venerated idols of youth society.

But unfortunately, the music industry ignored the predictions of scientists who foresaw the day when music would be easily sent over the Internet, like e-mail. The music industry, instead of laying the groundwork for how to earn money selling online, instead tried to sue upstart companies that offered music at a fraction of the cost of a CD. This was like trying to sweep back the ocean. This neglect is causing the present turmoil within the music industry.

(But the good thing is that unknown singers can now rise to the top, without having to face the de facto censorship of the big music companies. In the past, these music moguls could almost choose who the next rock star would be. So, in the future, the top musicians will be chosen more democratically, via a free-for-all involving market forces and technology, rather than by music business executives.)

Newspapers are also facing a similar dilemma. Traditionally, newspapers could rely on a steady stream of revenue from advertisers, especially in the classified ads section. The revenue stream came not so much from the purchase of the paper itself, but from the ad revenue those pages generated. But now we can download the day's news for free and advertise nationwide on a variety of online want-ads sites. As a consequence, newspapers around the country are shrinking in size and circulation.

But this process will continue only so far. There is so much noise on the Internet, with would-be prophets daily haranguing their audience and megalomaniacs trying to push bizarre ideas, that eventually people will cherish a new commodity: wisdom. Random facts do not correlate with wisdom, and in the future people will be tired of the rants of mad bloggers and will seek out respected sites that offer this rare commodity of wisdom.

As economist Hamish McRae has said, "In practice, the vast bulk of this 'information' is rubbish, the intellectual equivalent of junk mail." But he claims, "Good judgment will continue to be highly valued: successful financial analysts are, as a group, the best paid researchers in the world."

THE MATRIX

But what about Hollywood actors? Instead of becoming box-office celebrities and the talk of society, will actors find themselves on the unemployment line? Recently, there has been remarkable progress in computer animations of the human body, so that it appears nearly real. Animated characters now have 3-D features and shadowing. So will actors and actresses become obsolete anytime soon?

Probably not. There are fundamental problems modeling the human face by computer. Humans evolved an uncanny ability to differentiate one another's faces, since our survival depended on it. In a flash, we had to tell if someone was an enemy or a friend. Within seconds, we had to rapidly determine a person's age, sex, strength, and emotion. Those who could not do this simply did not survive to pass on their genes to the next generation. Hence, the human brain devotes a considerable amount of its processing power to reading people's faces. In fact, for most of our evolutionary history, before we learned how to speak, we communicated through gestures and body language, and a large part of our brain power was devoted to looking at subtle facial cues. But computers, which have a hard time recognizing simple objects around them, have even greater difficulty re-creating a realistic animated human face. Kids know immediately if the face they see on the movie screen is a real human or a computer simulation. (This goes back to the Cave Man Principle. Given a choice between seeing a live-action blockbuster action movie with our favorite actor or seeing a computer-animated cartoon action picture, we will still prefer the former.)

The body, by contrast, is much easier to model by computer. When Hollywood creates those realistic monsters and fantasy figures in the movies, they use a shortcut. An actor puts on a skintight suit that has sensors on its joints. As the actor moves or dances, the sensors send signals to a computer that then creates an animated figure performing the precise movements, as in the movie *Avatar*.

I once spoke at a conference sponsored by the Livermore National Laboratory, where nuclear weapons are designed, and at dinner sat next to someone who had worked on the movie *The Matrix*. He confessed that they had to use an enormous amount of computer time to create the dazzling special effects in that movie. One of the most difficult scenes, he said, required them to completely reconstruct an imaginary city as a helicopter flew overhead. With enough computer time, he said, he could create an entire fantasy city. But, he admitted, modeling a realistic human face was beyond his ability. This is because when a light beam hits the human face, it scatters in all directions, depending on its texture. Each particle of light has to be tracked by computer. Hence, each point of skin on a person's face has to be described by a complex mathematical function, which is a real headache for a computer programmer.

I remarked that this sounded very much like high-energy physics, my specialty. In our atom smashers, we create a powerful beam of protons that slams into a target, creating a shower of debris that scatters in all directions. We then introduce a mathematical function (called the form factor) that describes each particle.

Half jokingly, I asked if there was a relationship between the human face and high-energy particle physics? Yes, he replied. Computer animators use the same formalism used in high-energy physics to create the faces you see on the movie screen! I never realized that the arcane formulae that we theoretical physicists use may one day crack the problem of modeling the human face. So the fact that we can recognize the human face is similar to the way we physicists analyze subatomic particles!

FAR FUTURE (2070 TO 2010)

IMPACT ON CAPITALISM

These new technologies that we have been discussing in this book are so powerful that, by the end of the century, they are bound to have an impact on capitalism itself. The laws of supply and demand are the same, but the rise of science and technology has modified Adam Smith's capitalism in many ways, from the way that goods are distributed to the nature of wealth itself. Some of the more immediate ways in which capitalism has been affected are as follows:

- **Perfect capitalism**
  The capitalism of Adam Smith is based on the laws of supply and demand: prices are set when the supply for any good matches the demand. If an object is scarce and in demand, then its price rises. But the consumer and producer have only partial, imperfect understanding of supply and demand, and hence prices can vary widely from place to place. So the capitalism of Adam Smith was imperfect. But this will gradually change in the future.

  "Perfect capitalism" is when the producer and the consumer have infinite knowledge of the market, so that prices are perfectly determined. For example, in the future, consumers will scan the Internet via their contact lenses and have infinite knowledge of all comparative prices and performances. Already, one can scan the Internet to find the best airline fares. This will eventually apply to all products sold in the world. Whether through eyeglasses, wall screens, or cell phones, consumers will know everything about a product. Going through a grocery store, for example, you will scan the various products on display and, via the Internet in your contact lens, immediately evaluate if the product is a bargain or not. The advantage shifts to the consumers, because they will instantly know everything about a product—its history, its performance record, its price relative to others, and its strengths and liabilities.

The producer also has tricks up his sleeve, such as using data mining to understand the wants and needs of the consumer, and scanning the Internet for commodity prices. This removes much of the guesswork in setting prices. But in the main, it is the consumer who has the advantage, who instantly has comparative knowledge of any product, and who demands the cheapest price. The producer must then react to the constantly changing demands of the consumer.

- **Mass production to mass customization**

In the present system, goods are created by mass production. Henry Ford once famously said that the consumer could have the Model T in any color, as long as it's black. Mass production drastically lowered prices, replacing the inefficient, older system of guilds and handcrafted goods. The computer revolution will change all this.

Today, if a customer sees a dress of the perfect style and color but the wrong size, then there is no sale. But in the future, our precise 3-D measurements will be stored in our credit card or wallet. If a dress or other garment is the wrong size, you will e-mail your measurements to the factory and have it immediately produce one in the right size. In the future, everything will fit.

Mass customization today is impractical, since it is too costly to create a new product just for one consumer. But when everyone is hooked to the Internet, including the factory, custom-made objects can be manufactured at the same price as mass-produced items.

- **Mass technology as a utility**

When technologies become widely dispersed, such as electricity and running water, they eventually become utilities. With capitalism driving down prices and increasing competition, these technologies will be sold like utilities, that is, we don't care where they come from and we pay for them only when we want them. The same applies for computation. "Cloud computing," which relies heavily on the Internet for most computing functions, will gradually gain in popularity. Cloud computing reduces computation to a utility, something that we pay for only when we need it, and something that we don't think about when we don't need it.

This is different from the situation today, when most of us do our typing, word processing, or drawing on a desktop or laptop computer and then connect to the Internet when we want to search for information. In the future, we could gradually phase out the computer altogether and access all our information directly on the Internet, which then charges us for the time spent. So computation becomes a utility that is metered, like water and electricity. We will live in a world where our appliances, furniture, clothes, etc., are intelligent, and we will talk to them when we need specific services. Internet screens are hidden everywhere, and keyboards materialize whenever we need them. Function has replaced form, so, ironically, the computer revolution will eventually make the computer disappear into the clouds.

· **Targeting your customer**

Companies historically placed ads in newspapers, on radio, on TV, etc., often without the slightest idea of the impact the ads had. They could calculate the effectiveness of their ad campaign only by looking at upticks in sales. But in the future, companies will know almost immediately how many people have downloaded or viewed their products. If you are interviewed on an Internet radio site, for example, it is possible to determine precisely how many people have listened. This will allow companies to target their audience to tailor-made specifications.

(This, however, raises another question: the sensitive question of privacy, which will be one of the great controversies of the future. In the past, there were worries that the computer might make Big Brother possible. In George Orwell's novel *1984,* a totalitarian regime takes over the earth, unleashing a hellish future in which spies are everywhere, all freedoms are squashed, and life is an unending series of humiliations. At one point, the Internet might have evolved into such an all-pervasive spying machine. However, in 1989, after the breakup of the Soviet bloc, the National Science Foundation in effect opened it up, converting it from a primarily military device to one that networked universities and even commercial entities, eventually leading to the Internet explosion of the 1990s. Today, Big Brother is not possible. The real problem is "little brother," that is, nosy busybodies, petty criminals, tabloid newspapers, and even corporations that use data mining to find out our personal preferences. As we will discuss in the next chapter, this is a problem that will not go away but will evolve with time. More than likely, there will be an eternal cat-and-mouse game between software developers creating programs to protect our privacy and others creating programs to break it.)

FROM COMMODITY CAPITALISM TO INTELLECTUAL CAPITALISM

So far, we have asked only how technology is altering the way capitalism operates. But with all the turmoil created by the advances in high technology, what impact is this having on the nature of capitalism itself? All the turmoil that this revolution is creating can be summarized in one concept: the transition from commodity capitalism to intellectual capitalism.

Wealth in Adam Smith's day was measured in commodities. Commodity prices fluctuate, but on average commodity prices have been dropping steadily for the past 150 years. Today, you had breakfast that the king of England could not have had 100 years ago. Exotic delicacies from around the world are now routinely sold in supermarkets. The falling of commodity prices is due to a variety of factors, such as better mass production, containerization, shipping, communication, and competition.

(For example, today's high school students have a hard time understanding why Columbus risked life and limb to find a shorter trade route to the spices of the East. Why couldn't he simply go to the supermarket, they ask, and get some oregano? But in the days of Columbus, spices and herbs were extremely expensive. They were prized because they could mask the taste of rotting food, since there were no refrigerators in those days. At times, even kings and emperors had to eat rotten food at dinner. There were no refrigerated cars, containers, or ships to carry spices across the oceans.) That is why these commodities were so valuable that Columbus gambled his life to get them, although today they are sold for pennies.

What is replacing commodity capitalism is intellectual capitalism. Intellectual capital involves precisely what robots and AI cannot yet provide, pattern recognition and common sense.

As MIT economist Lester Thurow has said, "Today, knowledge and skills now stand alone as the only source of comparative advantage. . . . Silicon Valley and Route 128 are where they are simply because that is where the brainpower is. They have nothing else going for them."

Why is this historic transition rocking the foundation of capitalism? Quite simply, the human brain cannot be mass-produced. While hardware can be mass-produced and sold by the ton, the human brain cannot, meaning that common sense will be the currency of the future. Unlike with commodities, to create intellectual capital you have to nurture, cultivate, and educate a human being, which takes decades of individual effort.

As Thurow says, "With everything else dropping out of the competitive equation, knowledge has become the only source of long-run sustainable competitive advantage."

For example, software will become increasingly more important than hardware. Computer chips will be sold by the truckload as the price of chips continues to plunge, but software has to be created the old-fashioned way, by a human working with pencil and paper, sitting quietly in a chair. For example, the files stored in your laptop, which might contain valuable plans, manuscripts, and data, may be worth hundreds of thousands of dollars, but the laptop itself is worth only a few hundred. Of course, software can be easily copied and mass-produced, but the creation of new software cannot. That requires human thought.

According to UK economist Hamish McRae, "in 1991 Britain became the first country to earn more from invisible exports (services) than from visible ones."

While the share of the U.S. economy coming from manufacturing has declined dramatically over the decades, the sector that involves intellectual capitalism (Hollywood movies, the music industry, video games, computers, telecommunications, etc.) has soared. This shift from commodity capitalism to intellectual capitalism is a gradual one, starting in the last century, but it is accelerating every decade. MIT economist Thurow writes, "After correcting for general inflation, natural resource prices have fallen almost 60 percent from the mid-1970s to mid-1990s."

Some nations understand this. Consider the lesson of Japan in the postwar era. Japan has no great natural resources, yet its economy is among the largest in the world. The wealth of Japan today is a testament to the industriousness and unity of its people, rather than the wealth under its feet.

Unfortunately, many nations do not grasp this fundamental fact and do not prepare their citizens for the future, relying instead mainly on commodities. This means that nations that are rich in natural resources and do not understand this principle may sink into poverty in the future.

DIGITAL DIVIDE?

Some voices decry the information revolution, stating that we will have a widening chasm between the "digital rich" and the "digital poor," that is, those with access to computer power and those without. This revolution, they claim, will widen the fault lines of society, opening up new disparities of wealth and inequalities that could tear at the fabric of society.

But this is a narrow picture of the true problem. With computer power doubling every eighteen months, even poor children are getting access to computers. Peer pressure and cheap prices have encouraged computer and Internet use among poor children. In one experiment, funds were given to purchase a laptop for every classroom. Despite good intentions, the program was widely viewed as a failure. First, the laptop usually sat unused in a corner, because the teacher often did not know how to use it. Second, most of the students were already online with their friends and simply bypassed the classroom laptop.

The problem is not access. The real problem is jobs. The job market is undergoing a historic change, and the nations that will thrive in the future are those that take advantage of this.

For developing nations, one strategy is to use commodities to build a sound foundation, and then use that foundation as a stepping-stone to make the transition to intellectual capitalism. China, for example, has been successfully adopting this two-step process: the Chinese are building thousands of factories that produce goods for the world market, but they are using the profits to create a service sector built on intellectual capitalism. In the United States, 50 percent of the Ph.D. students in physics are foreign born (largely because the United States does not produce enough qualified students of its own). Of these foreign-born Ph.D. students, most are from China and India. Some of these students have returned to their native countries to create entirely new industries.

ENTRY-LEVEL JOBS

One casualty of this transition will be entry-level jobs. Every century has introduced new technologies that have created wrenching dislocations in the economy and people's lives. For example, in 1850, 65 percent of the American labor force worked on farms. (Today, only 2.4 percent does.) The same will be true in this century.

In the 1800s, new waves of immigrants flooded into the United States, whose economy was growing rapidly enough to assimilate them. In New York, for example, immigrants could find work in the garment industry or light manufacturing. Regardless of education level, any worker willing to do an honest day's work could find something to do in an expanding economy. It was like a conveyer belt that took immigrants from the ghettoes and slums of Europe and thrust them into the thriving middle class of America.

Economist James Grant has said, "The prolonged migration of hands and minds from the field to the factory, office and classroom is all productivity growth. . . . Technological progress is the bulwark of the modern economy. Then again, it has been true for most of the past 200 years."

Today, many of these entry-level jobs are gone. Moreover, the nature of the economy has changed. Many entry-level jobs have been sent overseas by corporations looking for cheaper labor. The old manufacturing job at the factory disappeared long ago.

But there is much irony in this. For years, many people demanded a level playing field, without favoritism or discrimination. But if jobs can be exported at the press of a button, the level playing field now extends to China and India. So entry-level jobs that used to act as conveyor belts to the middle class can now be exported elsewhere. This is fine for workers overseas, since they can benefit from the level playing field, but can cause inner cities to hollow out in the United States.

The consumer also benefits from this. Products and services become cheaper and production and distribution more efficient if there is global competition. Simply trying to prop up obsolete businesses and overpaid jobs creates complacency, waste, and inefficiency. Subsidizing failing industries only prolongs the inevitable, delays the pain of collapse, and actually makes thing worse.

There is another irony. Many high-paying, skilled service-sector jobs go unfilled for lack of qualified candidates. Often, the educational system does not produce enough skilled workers, so companies have to cope with a less-educated workforce. Corporations go begging for skilled workers whom the educational system often does not produce. Even in a depressed economy, there are jobs that go unfilled by skilled workers.

But one thing is clear. In a postindustrial economy, many of the old blue-collar factory jobs are gone for good. Over the years, economists have toyed with the idea of "reindustrializing America," until they realize that you cannot turn back the hands of time. The United States and Europe went through the transition from a largely industrial to a service economy decades ago, and this historic shift cannot be reversed. The heyday of industrialization has passed, forever.

Instead, efforts have to be made to reorient and reinvest in those sectors that maximize intellectual capitalism. This will be one of the most difficult tasks for governments in the twenty-first century, with no quick-and-easy solutions. On one hand, it means a major overhaul of the education system, so that workers can retrain and also so that high school students do not graduate into the unemployment lines. Intellectual capitalism does not mean jobs only for software programmers and scientists but in a broad spectrum of activities that involve creativity, artistic ability, innovation, leadership, and analysis—i.e., common sense. The workforce has to be educated to meet the challenges of the twenty-first century, not to duck them. In particular, science curricula have to be overhauled and teachers have to be retrained to become relevant for the technological society of the future. (It's sad that in America there is the old expression, "Those who can, do. Those who can't, teach.")

As MIT economist Lester Thurow has said, "Success or failure depends upon whether a country is making a successful transition to the man-made brainpower industries of the future—not on the size of any particular sector."

This means creating a new wave of innovative entrepreneurs who will create new industries and new wealth from these technological innovations. The energy and vitality of these people must be unleashed. They must be allowed to inject new leadership into the marketplace.

WINNERS AND LOSERS: NATIONS

Unfortunately, many countries are not taking this path, instead relying exclusively on commodity capitalism. But since commodity prices, on average, have been dropping for the past 150 years, their economies will eventually shrink with time, as the world bypasses them.

This process is not inevitable. Look at the examples of Germany and Japan in 1945, when their entire populations were near starvation, their cities were in ruins, and their governments had collapsed. In one generation, they were able to march to the front of the world economy. Look at China today, with its 8 to 10 percent galloping growth rate, reversing 500 years of economic decline. Once widely derided as the "sick man of Asia," in another generation it will join the ranks of the developed nations.

What distinguishes these three societies is that each was cohesive as a nation, had hardworking citizens, and made products that the world rushed to buy. These nations placed emphasis on education, on unifying their country and people, and on economic development.

As UK economist and journalist McRae writes, "The old motors of growth—land, capital, natural resources—no longer matter. Land matters little because the rise in agricultural yields has made it possible to produce far more food in the industrial world than it needs. Capital no longer matters because it is, at a price, almost infinitely available from the international markets for revenue-generating projects. . . . These quantitative assets, which have traditionally made countries rich, are being replaced by a series of qualitative features, which boil down to the quality, organization, motivation, and self-discipline of the people who live there. This is borne out by looking at the way the level of human skills is becoming more important in manufacturing, in private sector services, and in the public sector."

However, not every nation is following this path. Some nations are run by incompetent leaders, are culturally and ethnically fragmented to the point of dysfunction, and do not produce goods that the rest of the world wants. Instead of investing in education, they invest in huge armies and weapons to terrorize their people and maintain their privileges. Instead of investing in an infrastructure to speed up the industrialization of their country, they engage in corruption and keeping themselves in power, creating a kleptocracy, not a meritocracy.

Sadly, these corrupt governments have squandered much of the aid provided by the West, as small as it is. Futurists Alvin and Heidi Toffler note that between 1950 and 2000, more than $1 trillion in aid was given to poor nations by rich ones. But, they note, "we are told by the World Bank that nearly 2.8 billion people—almost half the population of the planet—still live on the equivalent of two dollars a day or less. Of these, some 1.1 billion survive in extreme or absolute poverty on less than one dollar."

The developed nations, of course, can do much more to alleviate the plight of developing nations rather than paying lip service to the problem. But after all is said and done, ultimately the main responsibility for development must come from wise leadership among the developing nations themselves. It goes back to that old saying, "Give me a fish, and I will eat for a day. Teach me how to fish, and I will eat forever." This means that instead of simply giving aid to developing nations, the stress should be on education and helping them develop new industries so they can become self-sufficient.

TAKING ADVANTAGE OF SCIENCE

Developing nations may be able to take advantage of the information revolution. They can, in principle, leapfrog past the developed nations in many areas. In the developed world, telephone companies had to tediously wire up every home or farm at great cost. But a developing nation does not have to wire up its country, since cell phone technology can excel in rural areas without any roads or infrastructure.

Also, developing nations have the advantage that they do not have to rebuild an aging infrastructure. For example, the subway systems of New York and London are more than a century old and badly in need of repairs. Today, renovating these creaky systems would cost more than building the original system itself. A developing nation may decide to create a subway system that is sparkling new with all the latest technology, taking advantage of vast improvements in metals, construction techniques, and technology. A brand-new subway system may cost much less than the systems of a century ago.

China, for example, was able to benefit from all the mistakes made in the West when building a city from the ground up. As a result, Beijing and Shanghai are being built at a fraction of the original cost of building a major city in the West. Today, Beijing is building one of the largest, most modern subway systems in the world, benefiting from all the computer technology created in the West, in order to serve an exploding urban population.

The Internet is another way for developing nations to take a shortcut to the future, bypassing all the mistakes made in the West, especially in the sciences. Previously, scientists in the developing world had to rely on a primitive postal system to deliver scientific journals, which usually arrived months to a year after publication, if they arrived at all. These journals were expensive and highly specialized, so that only the largest libraries could afford them. Collaborating with a scientist from the West was almost impossible. You had to be independently wealthy, or extremely ambitious, to obtain a position at a Western university to work under a famous scientist. Now it is possible for the most obscure scientist to obtain scientific papers less than a second after they are posted on the Internet, from almost anywhere in the world, for free. And, via the Internet, it is possible to collaborate with scientists in the West whom you have never met.

THE FUTURE IS UP FOR GRABS

The future is wide open. As we mentioned, Silicon Valley could become the next Rust Belt in the coming decades, as the age of silicon passes and the torch passes to the next innovator. Which nations will lead in the future? In the days of the Cold War, the superpowers were those nations that could wield military influence around the world. But the breakup of the Soviet Union has made it clear that in the future the nations that will rise to the top will be those that build their economies, which in turn depends on cultivating and nourishing science and technology.

So who are the leaders of tomorrow? The nations that truly grasp this fact. For example, the United States has maintained its dominance in science and technology in spite of the fact that U.S. students often score dead last when it comes to essential subjects like science and math. Proficiency test scores in 1991, for example, showed thirteen-year-old students in the United States ranking fifteenth in math and fourteenth in science, just above Jordanian students, who ranked eighteenth in both categories. Tests taken since then annually confirm these dismal numbers. (It should also be pointed out that this ranking corresponds roughly to the number of days that students were in school. China, which ranked number 1, averaged 251 days of instruction per year, while the United States averaged only 178 days per year.)

It seems like a mystery that, despite these awful numbers, the United States continues to do well internationally in science and technology, until you realize that much of the U.S. science comes from overseas, in the form of the "brain drain." The United States has a secret weapon, the H1B visa, the so-called genius visa. If you can show that you have special talents, resources, or scientific knowledge, you can jump ahead of the line and get an H1B visa. This has continually replenished our scientific ranks. Silicon Valley, for example, is roughly 50 percent foreign born, many coming from Taiwan and India. Nationwide, 50 percent of all Ph.D. students in physics are foreign born. At my university, the City University of New York, the figure is closer to 100 percent foreign born.

Some congressmen have tried to eliminate the H1B visa because, they claim, it takes jobs away from Americans, but they do not understand the true role that this visa plays. Usually, there are no Americans qualified to take the highest-level jobs in Silicon Valley, which we've seen often go unfilled as a consequence. This fact was apparent when former chancellor Gerhard Schroeder tried to pass a similar H1B visa immigration law for Germany, but the measure was defeated by those who claimed that this would take jobs away from native-born Germans. Again, the critics failed to understand that there are often no Germans to fill these high-level jobs, which then go unfilled. These H1B immigrants do not take away jobs, they create entire new industries.

But the H1B visa is only a stopgap measure. The United States cannot continue to live off foreign scientists, many of whom are beginning to return to China and India as their economies improve. So the brain drain is not sustainable. This means that the United States will eventually have to overhaul its archaic, sclerotic education system. At present, poorly prepared high school students flood the job market and universities, creating a logjam. Employers continually bemoan the fact that they have to take one year to train their new hires to bring them up to speed. And the universities are burdened by having to create new layers of remedial courses to compensate for the poor high school education system.

Fortunately, our universities and businesses eventually do a commendable job of repairing the damage done by the high school system, but this is a waste of time and talent. For the United States to remain competitive into the future, there have to be fundamental changes in the elementary and high school system.

To be fair, the United States still has significant advantages. I was once at a cocktail party at the American Museum of Natural History in New York and met a biotech entrepreneur from Belgium. I asked him why he left, given that Belgium has its own vigorous biotech industry. He said that in Europe, often you don't get a second chance. Since people know who you and your family are, if you make a mistake, you could be finished. Your mistakes tend to follow you, no matter where you are. But in the United States, he said, you can constantly reinvent yourself. People don't care who your ancestors were. They just care what you can do for them now, today. This was refreshing, he said, and one reason why other European scientists move to the United States.

LESSON OF SINGAPORE

In the West, there is the expression "The squeaky wheel gets the grease." But in the East, there is another expression: "The nail that sticks out gets hammered down." These two expressions are diametrically opposed to each other, but they capture some of the essential features of Western and Eastern thought.

In Asia, the students often have test scores that soar beyond those of their counterparts in the West. However, much of that learning is book learning and rote memorization, which will take you only to a certain level. To reach the higher levels of science and technology, you need creativity, imagination, and innovation, which the Eastern system does not nurture.

So although China may eventually catch up with the West when it comes to producing cheap factory-made copies of goods first manufactured in the West, it will lag for decades behind the West in the creative process of dreaming up new products and new strategies.

I once spoke at a conference in Saudi Arabia, where another featured speaker was Lee Kuan Yew, prime minister of Singapore from 1959 to 1990. He is something of a rock star among the developing nations, since he helped to forge the modern nation of Singapore, which ranks among the top nations in science. Singapore, in fact, is the fifth-richest nation in the world, if you calculate the per capita gross domestic product. The audience strained to hear every word from this legendary figure.

He reminisced about the early days after the war, when Singapore was viewed as a backwater port known primarily for piracy, smuggling, drunken sailors, and other unsavory activities. A group of his associates, however, dreamed of the day when this tiny seaport could rival the West. Although Singapore had no significant natural resources, its greatest resource was its own people, who were hardworking and semiskilled. His group embarked on a remarkable journey, taking this sleepy backwater nation and transforming it into a scientific powerhouse within one generation. It was perhaps one of the most interesting cases of social engineering in history.

He and his party began a systematic process of revolutionizing the entire nation, stressing science and education and concentrating on the high-tech industries. Within just a few decades, Singapore created a large pool of highly educated technicians, which made it possible for the country to become one of the leading exporters of electronics, chemicals, and biomedical equipment. In 2006, it produced 10 percent of the world's foundry wafer output for computers.

There have been a number of problems, he confessed, along the course of modernizing his nation. To enforce social order, they imposed draconian laws, outlawing everything from spitting on the street (punishable by whipping) to drug dealing (punishable by death). But he also noticed one important thing. Top scientists, he found, were eager to visit Singapore, yet only a handful stayed. Later, he found out one reason why: there were no cultural amenities and attractions to keep them in Singapore. This gave him his next idea: deliberately fostering all the cultural fringe benefits of a modern nation (ballet companies, symphony orchestras, etc.) so that top scientists would sink their roots in Singapore. Almost overnight, cultural organizations and events were springing up all over the country as a lure to keep the scientific elite anchored there.

Next, he also realized that the children of Singapore were blindly repeating the words of their teachers, not challenging the conventional wisdom and creating new ideas. He realized that the East would forever be trailing the West as long as it produced scientists who could only copy others. So he set into motion a revolution in education: creative students would be singled out and allowed to pursue their dreams at their own pace. Realizing that someone like a Bill Gates or a Steve Jobs would be crushed by Singapore's suffocating educational system, he asked schoolteachers to systematically identify the future geniuses who could revitalize the economy with their scientific imagination.

The lesson of Singapore is not for everyone. It is a small city-state, where a handful of visionaries could practice controlled nation building. And not everyone wants to be whipped for spitting on the street. However, it shows you what you can do if you systematically want to leap to the front of the information revolution.

CHALLENGE FOR THE FUTURE

I once spent some time at the Institute for Advanced Study at Princeton, and had lunch with Freeman Dyson. He began to reminisce about his long career in science and then mentioned a disturbing fact. Before the war, when he was a young university student in the UK, he found that the brightest minds of England were turning their backs on the hard sciences, like physics and chemistry, in favor of lucrative careers in finance and banking. While the previous generation was creating wealth, in the form of electrical and chemical plants and inventing new electromechanical machines, the next generation was indulging in massaging and managing other people's money. He lamented that it was a sign of the decline of the British Empire. England could not maintain its status as a world power if it had a crumbling scientific base.

Then he said something that caught my attention.

He remarked that he was seeing this for the second time in his life. The brightest minds at Princeton were no longer tackling the difficult problems in physics and mathematics but were being drawn into careers like investment banking. Again, he thought, this might be a sign of decay, when the leaders of a society can no longer support the inventions and technology that made their society great.

This is our challenge for the future.

People alive now are living in the midst of what may be seen as the most extraordinary three or four centuries in human history.

—JULIAN SIMON

Where there is no vision, the people perish.

—PROVERBS 29:18

8 FUTURE OF HUMANITY *Planetary Civilization*

In mythology, the gods lived in the divine splendor of heaven, far above the insignificant affairs of mere mortals. The Greek gods frolicked in the heavenly domain of Mount Olympus, while the Norse gods who fought for honor and eternal glory would feast in the hallowed halls of Valhalla with the spirits of fallen warriors. But if our destiny is to attain the power of the gods by the end of the century, what will our civilization look like in 2100? Where is all this technological innovation taking our civilization?

All the technological revolutions described here are leading to a single point: the creation of a planetary civilization. This transition is perhaps the greatest in human history. In fact, the people living today are the most important ever to walk the surface of the planet, since they will determine whether we attain this goal or descend into chaos. Perhaps 5,000 generations of humans have walked the surface of the earth since we first emerged in Africa about 100,000 years ago, and of them, the ones living in this century will ultimately determine our fate.

Unless there is a natural catastrophe or some calamitous act of folly, it is inevitable that we will enter this phase of our collective history. We can see this most clearly by analyzing the history of energy.

RANKING CIVILIZATIONS

When professional historians write history, they view it through the lens of human experience and folly, that is, through the exploits of kings and queens, the rise of social movements, and the proliferation of ideas. Physicists, by contrast, view history quite differently.

Physicists rank everything, even human civilizations, by the energy it consumes. When applied to human history, we see that for countless millennia, our energy was limited to 1/5 horsepower, the power of our bare hands, and hence we lived nomadic lives in small, wandering tribes, scavenging for food in a harsh, hostile environment. For eons, we were indistinguishable from the wolves. There were no written records, just stories handed down from generation to generation at lonely campfires. Life was short and brutish, with an average life expectancy of eighteen to twenty years. Your total wealth consisted of whatever you could carry on your back. Most of your life, you felt the gnawing pain of hunger. After you died, you left no trace that you had ever lived at all.

But 10,000 years ago, a marvelous event happened that set civilization into motion: the Ice Age ended. For reasons that we still do not understand, thousands of years of glaciation ended. This paved the way for the rise of agriculture. Horses and oxen were soon domesticated, which increased our energy to 1 horsepower. Now one person had the energy to harvest several acres of farmland, yielding enough surplus energy to support a rapidly expanding population. With the domestication of animals, humans no longer relied primarily on hunting animals for food, and the first stable villages and cities began to rise from the forests and plains.

The excess wealth created by the agricultural revolution spawned new, ingenious ways to maintain and expand this wealth. Mathematics and writing were created to count this wealth, calendars were needed to keep track of when to plant and harvest, and scribes and accountants were needed to keep track of this surplus and tax it. This excess wealth eventually led to the rise of large armies, kingdoms, empires, slavery, and ancient civilizations.

The next revolution took place about 300 years ago, with the coming of the Industrial Revolution. Suddenly, the wealth accumulated by an individual was not just the product of his hands and horse but the product of machines that could create fabulous wealth via mass production.

Steam engines could drive powerful machines and locomotives, so that wealth could be created from factories, mills, and mines, not just fields. Peasants, fleeing from periodic famines and tired of backbreaking work in the fields, flocked to the cities, creating the industrial working class. Blacksmiths and wagonmakers were eventually replaced by autoworkers. With the coming of the internal combustion engine, a person could now command hundreds of horsepower. Life expectancy began to grow, hitting forty-nine in the United States by the year 1900.

Finally, we are in the third wave, where wealth is generated from information. The wealth of nations is now measured by electrons circulating around the world on fiber-optic cables and satellites, eventually dancing across computer screens on Wall Street and other financial capitals. Science, commerce, and entertainment travel at the speed of light, giving us limitless information anytime, anywhere.

TYPE I, II, AND III CIVILIZATIONS

How will this exponential rise in energy continue into the coming centuries and millennia? When physicists try to analyze civilizations, we rank them on the basis of the energy they consume. This ranking was first introduced in 1964 by Russian astrophysicist Nikolai Kardashev, who was interested in probing the night sky for signals sent from advanced civilizations in space.

He was not satisfied with something as nebulous and ill defined as an "extraterrestrial civilization," so he introduced a quantitative scale to guide the work of astronomers. He realized that extraterrestrial civilizations may differ on the basis of their culture, society, government, etc., but there was one thing they all had to obey: the laws of physics. And from the earth, there was one thing that we could observe and measure that could classify these civilizations into different categories: their consumption of energy.

So he proposed three theoretical types: A Type I civilization is planetary, consuming the sliver of sunlight that falls on their planet, or about 10^{17} watts. A Type II civilization is stellar, consuming all the energy that their sun emits, or 10^{27} watts. A Type III civilization is galactic, consuming the energy of billions of stars, or about 10^{37} watts.

The advantage of this classification is that we can quantify the power of each civilization rather than make vague and wild generalizations. Since we know the power output of these celestial objects, we can put specific numerical constraints on each of them as we scan the skies.

Each type is separated by a factor of 10 billion: a Type III civilization consumes 10 billion times more energy than a Type II civilization (because there are roughly 10 billion or more stars in a galaxy), which in turn consumes 10 billion times more energy than a Type I civilization.

According to this classification, our present-day civilization is Type 0. We don't even rate on this scale, since we get our energy from dead plants, that is, from oil and coal. (Carl Sagan, generalizing this classification, tried to get a more precise estimate of where we ranked on this cosmic scale. His calculation showed that we are actually a Type .7 civilization.)

On this scale, we can also classify the various civilizations we see in science fiction. A typical Type I civilization would be that of Buck Rogers or Flash Gordon, where an entire planet's energy resources have been developed. They can control all planetary sources of energy, so they might be able to control or modify the weather at will, harness the power of a hurricane, or have cities on the oceans. Although they roam the heavens in rockets, their energy output is still largely confined to a planet.

A Type II civilization might include *Star Trek*'s United Federation of Planets (without the warp drive), able to colonize about 100 nearby stars. Their technology is barely capable of manipulating the entire energy output of a star.

A Type III civilization may be the Empire in the *Star Wars* saga, or perhaps the Borg in the *Star Trek* series, both of which have colonized large portions of a galaxy, embracing billions of star systems. They can roam the galactic space lanes at will.

(Although the Kardashev scale is based on planets, stars, and galaxies for its classification, we should point out the possibility of a Type IV civilization, which derives its energy from extragalactic sources. The only known energy source beyond our galaxy is dark energy, which makes up 73 percent of the matter and energy of the known universe, while the world of stars and galaxies makes up only 4 percent of the universe. A possible candidate for a Type IV civilization might be the godlike Q in the *Star Trek* series, whose power is extragalactic.)

We can use this classification to calculate when we might achieve each of these types. Assume that world civilization grows at the rate of 1 percent each year in terms of its collective GDP. This is a reasonable assumption when we average over the past several centuries. According to this assumption, it takes roughly 2,500 years to go from one civilization to the next. A 2 percent growth rate would give a transition period of 1,200 years.

But we can also calculate how long it would take for our planet to attain Type I classification. In spite of economic recessions and expansions, booms and busts, we can mathematically estimate that we will attain Type I status in about 100 years, given an average rate of our economic growth.

FROM TYPE 0 TO TYPE I

We see evidence of this transition from Type 0 to Type I every time we open a newspaper. Many of the headlines can be traced to the birth pangs of a Type I civilization being born right in front of our eyes.

- The Internet is the beginning of a Type I planetary telephone system. For the first time in history, a person on one continent can effortlessly exchange unlimited information with someone on another continent. In fact, many people already feel they have more in common with someone on the other side of the world than with their next-door neighbor. This process will only accelerate as nations lay even more fiber-optic cables and launch more communications satellites. This process is also unstoppable. Even if the president of the United States tried to ban the Internet, he would be met only with laughter. There are almost a billion personal computers in the world today, and roughly a quarter of humanity has been on the Internet at least once.
- A handful of languages, led by English, followed by Chinese, are rapidly emerging as the future Type I language. On the World Wide Web, for example, 29 percent of visitors log on in English, followed by 22 percent in Chinese, 8 percent in Spanish,

6 percent in Japanese, and 5 percent in French. English is already the de facto planetary language of science, finance, business, and entertainment. English is the number-one second language on the planet. No matter where I travel, I find that English has emerged as the lingua franca. In Asia, for example, when Vietnamese, Japanese, and Chinese are in a meeting, they use English to communicate. Currently, there are about 6,000 languages being spoken on earth, and 90 percent of them are expected to become extinct in the coming decades, according to Michael E. Krauss, formerly of the University of Alaska's Native Language Center. The telecommunications revolution is accelerating this process, as people living in even the most remote regions of the earth are exposed to English. This will also accelerate economic development as their societies are further integrated into the world economy, thereby raising living standards and economic activity.

Some people will bemoan the fact that some ancestral languages will no longer be spoken. But on the other hand, the computer revolution will guarantee that these languages are not lost. Native speakers will add their language and their culture to the Internet, where they will last forever.

- We are witnessing the birth of a planetary economy. The rise of the European Union and other trade blocs represents the emergence of a Type I economy. Historically, the peoples of Europe have fought blood feuds with their neighbors for thousands of years. Even after the fall of the Roman Empire, these tribes would continue to slaughter one another, eventually becoming the feuding nations of Europe. Yet today, these bitter rivals have suddenly banded together to form the European Union, representing the largest concentration of wealth on the planet. The reason these nations have abruptly put aside their famous rivalries is to compete with the economic juggernaut of nations that signed the North American Free Trade Agreement (NAFTA). In the future, we will see more economic blocs forming, as nations realize that they cannot remain competitive unless they join lucrative trading blocs.

We see graphic evidence of this when analyzing the great recession of 2008. Within a matter of days, the shock waves emanating from Wall Street rippled through the financial halls of London, Tokyo, Hong Kong, and Singapore. Today, it is impossible to understand the economics of a single nation without understanding the trends affecting the world economy.

- We are seeing the rise of a planetary middle class. Hundreds of millions of people in China, India, and elsewhere are entering its ranks, which is perhaps the greatest social upheaval in the last half century. This group is savvy about cultural, educational, and economic trends affecting the planet. The focus of this planetary middle class is not wars, religion, or strict moral codes, but political and social stability and consumer goods. The ideological and tribal passions that might have gripped their ancestors mean little to them if their goal is to have a suburban house with two cars. While their ancestors might have celebrated the day their sons went off to war, one of their main concerns now is to get them into a good college. And for people who enviously watch other people rise, they will wonder when their time will come. Kenichi Ohmae, a former senior partner of McKinsey & Company, writes, "People will inevitably start to look around them and ask why they cannot have what others have. Equally important, they will start to ask why they were not able to have it in the past."

- The economy, not weapons, is the new criterion for a superpower. The rise of the EU and NAFTA underscores an important point: with the end of the Cold War, it is clear that a world power can maintain its dominant position mainly through economic might. Nuclear wars are simply too dangerous to fight, so it is economic might that will largely determine the destiny of nations. One contributing factor to the collapse of the Soviet Union was the economic stress of competing militarily with the United States. (As the advisers to President Ronald Reagan once commented, the strategy of the United States was to spend Russia into a depression, that is, increase U.S. military expenditure so that the Russians, with an economy less than half the size of the United States', would have to starve their own people to keep up.) In the future, it is clear that a superpower can maintain its status only through economic might, and that in turn stems from science and technology.

- A planetary culture is emerging, based on youth culture (rock and roll and youth fashion), movies (Hollywood blockbusters), high fashion (luxury goods), and food (mass-market fast-food chains). No matter where you travel, you can find evidence of the same cultural trends in music, art, and fashion. For example, Hollywood carefully factors in global appeal when it estimates the success of a potential blockbuster movie. Movies with cross-cultural themes (such as action or romance), packed with internationally recognized celebrities, are the big moneymakers for Hollywood, evidence of an emerging planetary culture.

We saw this after World War II when, for the first time in human history, an entire generation of young people possessed enough disposable income to alter the prevailing culture. Formerly, children were sent into the fields to toil with their parents once they hit puberty. (This is the origin of the three-month summer vacation. During the Middle Ages, children were required to do backbreaking work in the fields during summer as soon as they were of age.) But with rising prosperity, the postwar baby boom generation left the fields to head to the streets. Today, we see the same pattern taking place in country after country, as economic development empowers youth with ample disposable incomes. Eventually, as most of the people of the world enter the middle class, rising incomes will filter down to their youth, fueling a perpetuation of this planetary youth culture.

Rock and roll, Hollywood movies, etc., are in fact prime examples of how intellectual capitalism is replacing commodity capitalism. Robots for decades to come will be incapable of creating music and movies that can thrill an international audience.

This is also happening in the world of fashion, where a handful of brand names are extending their reach worldwide. High fashion, once reserved for the aristocracy and the extremely wealthy, is rapidly proliferating around the world as more people enter the middle class and aspire to some of the glamour of the rich. High fashion is no longer the exclusive province of the privileged elite.

But the emergence of a planetary culture does not mean that local cultures or customs will be wiped out. Instead, people will be bicultural. On one hand, they will keep their local cultural traditions alive (and the Internet guarantees that these regional customs will survive forever). The rich cultural diversity of the world will continue to thrive into the future. In fact, certain obscure features of local culture can spread around the world via the Internet, gaining them a worldwide audience. On the other hand, people will be fluent in the changing trends that affect global culture. When people communicate with those from another culture, they will do so via the global culture. This has already happened to many of the elites on the planet: they speak the local language and obey local customs but use English and follow international customs when dealing with people from other countries. This is the model for the emerging Type I civilization. Local cultures will continue to thrive, coexisting side by side with the larger global culture.

- The news is becoming planetary. With satellite TV, cell phones, the Internet, etc., it becomes impossible for one nation to completely control and filter the news. Raw footage is emerging from all parts of the world, beyond the reach of censors. When wars or revolutions break out, the stark images are broadcast instantly around the world as they happen in real time. In the past, it was relatively easy for the Great Powers of the nineteenth century to impose their values and manipulate the news. Today, this is still possible, but on a much reduced basis because of advanced technology. Also, with rising education levels around the world, there is a much larger audience for world news. Politicians today have to include world opinion when they think about the consequences of their actions.

- Sports, which in the past were essential in forging a tribal and then a national identity, are now forging a planetary identity. Soccer and the Olympics are emerging to dominate planetary sports. The 2008 Olympics, for example, were widely interpreted as a coming-out party for the Chinese, who wanted to assume their rightful cultural position in the world after centuries of isolation. This is also an example of the Cave Man Principle, since sports are High Touch but are entering the world of High Tech.

- Environmental threats are also being debated on a planetary scale. Nations realize that the pollution they create crosses national boundaries and hence can precipitate an international crisis. We first saw this when a gigantic hole in the ozone layer opened over the South Pole. Because the ozone layer prevents harmful UV and X-rays from the sun from reaching the ground, nations banded together to limit the production and consumption of chlorofluorocarbons used in refrigerators and industrial systems. The Montreal Protocol was signed in 1987 and successfully decreased the use of the ozone-depleting chemicals. Building on this international success, most nations adopted the Kyoto Protocol in 1997 to address the threat of global warming, which is an even greater threat to the environment of the planet.
- Tourism is one of the fastest-growing industries on the planet. During most of human history, it was common for people to live out their entire lives within a few miles of their birthplace. It was easy for unscrupulous leaders to manipulate their people, who had little to no contact with other peoples. But today, one can go around the globe on a modest budget. Backpacking youths of today who stay in budget youth hostels around the world will become the leaders of tomorrow. Some people decry the fact that tourists have only the crudest understanding of local cultures, histories, and politics. But we have to weigh that against the past, when contact between distant cultures was almost nonexistent, except during times of war, often with tragic results.
- Likewise, the falling price of intercontinental travel is accelerating contact between diverse peoples, making wars more difficult to wage and spreading the ideals of democracy. One of the main factors that whipped up animosity between nations was misunderstanding between people. In general, it is quite difficult to wage war on a nation you are intimately familiar with.
- The nature of war itself is changing to reflect this new reality. History has shown that two democracies almost never wage war against each other. Almost all wars of the past have been waged between nondemocracies, or between a democracy and a nondemocracy. In general, war fever can be easily whipped up by

demagogues who demonize the enemy. But in a democracy, with a vibrant press, oppositional parties, and a comfortable middle class that has everything to lose in a war, war fever is much more difficult to cultivate. It is hard to whip up war fever when there is a skeptical press and mothers who demand to know why their children are going to war.

There will still be wars in the future. As the Prussian military theorist Carl von Clausewitz once said, "War is politics by other means." Although we will still have wars, their nature will change as democracy is spread around the world.

(There is another reason why wars are becoming more difficult to wage as the world becomes more affluent and people have more to lose. Political theorist Edward Luttwak has written that wars are much more difficult to wage because families are smaller today. In the past, the average family had ten or so children; the eldest inherited the farm, while the younger siblings joined the church, the military, or sought their fortunes elsewhere. Today, when a typical family has an average of 1.5 children, there is no more surplus of children to easily fill the military and the priesthood. Hence, wars will be much more difficult to wage, especially between democracies and third-world guerrillas.)

- Nations will weaken but will still exist in 2100. They will still be needed to pass laws and fix local problems. However, their power and influence will be vastly decreased as the engines of economic growth become regional, then global. For example, with the rise of capitalism in the late 1700s and early 1800s, nations were needed to enforce a common currency, language, tax laws, and regulations concerning trade and patents. Feudal laws and traditions, which hindered the advance of free trade, commerce, and finance, were quickly swept away by national governments. Normally, this process might take a century or so, but we saw an accelerated version of this when Otto von Bismarck, the Iron Chancellor, forged the modern German state in 1871. In the same way, this march toward a Type I civilization is changing the nature of capitalism, and economic power is gradually shifting from national governments to regional powers and trade blocs.

This does not necessarily mean a world government. There are many ways a planetary civilization could exist. It is clear that national governments will lose relative power, but what power will fill the vacuum will depend on many historical, cultural, and national trends that are hard to predict.

- Diseases will be controlled on a planetary basis. In the ancient past, virulent diseases were actually not so dangerous because the human population was very low. The incurable Ebola virus, for example, is probably an ancient disease that infected just a few villages over thousands of years. But the rapid expansion of civilization into previously uninhabited areas and the rise of cities mean that something like Ebola has to be monitored very carefully.

   When the population of cities hit several hundred thousand to a million, diseases could spread rapidly and create genuine epidemics. The fact that the Black Plague killed perhaps half the European population was an indication, ironically, of progress, because populations had reached critical mass for epidemics and shipping routes connected ancient cities around the world.

   The recent outbreak of the H1N1 flu is thus a measure of our progress as well. Perhaps originating in Mexico City, the disease spread quickly around the globe via jet travel. More important, it took only a matter of months for the nations of the world to sequence the genes of the virus and then create a vaccine for it that was available to tens of millions of people.

TERRORISM AND DICTATORSHIPS

There are groups, however, that instinctively resist the trend toward a Type I planetary civilization, because they know that it is progressive, free, scientific, prosperous, and educated. These forces may not be conscious of this fact and cannot articulate it, but they are in effect struggling against the trend toward a Type I civilization. These are:

- Islamic terrorists, who would prefer to go back a millennium, to the eleventh century, rather than live in the twenty-first century.

They cannot frame their discontent in this fashion, but, judging from their own statements, they prefer to live in a theocracy where science, personal relations, and politics are all subject to strict religious edicts. (They forget that, historically, the greatness and scientific and technological prowess of the Islamic civilization were matched only by its tolerance of new ideas. These terrorists do not understand the true source of the greatness of the Islamic past.)

• Dictatorships that depend on keeping their people ignorant of the wealth and progress of the outside world. One striking example was the demonstrations that gripped Iran in 2009, where the government tried to suppress the ideas of the demonstrators, who were using Twitter and YouTube in their struggle to carry their message to the world.

In the past, people said that the pen was mightier than the sword. In the future, it will be the chip that is mightier than the sword.

One of the reasons the people of North Korea, a horribly impoverished nation, do not rebel is because they are denied all contact with the world, whose people, they believe, are also starving. In part, not realizing that they do not have to accept their fate, they endure incredible hardship.

TYPE II CIVILIZATIONS

By the time a society attains Type II status thousands of years into the future, it becomes immortal. Nothing known to science can destroy a Type II civilization. Since it will have long mastered the weather, ice ages can be avoided or altered. Meteors and comets can be also be deflected. Even if their sun goes supernova, the people will be able to flee to another star system, or perhaps prevent their star from exploding. (For example, if their sun turns into a red giant, they might be able swing asteroids around their planet in a slingshot effect in order to move their planet farther from the sun.)

One way in which a Type II civilization may be able to exploit the entire energy output of a star is to create a gigantic sphere around it that absorbs all the sunlight of the star. This is called a Dyson sphere.

A Type II civilization will probably be at peace with itself. Since space travel is so difficult, it will have remained a Type I civilization for centuries, plenty of time to iron out the divisions within their society. By the time a Type I civilization reaches Type II status, they will have colonized not just their entire solar system but also the nearby stars, perhaps out to several hundred light-years, but not much more. They will still be restricted by the speed of light.

TYPE III CIVILIZATIONS

By the time a civilization reaches Type III status, it will have explored most of the galaxy. The most convenient way to visit the hundreds of billions of planets is to send self-replicating robot probes throughout the galaxy. A von Neumann probe is a robot that has the ability to make unlimited copies of itself; it lands on a moon (since it is free of rust and erosion) and makes a factory out of lunar dirt, which creates thousands of copies of itself. Each copy rockets off to other distant star systems and makes thousands more copies. Starting with one such probe, we quickly create a sphere of trillions of these self-replicating probes expanding at near the speed of light, mapping out the entire Milky Way galaxy in just 100,000 years. Since the universe is 13.7 billion years old, there is plenty of time in which these civilizations may have risen (and fallen). (Such rapid, exponential growth is also the mechanism by which viruses spread in our body.)

There is another possibility, however. By the time a civilization has reached Type III status, its people have enough energy resources to probe the "Planck energy," or 10^{19} billion electron volts, the energy at which space-time itself become unstable. (The Planck energy is a quadrillion times larger than the energy produced by our largest atom smasher, the Large Hadron Collider outside Geneva. It is the energy at which Einstein's theory of gravity finally breaks down. At this energy, it is theorized that the fabric of space-time will finally tear, creating tiny portals that might lead to other universes, or other points in space-time.) Harnessing such vast energy would require colossal machines on an unimaginable scale, but if successful they might make possible shortcuts through the fabric of space and time, either by compressing space or by passing through wormholes. Assuming that they can overcome a number of stubborn theoretical and practical obstacles (such as harnessing sufficient positive and negative energy and removing instabilities), it is conceivable that they might be able to colonize the entire galaxy.

This has prompted many people to speculate about why they have not visited us. Where are they? the critics ask.

One possible answer is that perhaps they already have, but we are too primitive to notice. Self-replicating von Neumann probes would be the most practical way of exploring the galaxy, and they do not have to be huge. They might be just a few inches long, because of revolutionary advances in nanotechnology. They might be in plain view, but we don't recognize them because we are looking for the wrong thing, expecting a huge starship carrying aliens from outer space. More than likely, the probe will be fully automatic, part organic and part electronic, and will not contain any space aliens at all.

And when we do eventually meet the aliens from space, we may be surprised, because they might have long ago altered their biology using robotics, nanotechnology, and biotechnology.

Another possibility is that they have self-destructed. As we mentioned, the transition from Type 0 to Type I is the most dangerous one, since we still have all the savagery, fundamentalism, racism, and so on of the past. It is possible that one day, when we visit the stars, we may find evidence of Type 0 civilizations that failed to make the transition to Type I (for example, their atmospheres may be too hot, or too radioactive, to support life).

SETI (SEARCH FOR EXTRATERRESTRIAL INTELLIGENCE)

At the present time, the people of the world are certainly not conscious of the march toward a Type I planetary civilization. There is no collective self-awareness that this historic transition is taking place. If you take a poll, some people might be vaguely aware of the process of globalization, but beyond that there is no conscious awareness that we are headed to a specific destination.

All this might suddenly change if we find evidence of intelligent life in outer space. Then, we would immediately be aware of our technological level in relation to this alien civilization. Scientists in particular would be intensely interested in which types of technologies this alien civilization has mastered.

Although one cannot know for sure, probably within this century we will detect an advanced civilization in space, given the rapid advances in our technology.

Two trends have made this possible. First is the launching of satellites specifically designed to find small, rocky extrasolar planets, the COROT and Kepler satellites. The Kepler is expected to identify up to 600 small, earthlike planets in space. Once these planets have been identified, the next step is to focus our search for intelligent emissions from these planets.

In 2001, Microsoft billionaire Paul Allen began donating funds, now more than $30 million, to jump-start the stalled SETI program. This will vastly increase the number of radio telescopes at the Hat Creek installation, located north of San Francisco. The Allen Telescope Array, when fully operational, will have 350 radio telescopes, making it the most advanced radio telescope facility in the world. While in the past astronomers have scanned little more than 1,000 stars in their search for intelligent life, the new Allen Array will increase that number by a factor of 1,000, to a million stars.

Although scientists have been searching vainly for signals from advanced civilizations for almost fifty years, only recently have these two developments given a much-needed boost to the SETI program. Many astronomers believe that there was simply too little effort and too few resources devoted to this project. With this influx of new resources and new data, the SETI program is becoming a serious scientific project.

It is conceivable that we may, within this century, detect signals from an intelligent civilization in space. (Seth Shostak, the director of the SETI Institute in the Bay Area, told me that within twenty years, he expects to make contact with such a civilization. That may be too optimistic, but it is safe to say that within this century it would be strange if we did not detect signals from another civilization in space.)

If signals are found from an advanced civilization, it could be one of the most significant milestones in human history. Hollywood movies love to describe the chaos this event might unleash, with prophets telling us that the end is near, with crazy religious cults going into overtime, etc.

The reality, however, is more mundane. There will be no need for immediate panic, since this civilization may not even know that we are eavesdropping on their conversations. And if they did, direct conversations between them and us would be difficult, given their enormous distance from us. First, it may take months to years to fully decode the message, and then to rank this civilization's technology, to see if it fits the Kardashev classification. Second, direct communication with them will probably be unlikely, since the distance to this civilization will be many light-years away, too far for any direct contact. So we will be able only to observe this civilization, rather than carry on any conversation. There will be an effort to build gigantic radio transmitters that can send messages back to the aliens. But in fact, it may take centuries before any two-way communication is possible with this civilization.

NEW CLASSIFICATIONS

The Kardashev classification was introduced in the 1960s, when physicists were concerned about energy production. However, with the spectacular rise of computer power, attention turned to the information revolution, where the number of bits processed by a civilization became as relevant as its energy production.

One can imagine, for example, an alien civilization on a planet where computers are impossible because their atmosphere conducts electricity. In this case, any electrical device will soon short-circuit, creating sparks, so that only the most primitive forms of electrical appliances are possible.

Any large-scale dynamo or computer would quickly burn out. We can imagine that such a civilization might eventually master fossil fuels and nuclear energy, but their society would be unable to process large amounts of information. It would be difficult for them to create an Internet or a planetary telecommunications system, so their economy and scientific progress would be stunted. Although they would be able to rise up the Kardashev scale, it would be very slow and painful without computers.

Therefore, Carl Sagan introduced another scale, based on information processing. He devised a system in which the letters of the alphabet, from A to Z, correspond to information. A Type A civilization is one that processes only a million pieces of information, which corresponds to a civilization that has only a spoken language but not a written one. If we compile all the information that has survived from ancient Greece, which had a flourishing written language and literature, it is about a billion bits of information, making it a Type C civilization. Moving up the scale, we can then estimate the amount of information that our civilization processes. An educated guess puts us at a Type H civilization. So therefore, the energy and information processing of our civilization yields a Type .7 H civilization.

In recent years, another concern has arisen: pollution and waste. Energy and information are not enough to rank a civilization. In fact, the more energy a civilization consumes and the more information it spews out, the more pollution and waste it might produce. This is not an academic question, since the waste from a Type I or II civilization might be enough to destroy it.

A Type II civilization, for example, consumes all the energy that is produced by a star. Let us say that its engines are 50 percent efficient, meaning that half the waste it produces is in the form of heat. This is potentially disastrous, because it means that the temperature of the planet will rise until it melts! Think of billions of coal plants on such a planet, belching huge amounts of heat and gases that heat the planet to the point that life is impossible.

Freeman Dyson, in fact, once tried to find Type II civilizations in outer space by searching for objects that emit primarily infrared radiation, rather than X-rays or visible light. This is because a Type II civilization, even if it wanted to hide its presence from prying eyes by creating a sphere around itself, would inevitably produce enough waste heat so that it would glow with infrared radiation. Therefore he suggested that astronomers search for star systems that produce mainly infrared light. (None, however, were found.)

But this raises the concern that any civilization that lets its energy grow out of control may commit suicide. We see, therefore, that energy and information are not enough to ensure the survival of the civilization as it moves up the scale. We need a new scale, one that takes efficiency, waste heat, and pollution into account. A new scale that does is based on another concept, called entropy.

RANKING CIVILIZATIONS BY ENTROPY

Ideally, what we want is a civilization that grows in energy and information, but does so wisely, so that its planet does not become unbearably hot or deluged with waste.

This was graphically illustrated in the Disney movie *Wall-E*, where in the distant future we have so polluted and degraded the earth that we simply left the mess behind and lead self-indulgent lives in luxury cruise ships drifting in outer space.

Here's where the laws of thermodynamics become important. The first law of thermodynamics simply says that you can't get something for nothing, i.e., there is no free lunch. In other words, the total amount of matter and energy in the universe is constant. But as we saw in Chapter 3, the second law is the most interesting and, in fact, may eventually determine the fate of an advanced civilization. Simply put, the second law of thermodynamics says that the total amount of entropy (disorder or chaos) always increases. This means that all things must pass; objects must rot, decay, rust, age, or fall apart. (We never see total entropy *decrease*. For example, we never see fried eggs leap from the frying pan and back into the shell. We never see sugar crystals in a cup of coffee suddenly unmix and jump into your spoon. These events are so exceedingly rare that the word "unmix" does not exist in the English—or any other—language.)

So if civilizations of the future blindly produce energy as they rise to a Type II or III civilization, they will create so much waste heat that their home planet will become uninhabitable. Entropy, in the form of waste heat, chaos, and pollution, will essentially destroy their civilization. Similarly, if they produce information by cutting down entire forests and generating mountains of waste paper, the civilization will be buried in its own information waste.

So we have to introduce yet another scale to rank civilizations. We have to introduce two new types of civilizations. The first is an "entropy conserving" civilization, one that uses every means at its disposal to control excess waste and heat. As its energy needs continue to grow exponentially, it realizes that its energy consumption may change the planetary environment, making life impossible. The total disorder or entropy produced by an advanced civilization will continue to soar; that is unavoidable. But local entropy can decrease on their planet if they use nanotechnology and renewable energy to eliminate waste and inefficiency.

The second civilization, an "entropy wasteful" civilization, continues to expand its energy consumption without limit. Eventually, if the home planet becomes uninhabitable, the civilization might try to flee its excesses by expanding to other planets. But the cost of creating colonies in outer space will limit its ability to expand. If its entropy grows faster than its ability to expand to other planets, then it will face disaster.

FROM MASTERS OF NATURE TO CONSERVATORS OF NATURE

As we mentioned earlier, in ancient times we were passive observers of the dance of nature, gazing in wonder at all the mysteries around us. Today, we are like choreographers of nature, able to tweak the forces of nature here and there. And by 2100, we will become masters of nature, able to move objects with our minds, controlling life and death, and reaching for the stars.

But if we become masters of nature, we will also have to become conservators of nature. If we let entropy increase without limit, we will inevitably perish by the laws of thermodynamics. A Type II civilization, by definition, consumes as much energy as a star, and hence the surface temperature of the planet will be scorching hot if entropy is allowed to grow unabated. But there are ways to control entropy growth.

For example, when we visit a museum and see the huge steam engines of the nineteenth century, with their enormous boilers and carloads of black coal, we see how inefficient they were, wasting energy and generating enormous amounts of heat and pollution. If we compare them to a silent, sleek electric train, we see how much more efficiently we use energy today. The need for gigantic coal-burning power plants, belching huge amounts of waste heat and pollution into the air, can be vastly reduced if people's appliances are energy efficient via renewable energy and miniaturization. Nanotechnology gives us the opportunity to reduce waste heat even further as machines are miniaturized to the atomic scale.

Also, if room temperature superconductors are found in this century, it means a complete overhaul of our energy requirements. Waste heat, in the form of friction, will be greatly reduced, increasing the efficiency of our machines. As we mentioned, the majority of our energy consumption, especially transportation, goes into overcoming friction. That is why we put gasoline into our gas tanks, even though it would take almost no energy to move from California to New York if there were no friction. One can imagine that an advanced civilization will be able to perform vastly more tasks with less energy than we use today. This means that we might be able to put numerical limits on the entropy produced by an advanced civilization.

MOST DANGEROUS TRANSITION

The transition between our current Type 0 civilization and a future Type I civilization is perhaps the greatest transition in history. It will determine whether we will continue to thrive and flourish, or perish due to our own folly. This transition is extremely dangerous because we still have all the barbaric savagery that typified our painful rise from the swamp. Peel back the veneer of civilization, and we still see the forces of fundamentalism, sectarianism, racism, intolerance, etc., at work. Human nature has not changed much in the past 100,000 years, except now we have nuclear, chemical, and biological weapons to settle old scores.

However, once we make the transition to a Type I civilization, we will have many centuries to settle our differences. As we saw in earlier chapters, space colonies will continue to be extremely expensive into the future, so it is unlikely that a significant fraction of the world's population will leave to colonize Mars or the asteroid belt. Until radically new rocket designs bring down the cost or until the space elevator is built, space travel will continue to be the province of governments and the wealthy. For the majority of the earth's population, this means that they will remain on the planet as we attain Type I status. This also means that we will have centuries to work out our differences as a Type I civilization.

THE SEARCH FOR WISDOM

We live in exciting times. Science and technology are opening up worlds to us that we could previously only dream about. When looking at the future of science, with all its challenges and dangers, I see genuine hope. We will discover more about nature in the coming decades than in all human history combined—many times over.

But it wasn't always that way.

Consider the words of Benjamin Franklin, America's last great scientist/statesman, when he made a prediction not just about the next century but about the next thousand years. In 1780, he noted with regret that men often acted like wolves toward one another, mainly because of the grinding burden of surviving in a harsh world.

He wrote:

It is impossible to imagine the height to which may be carried, in a thousand years, the power of man over matter. We may perhaps learn to deprive large masses of their gravity, and give them absolute levity, for the sake of easy transport. Agriculture may diminish its labor and double its produce; all diseases may by sure means be prevented or cured, not excepting even that of old age, and our lives lengthened at pleasure even beyond the antediluvian standard.

He was writing at a time when peasants were scratching a bleak existence from the soil, when ox-drawn carts brought rotting produce to the market, when plagues and starvation were a fact of life, and only the lucky few lived beyond the age of forty. (In London in 1750, two-thirds of children died before they reached the age of five.) Franklin lived in a time when it appeared hopeless that one day we might be able to solve these age-old problems. Or, as Thomas Hobbes wrote in 1651, life was "solitary, poor, nasty, brutish, and short."

But today, well short of Franklin's thousand years, his predictions are coming to pass.

This faith—that reason, science, and intellect would one day free us of the oppression of the past—was echoed in the work of the Marquis de Condorcet's 1795 *Sketch for a Historical Picture of the Progress of the Human Mind,* which some claim is the most accurate prediction of future events ever written. He made a wide variety of predictions, all of which were quite heretical, but all of which came true. He predicted that the colonies of the New World would eventually break free from Europe and then advance rapidly by benefiting from the technology of Europe. He predicted the end of slavery everywhere. He predicted that farms would greatly increase the amount and quality of the food they produced per acre. He predicted that science would increase rapidly and benefit mankind. He predicted that we would be free of the grind of daily life and have more leisure time. He predicted that birth control would one day be widespread.

In 1795, it seemed hopeless that these predictions would be fulfilled.

Benjamin Franklin and the Marquis de Condorcet both lived in a time when life was short and brutal and science was still in its infancy. Looking back at these predictions, we can fully appreciate the rapid advances made in science and technology, which created enough bounty and wealth to lift billions out of the savagery of the past. Looking back at the world of Franklin and Condorcet, we can appreciate that, of all the creations of humanity, by far the most important has been the creation of science. Science has taken us from the depths of the swamp and lifted us to the threshold of the stars.

But science does not stand still. As we mentioned earlier, by 2100, we shall have the power of the gods of mythology that we once worshipped and feared. In particular, the computer revolution should give us the ability to manipulate matter with our minds, the biotech revolution should give us the ability to create life almost on demand and extend our life span, and the nanotech revolution may give us the ability to change the form of objects and even create them out of nothing. And all this may eventually lead to the creation of a planetary Type I civilization. So the generation now alive is the most important ever to walk the surface of the earth, for we will determine if we will reach a Type I civilization or fall into the abyss.

But science by itself is morally neutral. Science is like a double-edged sword. One side of the sword can cut against poverty, disease, and ignorance. But the other side of the sword can cut against people. How this mighty sword is wielded depends on the wisdom of its handlers.

As Einstein once said, "Science can only determine what is, but not what shall be; and beyond its realm, value judgments remain indispensable." Science solves some problems, only to create others, but on a higher level.

We saw the raw, destructive side of science during World Wars I and II. The world witnessed in horror how science could bring on ruin and devastation on a scale never seen before, with the introduction of poison gas, the machine gun, firebombings of entire cities, and the atomic bomb. The savagery of the first part of the twentieth century unleashed violence almost beyond comprehension.

But science also allowed humanity to rebuild and rise above the ruin of war, creating even greater peace and prosperity for billions of people. So the true power of science is that it enables us and empowers us—giving us more options. Science magnifies the innovative, creative, and enduring spirit of humanity, as well as our glaring deficiencies.

KEY TO THE FUTURE: WISDOM

The key, therefore, is to find the wisdom necessary to wield this sword of science. As the philosopher Immanuel Kant once said, "Science is organized knowledge. Wisdom is organized life." In my opinion, wisdom is the ability to identify the crucial issues of our time, analyze them from many different points of view and perspectives, and then choose the one that carries out some noble goal and principle.

In our society, wisdom is hard to come by. As Isaac Asimov once said, "The saddest aspect of society right now is that science gathers knowledge faster than society gathers wisdom." Unlike information, it cannot be dispensed via blogs and Internet chatter. Since we are drowning in an ocean of information, the most precious commodity in modern society is wisdom. Without wisdom and insight, we are left to drift aimlessly and without purpose, with an empty, hollow feeling after the novelty of unlimited information wears off.

But where does wisdom come from? In part, wisdom comes from reasoned and informed democratic debate from opposing sides. This debate is often messy, unseemly, and always raucous, but out of the thunder and smoke emerges genuine insight. In our society, this debate emerges in the form of democracy. As Winston Churchill once observed, "Democracy is the worst form of government, except for all the others that have been tried from time to time."

So democracy is not easy. You have to work at it. George Bernard Shaw once said, "Democracy is a device that ensures we shall be governed no better than we deserve."

Today, the Internet, with all its faults and excesses, is emerging as a guardian of democratic freedoms. Issues that were once debated behind closed doors are now being dissected and analyzed on a thousand Web sites.

Dictators live in fear of the Internet, terrified of what happens when their people rise up against them. So today, the nightmare of *1984* is gone, with the Internet changing from an instrument of terror into an instrument of democracy.

Out of the cacophony of debate emerges wisdom. But the surest way to enhance vigorous, democratic debate is through education, for only an educated electorate can make decisions on technologies that will determine the fate of our civilization. Ultimately, the people will decide for themselves how far to take this technology, and in what directions it should develop, but only an informed, educated electorate can make these decisions wisely.

Unfortunately, many are woefully ignorant of the enormous challenges that face us in the future. How can we generate new industries to replace the old ones? How will we prepare young people for the job market of the future? How far should we push genetic engineering in humans? How can we revamp a decaying, dysfunctional educational system to meet the challenges of the future? How can we confront global warming and nuclear proliferation?

The key to a democracy is an educated, informed electorate that can rationally and dispassionately discuss the issues of the day. The purpose of this book is to help start the debate that will determine how this century unfolds.

FUTURE AS A FREIGHT TRAIN

In summary, the future is ours to create. Nothing is written in stone. As Shakespeare wrote in *Julius Caesar,* "The fault, dear Brutus, is not in our stars, but in ourselves. . . ." Or, as Henry Ford once said, perhaps less eloquently, "History is more or less bunk. It's tradition. We don't want tradition. We want to live in the present and the only history that is worth a tinker's damn is the history we make today."

So the future is like a huge freight train barreling down the tracks, headed our way. Behind this train is the sweat and toil of thousands of scientists who are inventing the future in their labs. You can hear the whistle of the train. It says: biotechnology, artificial intelligence, nanotechnology, and telecommunications. However, the reaction of some is to say, "I am too old. I can't learn this stuff. I will just lie down and get run over by the train." However, the reaction of the young, the energetic, and the ambitious is to say, "Get me on that train! This train represents my future. It is my destiny. Get me in the driver's seat."

Let us hope that the people of this century use the sword of science wisely and with compassion.

But perhaps to better understand how we might live in a planetary civilization, it may be instructive to live out a day in the year 2100, to see how these technologies will affect our daily lives as well as our careers and our hopes and dreams.

From Aristotle to Thomas Aquinas, perfection meant wisdom rooted in experience and in the relationships by which the moral life is learned through example. Our perfection lies not in gene enhancement, but in the enhancement of character.

—STEVEN POST

9 A DAY IN THE LIFE IN 2100

JANUARY 1, 2100, 6:15 A.M.

After a night of heavy partying on New Year's Eve, you are sound asleep.

Suddenly, your wall screen lights up. A friendly, familiar face appears on the screen. It's Molly, the software program you bought recently. Molly announces cheerily, "John, wake up. You are needed at the office. In person. It's important."

"Now wait a minute, Molly! You've got to be kidding," you grumble. "It's New Year's Day, and I have a hangover. What could possibly be so important anyway?"

Slowly you drag yourself out of bed and reluctantly head off to the bathroom. While washing your face, hundreds of hidden DNA and protein sensors in the mirror, toilet, and sink silently spring into action, analyzing the molecules you emit in your breath and bodily fluids, checking for the slightest hint of any disease at the molecular level.

Leaving the bathroom, you wrap some wires around your head, which allow you to telepathically control your home: you mentally raise the temperature of the apartment, turn on some soothing music, tell the robotic cook in your kitchen to make breakfast and brew some coffee, and order your magnetic car to leave the garage and be ready to pick you up. As you enter the kitchen, you see the mechanical arms of the robotic cook preparing eggs just the way you like them.

Then you put in your contact lenses and connect to the Internet. Blinking, you see the Internet as it shines onto the retina of your eye. While drinking hot coffee, you start scanning the headlines that flash in your contact lenses.

- The outpost on Mars is requesting more supplies. Winter on Mars is fast approaching. If the settlers are going to complete the next stage in colonization, they need more resources from earth to handle the freezing-cold weather. The plan is to start the first phase of terraforming Mars by raising its surface temperature.
- The first starships are ready for launch. Millions of nanobots, each about the size of a pinhead, will be fired from the moon base, whip around Jupiter using its magnetic field, and head off to a nearby star. It will take years, however, before a handful of these nanobots reach their destination in another star system.
- Yet another extinct animal is going to join the local zoo. This time, it's a rare saber-toothed tiger, brought back via DNA found frozen in the tundra. Because the earth has been heating up, DNA from more and more extinct animals has been recovered and then cloned to fill zoos around the world.
- The space elevator, after years of hauling freight into space, is now allowing a limited number of tourists into outer space. The cost of space travel has already plummeted in recent years by a factor of 50 since the space elevator opened.
- The oldest fusion plants are now almost fifty years old. The time is coming to begin decommissioning some of them and building new ones.
- Scientists are carefully monitoring a new lethal virus that suddenly sprang out of the Amazon. So far, it seems confined to a small

area, but there is no known cure. Teams of scientists are frantically sequencing its genes to learn its weak spots and how to fight it.

Suddenly, one item catches your eye:

- A large leak has unexpectedly been detected in the dikes surrounding Manhattan. Unless the dikes are repaired, the entire city could be submerged, like scores of other cities in the past.

"Uh-oh," you say to yourself. "So that's why the office called and woke me up."

You skip breakfast, dress, and dash out the door. Your car, which drove itself out of the garage, is waiting for you outside. You telepathically order the car to take you to your office as quickly as possible. The magnetic car instantly accesses the Internet, the GPS, and billions of chips hidden in the road that constantly monitor traffic.

Your magnetic car takes off silently, floating on a cushion of magnetism created by the superconducting pavement. Molly's face suddenly appears on the windshield of your car. "John, the latest message from your office says for you to meet everyone in the conference room. Also, you have a video message from your sister."

With the car driving itself, you have time to scan the video mail left by your sister. Her image appears in your wristwatch and says, "John, remember this weekend we have a birthday party for Kevin, who is now six. You promised to buy him the latest robot dog. And, by the way, are you seeing anyone? I was playing bridge on the Internet, and met someone you might like."

"Uh-oh," you say to yourself.

You love cruising in your magnetic car. There are no bumps or potholes to worry about, since it's hovering over the road. Best of all, you rarely need to fuel it up, since there is almost no friction to slow it down. (It's hard to believe, you muse to yourself, that there was an energy crisis in the early part of the century. You shake your head, realizing that most of that energy was wasted in overcoming friction.)

You remember when the superconducting highway first opened. The media lamented that the familiar age of electricity was coming to a close, ushering in the new age of magnetism. Actually, you don't miss the age of electricity one bit. Glancing outside, seeing sleek cars, trucks, and trains whizzing past you in the air, you realize that magnetism is the way to go, and saves money in the process.

Your magnetic car now cruises past the city dump. You see that most of the junk is computer and robot parts. With chips costing almost nothing, even less than water, obsolete ones are piling up in city dumps around the world. There is talk about using chips as landfill.

THE OFFICE

Finally, you reach your office building, the headquarters of a major construction company. As you enter, you hardly notice that a laser is silently checking your iris and identifying your face. No more need for plastic security cards. Your identity is your body.

The conference room is nearly empty, with only a few coworkers sitting around the table. But then, in your contact lens, the 3-D images of the participants begin to rapidly materialize around the table. Those who cannot come to the office are here holographically.

You glance around the room. Your contact lens identifies all the people sitting at the table, displaying their biographies and backgrounds. Quite a few big shots here, you notice. You make a mental note of the important people attending.

The image of your boss suddenly materializes in his chair. "Gentlemen," he announces, "as you've probably heard, the dikes around Manhattan have suddenly begun to leak. It's serious, but we caught it in time, so there is no danger of collapse. Yet, unfortunately, the robots we have sent down to repair the dikes have failed."

Instantly, the lights dim, and you are completely surrounded by the 3-D image of the underwater dike. You are completely immersed in the water, the image of the dike with a huge crack staring you in the face.

As the image rotates, you can see precisely where the leak has occurred. You can see a large, strange gash in the dike that catches your attention. "Robots are not enough," your boss continues. "This is a type of leak that is not part of their programming. We need to send experienced people down there who can size up the situation and improvise. I don't have to remind you that if we fail, New York could suffer the same fate as other great cities, some now underwater."

A shudder goes through the group. Everyone knows the names of the great cities that had to be abandoned as sea levels rose. Although renewable technologies and fusion power displaced fossil fuels many decades ago as the main source of the planet's energy, people are still suffering from the carbon dioxide that was already released into the atmosphere in the first part of the last century.

After much discussion, it is decided to send the human-controlled robot repair crew. This is where you come in. You helped to design these robots. Trained human workers are placed in pods, where electrodes are fitted around their heads. Their brain signals allow them to make telepathic contact with robots. From their pods, the workers can see and feel everything that the robots see and feel. It's just like being there in person, except in a new superhuman body.

You are justifiably proud of your work. These telepathically controlled robots have proven their worth many times over. The moon base is largely controlled by human workers, who lie comfortably and safely in their pods on earth. But since it takes about a second for a radio signal to reach the moon, it also means that these workers have to be trained to adjust for this time delay.

(You would have loved to put your robots on the Mars base, too. But since it takes up to twenty minutes for a signal to reach Mars and twenty minutes to come back, communicating with robots on Mars would be too difficult, it was decided. Alas, for all our progress, there is one thing you cannot adjust: the speed of light.)

But something is still bothering you at the meeting.

Finally, you summon the nerve to interrupt your boss. "Sir, I hate to say this, but looking at the leak in the dike, the crack looks suspiciously like a mark left by one of our own robots."

A loud murmur immediately fills the room. You can hear the rising chorus of objections: "Our own robot? Impossible. Preposterous. It's never happened before," people protest.

Then your boss quiets the room and responds solemnly. "I was afraid someone would raise this issue, so let me say that this is a matter of great importance, which has to be kept strictly confidential. This information must not leave this room, until we issue our own press release. Yes, the leak was caused by one of own robots that suddenly went out of control."

Pandemonium breaks out in the meeting. People are shaking their heads. How can this be?

"Our robots have had a perfect record," your boss insists. "Absolutely spotless. Not a single robot has caused any harm, ever. Their fail-safe mechanisms have proven effective again and again. We stand by that record. But as you know, our latest generation of advanced robots use quantum computers, which are the most powerful available, even approaching human intelligence. Yes, human intelligence. And in the quantum theory, there is always a small but definite probability that something wrong will happen. In this case, go berserk."

You slump back into your chair, overwhelmed by the news.

HOME AGAIN

It has been a very long day, first organizing the robot repair crew to fix the leak, and then helping to deactivate all experimental robots that use quantum computers, at least until this issue is finally resolved. You finally arrive back home again. You are exhausted. Just as you sink comfortably into your sofa, Molly appears on the wall screen. "John, you have an important message from Dr. Brown."

Dr. Brown? What does your robot doctor have to say?

"Put him on screen," you say to Molly. Your doctor appears on the wall screen. "Dr. Brown" is so realistic that you sometimes forget that he is just a software program.

"Sorry to bother you, John, but there is something I have to bring to your attention. Remember your skiing accident last year, the one that almost killed you?"

How could you forget? You still cringe when you remember how you plowed into a tree while skiing in what is left of the Alps. Since most of the Alpine snow has already melted, you had to choose an unfamiliar resort at a very high altitude. Unaccustomed to the terrain, you accidentally tumbled down the slope and slammed into a bunch of trees at forty miles per hour. Ouch!

Dr. Brown continues, "My records show that you were knocked unconscious, suffering a concussion and massive internal injuries, but your clothes saved your life."

Although you were unconscious, your clothes automatically called for an ambulance, uploaded your medical history, and located your precise coordinates. Then at the hospital robots performed microsurgery to stop the bleeding, sew up tiny ruptured blood vessels, and patch up other damage.

"Your stomach, liver, and intestines were damaged beyond repair," Dr. Brown reminds you. "Luckily, we could grow a new set of organs for you just in time."

Suddenly, you feel a little bit like a robot yourself, with so much of your body made from organs grown in a tissue factory.

"You know, John, my records also show that you could have replaced your shattered arm with a fully mechanical one. The latest robot arm would have increased the strength in your arm by a factor of five. But you declined."

"Yes," you reply, "I guess I'm still an old-fashioned guy. I'll take flesh over steel any day," you say.

"John, we have to do a periodic checkup on your new organs. Pick up your MRI scanner and slowly pass it over your stomach area."

You go to the bathroom and pick up a small device, about the size of cell phone, and slowly pass it over your organs. Immediately on the wall screen, you can see the 3-D image of your internal organs lighting up.

"John, we are going to analyze these images to see how your body is healing. By the way, this morning the DNA sensors in your bathroom detected cancer growing in your pancreas."

"Cancer?" You suddenly straighten up. You are puzzled. "But I thought cancer was cured years ago. No one even talks about it much anymore. How can I have cancer?"

"Actually, scientists never cured cancer. Let's just say that we are in a truce with cancer, a stalemate. There are too many kinds of cancer. Like the common cold. We never cured that, either. We simply keep it at bay. I've ordered some nanoparticles to zap those cancer cells. There are only a few hundred of them. Just routine. But without this intervention, you would probably die in about seven years," he deadpans.

"Oh, that's a relief," you say to yourself.

"Yes, today we can spot cancer years before a tumor forms," says Dr. Brown.

"Tumor? What's that?"

"Oh, that's an old-fashioned word for a type of advanced cancer. It's pretty much disappeared from the langauge. We never see them anymore," adds Dr. Brown.

Then you realize that in all this excitement, you forgot that your sister threatened to set you up with someone. You call up Molly again.

"Molly, I am not doing anything this weekend, so can you find a date for me? You know the kind of person I like."

"Yes, your preferences are programmed in my memory. Wait a minute while I scan the Internet." After a minute, Molly displays the profiles of promising candidates who are also sitting in front of their wall screens, asking the same question.

After scanning the candidates, you finally select one who appeals to you. This person, called Karen, somehow looks special, you think to yourself. "Molly, send Karen a polite message, asking her if she is available this weekend. There's a new restaurant that just opened that I want to try."

Molly then sends Karen your profile in a video mail.

That night, you relax by having some of your coworkers come over for beer and to watch some football. Your friends could have watched the game by appearing in your living room via holographic images, but somehow, cheering for the home team is more enjoyable with your friends joining in the excitement in person. You smile, imagining that this is probably how it was thousands of years ago, when cavemen had to bond with one another.

Suddenly, the entire living room is illuminated, and it appears as if you are right on the football field, at the 50-yard line. As the quarterback makes a forward pass, you stand right next to him. The game is being played all around you.

During halftime, you and your friends begin sizing up the players. Over beer and popcorn, you hotly debate who trains the most, practices the hardest, has the best coaches, and has the best gene therapist. Your home team, you all agree, has the best geneticist in the league, with the best genes that money can buy.

After your friends have left, you still are too excited to go to sleep. So you decide to play a quick game of poker before turning in.

"Molly," you ask, "it's late, but I want to set up a game of poker. I'm feeling lucky. Someone must be awake in England, China, India, or Russia who might want to play a few hands right now."

"No problem," says Molly. A number of promising faces appear on the screen. As the 3-D images of each player materialize in your living room, you relish the idea of seeing who can bluff the best. It's funny, you say to yourself, that you are more familiar with people in distant countries, thousands of miles away, than with your next-door neighbors. National boundaries don't mean much these days.

Finally, just before you finally turn in, Molly interrupts you again, appearing in the bathroom mirror.

"John, Karen accepted your invitation. Everything is set for this weekend. I will make a reservation at that new restaurant. Do you want to see the profile that she wrote about herself? Do you want me to scan the Internet to verify the accuracy of her profile? People have been known to . . . ah . . . lie about their profiles."

"No," you say. "Let's keep it a surprise for the weekend." After that poker game, you feel lucky again.

THE WEEKEND

It's the weekend now, and time to go shopping and buy a present for Kevin. "Molly, put the mall on the screen."

The mall suddenly appears on the wall screen. You wave your arms and fingers, and the image on the wall screen traces a path through the mall. You take a virtual tour until you arrive at the image of the toy store. Yes, they have exactly the toy robot pets you want. You telepathically order the car to take you to the mall. (You could have ordered the toy online. Or you could have had the blueprints e-mailed to you, and then had your fabricator materialize the toy at home from scratch using programmable matter. But it's always good to get out of the apartment and shop once in a while.)

Cruising in your magnetic car, you look outside and see people taking a walk. It's such a nice day. You also see robots of all sorts. Robots to walk the dog. Robot clerks, cooks, receptionists, and pets. It seems that every task that is dangerous, repetitive, or requires only the simplest human interaction is being duplicated by robots. In fact, robots are now big business. All around you, you see ads for anyone who can repair, service, upgrade, or build robots. Anyone in the field of robotics has a bright future. The robot business is bigger than the automobile industry of the last century. And most of the robots, you realize, are hidden from view, silently repairing the city's infrastructure and maintaining essential services.

When you reach the toy store, a robot clerk greets you at the entrance. "Can I help you," it says.

"Yes, I want to buy a robot dog."

You look over the latest robot dogs. Amazing what these pet robots can do, you say to yourself. They can play, run, fetch, do anything a dog can do. Everything but pee on the carpet. Maybe that's why parents buy them for their kids, you muse.

Then you say to the robot clerk, "I'm buying a robot pet for my six-year-old nephew. He's a very intelligent, hands-on type of kid. But he's also sometimes shy and quiet. What kind of dog might help bring him out of his shell?"

The robot replies, "I am sorry, sir. That is outside my programming. Perhaps I can interest you in a space toy?"

You forgot that robots, no matter how versatile, have a long way to go before they understand human behavior.

Then you go to the men's department store. Time to replace that ratty old outfit of yours if you want to impress your date. You put on some designer suits. They all look stylish, but they are all the wrong size. You are disappointed. But then you take out your credit card, which contains all your precise 3-D measurements. Your data is fed into the computer, and then a new suit is being cut at a factory and will soon be delivered to your door. A perfect fit every time.

Last, you go to the supermarket. You scan all the chips hidden in each plastic item in the market, and then in your contact lenses you compare prices to see which store in the city has the cheapest, best products. No more guesswork about who has the lowest prices.

THE DATE

You have been looking forward to this date all week. Preparing to meet Karen, you are surprised that you feel like a schoolboy again. You decide that if you are going to invite her to your apartment after dinner, you will have to do some serious remodeling of your worn-out furniture. Fortunately, most of the kitchen counters and living room furniture is made of programmable matter.

"Molly," you say, "can you show me the catalog of new kitchen counters and furniture offered by the manufacturer? I want to reprogram the furniture. It looks so old."

Soon, the pictures of the latest furniture designs flash on the screen.

"Molly, please download the blueprints for this kitchen counter, that sofa, and this table, and then install them, please."

While you are getting ready for your date, Molly downloads and installs the blueprints. Instantly, the kitchen countertop, living room sofa, and table begin to dissolve, turn into something that looks like putty, and then gradually re-form into the new shapes. In an hour, your apartment looks brand-new. (Recently, you were scanning the real estate section on the Internet, and noticed that houses made of programmable matter were becoming quite fashionable. In fact, at your engineering company, there are ambitious plans to create an entire city in the desert totally out of programmable matter. Push a button, and—poof!—an instant city.)

Your apartment still looks a bit drab, you decide. You wave your hand, and the pattern and color of the wallpaper change immediately. Having intelligent wallpaper certainly beats having to repaint the walls, you say to yourself.

You grab some flowers along the way, and finally pick up your date. You are pleasantly surprised. You hit it off right away. Something is clicking.

Over dinner, you find out that Karen is an artist. Normally, she jokes, she might be penniless, starving, and selling paintings along the sidewalk for a pittance. Instead, she is a very successful Web designer. In fact, she has her own company. Everyone, it seems, wants the very latest designs for the Web. There is such a huge demand for creative art.

She draws some circles in the air with her fingers, and some of her animations appear in thin air. "Here are some of my latest creations," she says proudly.

You comment, "You know, as an engineer, I work with robots all day long. Some are pretty advanced, but they can also act pretty stupid at times. What about your field? Are robots making inroads?"

"Absolutely not," she protests. Karen says she works exclusively with creative people, where the most prized commodity is imagination, something the most advanced robots lack.

"I may be old-fashioned, but in my field, we use robots only to make copies or do clerical work," she says proudly. "I would like to see the day when robots can do something really original, like tell a joke, write a novel, or compose a symphony."

That hasn't happened yet, but it might, you think to yourself.

While she is talking, a question crosses your mind. How old is she? Since the aging process was medically slowed years ago, people can be any age. Her Web site did not say how old she was. But she does not look a day over twenty-five.

After taking her home, you begin to idly daydream. What would it be like to live with a person like her? To spend the rest of your life with her? But there is something that is bothering you. It's been nagging at you all day.

You face the wall screen and say, "Molly, please call Dr. Brown for me." You are suddenly grateful that robotic doctors do house calls any time of the day. And they never complain or bellyache. It's not part of their programming.

Dr. Brown's image instantly appears on the wall screen. "Is there something bothering you, son?" he asks in a fatherly way.

"Doc, I have to ask you a question that's been troubling me lately."

"Yes, what is that?" asks Dr. Brown.

"Doc," you say, "how long do you think I'll live?"

"You mean what is your life expectancy? Well, we don't really know. Your records say you are seventy-two years old, but biologically your organs are more like thirty years old. You were part of the first generation to be genetically reprogrammed to live longer. You chose to stop aging at around thirty. Not enough of your generation has died yet, so we have no data to work with. So we have no way of knowing how long you will live."

"Then do you think I will live forever?" you ask.

"And be immortal?" Dr. Brown frowns. "No, I don't think so. There is a big difference between someone who lives forever and someone who has a life span so long that it hasn't been measured yet."

"But if I don't age," you protest, "then how am I supposed to know when to get . . ." You stop yourself in midsentence. "Ah, okay . . . you see, I just met someone, ah, special, and, assuming I want to plan a life with her, how do I adjust the stages in my life to hers? If my generation hasn't lived long enough to die," you continue, "then how am I supposed to know when to get married, have kids, and plan for retirement? You know, how do I set the milestones in my life?"

"I don't know the answer to that. You see, the human race is now a guinea pig of some sort," says Dr. Brown. "I'm sorry, John. You are in uncharted waters here."

NEXT FEW MONTHS

The next few months are a wonderful surprise for you and Karen. You take her to the virtual reality parlor, and have great fun living out silly, imaginary lives. Like being a kid again. You enter a vacant chamber. The software of a virtual world is beamed into your contact lenses, and the scenery instantly changes. In one program, you are fleeing dinosaurs, but everywhere you run, another dinosaur pops out of the bushes. In another program, you are battling space aliens or pirates trying to board your ship. In another, you decide to change species and morph into two eagles that are soaring in the air. And in another program, you are basking on a romantic South Sea island, or dancing in the moonlight with music gently floating in the air.

After a while, you and Karen want to try something new. Instead of living out imaginary lives, you decide to lead real ones. So, when you both have vacation time together, you decide to take a whirlwind tour through Europe.

You say to the wall, "Molly, Karen and I want to plan a European vacation. A real one. Please check on flights, hotels, and any specials. Then list possible shows or events that may interest us. You know our tastes." In a few minutes, Molly has prepared a detailed itinerary.

Later, when walking through the ruins of the Roman Forum, you can see the Roman Empire resurrected in your contact lenses. Passing by the scattered columns, stones, and debris, you gaze on the might that was once Imperial Rome at the height of its glory.

And shopping is a delight, even when bargaining in the local shops in Italian. You can clearly see the translations appearing beneath the person you are talking to. And no more guidebooks and clumsy maps. Everything is in your contact lens.

At night, gazing at the night sky over Rome, you can clearly see the stars arranged into constellations in your contact lens. Glancing across the sky, you can see magnified images of the rings of Saturn, soaring comets, beautiful gas clouds, and exploding stars.

One day, Karen finally reveals a secret, her true age. It's sixty-one. Somehow, that doesn't seem very important anymore.

"So, Karen, do you feel happier now that we live so long?"

"Yes, yes!" she replies immediately. "You know, my grandmother lived in a time when women got married, had a family, and maybe squeezed in a career. But I like to feel that I've been reincarnated three times, with three careers, and never looked back. First, I was a tour guide in several countries, traveling the world. It was a wonderful life. Tourism is such a huge industry, with plenty of jobs. But later, I wanted to do something more relevant. So I became a lawyer, defending cases and people I cared about. And then, I decided to indulge my artistic side and started my Web design company. And you know something? I'm proud to say I never used a robot. No robot can be a personal tour guide, win a case in court, or produce beautiful artwork."

Time will tell, you think to yourself.

"And are you planning a fourth career?" you ask.

"Well, maybe if something better comes along." She smiles at you.

"Karen," you finally say, "if we stop aging, then how do you know when the best time is for, you know, well, getting married, having kids, and raising a family? The biological clock went out the window decades ago. So I was thinking, maybe it's time to settle down and have a family."

"You mean have children?" Karen says, a bit surprised. "That's something that I haven't considered seriously. Well, until now, that is. It all depends on if the right man comes along," she says, as she smiles mischievously at you.

Later, you and Karen discuss marriage, and what name you might choose for a kid, and also what genes you want the kid to have.

You go to the wall screen and say, "Molly, can you give me the list of the latest genes that have been approved by the government?" As you scan the list, you see the various genes for hair color, eye color, height, build, and even some personality traits that are now being offered. The list seems to grow every year. You also see the long list of hereditary diseases that can be cured. Since cystic fibrosis has run in your family for centuries, it is a relief not to have to worry about that anymore.

Scanning the list of approved genes, you feel that you are not just a future parent, but some sort of god, creating a child to order in your image.

Then Molly says, "There is a program that can analyze a baby's DNA and then give a reasonable approximation of its future face, body shape, and personality. Do you want to download this program and see what your child might look like in the future?"

"No," you say. "Some things should be left as a mystery."

A YEAR LATER

Karen is now pregnant, but her doctors reassure her that there is no danger in taking a ride on the space elevator, which is now open to tourists.

"You know," you admit to Karen, "as a kid, I always wanted to go into outer space. You know, do the astronaut thing. But one day I thought about sitting on top of millions of gallons of volatile rocket fuel that could explode with a single spark. Then my enthusiasm for space travel began to cool a bit. But the space elevator is different. Clean, safe, with no mess. That's the way to go."

As you and Karen get into the elevator, you see the operator push what looks like the Up button. You half expect to see the lingerie department. Instead, you feel yourself soaring into outer space. You feel the slow acceleration as you rapidly rise into the air. The gauge on the elevator reads, "10 miles, 20 miles, 30 miles . . ."

Outside, you see the scenery changing, second by second. One moment, you are staring at fluffy clouds passing by as you soar into the atmosphere. Then the sky changes from blue to purple to a deep black, and finally you see the stars surrounding you in all their splendor. You begin to make out the constellations as you have never seen them before, blazing away in the distance. The stars are not twinkling, as they appear from the earth, but staring brightly, as they have for billions of years.

The elevator slowly comes to a stop about 100 miles from the surface of the earth. From space, you see a dazzling sight that you previously saw only in pictures.

Looking down, you suddenly see the earth in an entirely new light. You see the oceans, the continents, and the lights of megacities that shine into outer space.

From space, the earth appears so serene that it's hard to believe people once spilled blood fighting wars over silly borders. These nations still exist, but they seem so quaint, less relevant today, in an age when communication is instantaneous and ubiquitous.

As Karen puts her head on your shoulder, you begin to realize that you are witnessing the birth of a new planetary civilization. And your kid will be among the first citizens of this new civilization.

And then you take out an old, worn book from your back pocket, and read to her the words of someone who died more than 100 years ago. It reminds you of the challenges facing humanity before it attains a planetary civilization.

Mahatma Gandhi once wrote:

The Roots of Violence:

Wealth without work,
Pleasure without conscience,
Knowledge without character,
Commerce without morality,
Science without humanity,
Worship without sacrifice,
Politics without principles.

NOTES

(The authors' names refer to the books listed in Recommended Reading.)

INTRODUCTION

6   *"In his newspapers of January 1, 1900"*: Rhodes, pp. 29–30.
7   *"It will be as common for the citizen"*: www.learner.org/workshops/primarysources/corporations/docs/.
7   *"Everything that can be invented"*: quoted in Canton, p. 247.
7   *"Who the hell wants to hear actors talk?"*: quoted in Canton, p. 247.
7   *"I think there is a world market"*: quoted in Canton, p. 247.
7   *"It is now definitely established"*: Cornish, p. 149. See also: "The Facts that Got Away," *New York Times*, November 14, 2001.

1. FUTURE OF THE COMPUTER: MIND OVER MATTER

21   *"Where a calculator like the ENIAC"*: *Popular Mechanics*, quoted in Kurzweil, p. 56. See also: Andrew Hamilton, "Brains That Click," *Popular Mechanics*, March 1940, p. 258.
23   *"Technology [is] the knack"*: Rhodes, p. 206.
26   *"Those components will eventually include"*: Babak A. Parvie, "Augmented Reality in a Contact Lens," *IEEE Spectrum*, September 2009, www.spectrum.ieee.org/biomedical/bionics/augmented-reality-in-a-contact-lens/0.
53   *"There's some physiological evidence"*: Gary Stix, "Jacking into the Brain—Is the Brain the Ultimate Computer Interface?" *Scientific American*, November 2008, pp. 56–61.
54   *"It's like being an astronomer"*: Jeff Wise, "Thought Police: How Brain Scans Could Invade Your Private Life," *Popular Mechanics*, October 15, 2007, www.popularmechanics.com/science/health/neuroscience/4226614.
55   *"possible to identify, from a large set of completely novel natural images"*: *New Scientist*, October 15, 2008, issue 2678.
58   *"Can we tap into the thoughts of others"*: David Baltimore, "How Biology Became Information Science," in Denning, pp. 53–54.
58   *"I am told"*: Ibid., p. 54.
60   *"Perhaps something like the* Star Trek *tricorder"*: Bernhard Blümich, "The Incredible Shrinking Scanner: MRI-like Machine Becomes Portable," *Scientific American*, November 2008, p. 68.

2. FUTURE OF AI: RISE OF THE MACHINES

66   *"Scientists Worry Machines May Outsmart Man"*: John Markoff, *New York Times*, July 25, 2009, p. A1, www.nytimes.com/2009/07/26/science/26robot.html?scp= 1&sq=Scientists Worry Machines May Outsmart Man&st=cse.

66   *"Technologists are providing"*: Ibid.

68   *"just at the stage where they're robust"*: Kaku, p. 75.

69   *"Machines will be capable, within twenty years"*: Crevier, p. 109.

69   *"It's as though a group of people"*: Paul W. Abrahams, "A World Without Work," in Denning and Metcalfe, p. 136.

70   *"Today, you can buy chess programs for $49"*: Richard Strozzi Heckler, "Somatics in Cyberspace," in Denning, p. 281.

73   *"To this day, AI programs"*: Sheffield et al., p. 30.

74   *"100 million things, about the number a typical person knows"*: Kurzweil, p. 267.

77   *In 2006, it was estimated that there were 950,000 industrial robots*: World Robotics 2007, IFR Statistical Department (Frankfurt: International Federation of Robotics, 2007).

87   *"Discovering how the brain works"*: Fred Hapgood, "Reverse Engineering the Brain," *Technology Review*, July 11, 2006, www.technologyreview.com/read_article .aspx?id=17111.

88   *He was in a semiconscious state for several weeks:* John M. Harlow, M.D., "Passage of an Iron Rod Through the Head," *Journal of Neuropsychiatry and Clinical Neurosciences* 11, May 1999, pp. 281–83, www.neuro.psychiatryonline.org/cgi/ content/full/11/2/281.

91   *"It is not impossible to build a human brain"*: Jonathan Fildes, "Artificial Brain '10 Years Away,' " BBC News, July 22, 2009, http://news.bbc.co.uk/2/hi/8164060.stm.

92   *"It's not a question of years"*: Jason Palmer, "Simulated Brain Closer to Thought," BBC News, April 22, 2009, http://news.bbc.co.uk/2/hi/sci/tech/8012496.stm.

92   *"This is a Hubble Telescope of the mind . . . it's inevitable"*: Douglas Fox, "IBM Reveals the Biggest Artificial Brain of All Time," *Popular Mechanics*, December 18, 2009, www.popularmechanics.com/technology/engineering/extreme-machines/ 4337190.

94   *"After we solve this"*: Sally Adee, "Reverse Engineering the Brain," *IEEE Spectrum*, June 2008, http://spectrum.ieee.org/biomedical/ethics/reverse-engineering-the -brain/0.

96   *"Within thirty years"*: Vernor Vinge, "What Is the Singularity?" paper presented at the VISION-21 Symposium sponsored by NASA Lewis Research Center and the Ohio Aerospace Institute, March 30–31, 1993. A slightly changed version appeared in *Whole Earth Review*, Winter 1993, http://mindstalk.net/vinge/vinge-sing.html.

96   *"I'd be very surprised if anything remotely like this happened"*: Tom Abate, "Smarter Than Thou? Stanford Conference Ponders a Brave New World with Machines More Powerful Than Their Creators," *San Francisco Chronicle*, May 12, 2006, http://articles.sfgate.com/2006–05–12/business/17293318_1_ray-kurzweil-machines -artificial-intelligence.

96   *"If you could blow the brain up"*: Kurzweil, p. 376.

96   *Philosopher David Chalmers has even catalogued:* http://consc.net/mindpapers.com.

101   *"life may seem pointless if we are fated"*: Sheffield, p. 38.

101   *"One conversation centered"*: Kurzweil, p. 10.

103   *"It's not going to be an invasion"*: Abate, *San Francisco Chronicle*, May 12, 2006.

103 *"intelligent design for the IQ 140 people":* Brian O'Keefe, "The Smartest (or the Nuttiest) Futurist on Earth," *Fortune,* May 2, 2007, http://money.cnn.com/magazines/fortune/fortune_archive/2007/05/14/100008848/.

103 *"It's as if you took a lot of good food":* Greg Ross, "An Interview with Douglas R. Hofstadter," *American Scientist,* January 2007, www.americanscientist.org/bookshelf/pub/douglas-r-hofstadter.

105 *"will evolve into socially intelligent beings":* P. W. Singer, "Gaming the Robot Revolution," *Slate,* May 21, 2009, www.slate.com/id/2218834/.

107 *"When I was a kid":* Rodney A. Brooks, "Making Living Systems," in John Brockman, ed., *Science at the Edge: Conversations with the Leading Scientific Thinkers of Today* (New York: Sterling, 2008), p. 250.

108 *"My prediction is that by the year 2100":* Rodney A. Brooks, "Flesh and Machines," in Denning, p. 63.

109 *"At Little League games":* Pam Belluck, "Burst of Technology Helps Blind to See," *New York Times,* September 27, 2009, p. A1, www.nytimes.com/2009/09/27/health/research/27eye.html?_r=1&scp=1&sq="burst of technology"&st=cse.

110 *"It's great. I have a feeling":* BBC-TV, October 18, 2009.

110 *"Over the next ten to twenty years . . . wireless Internet":* Rodney A. Brooks, "The Merger of Flesh and Machines," in John Brockman, ed., *The Next Fifty Years* (New York: Vintage, 2002), p. 189.

111 *"Fifty years from now . . . Darwinian evolution":* Ibid., pp. 191–92.

115 *"When I try to think of what I might gain":* Stock, p. 23.

3. FUTURE OF MEDICINE: PERFECTION AND BEYOND

122 *"Biology is today an information science":* David Baltimore, "How Biology Became an Information Science," in Denning, p. 43.

123 *"You have to have a strong stomach":* Nicholas Wade, "Cost of Decoding a Genome Is Lowered," *New York Times,* August 10, 2009, p. D3, www.nytimes.com/2009/08/11/science/11gene.html.

128 *"Embryonic stem cells represent":* Jeanne Lenzer, "Have We Entered the Stem Cell Era?" *Discover,* November 2009, p. 33, http://discovermagazine.com/2009/nov/14-have-we-entered-the-stem-cell-era/article_view?b_start:int=1&-C=.

129 *"It's gorgeous":* Ibid.

133 *By 2001, there were more than 500:* Stock, p. 5.

133 *But there have been setbacks:* Ibid., p. 36.

135 *"What we are seeing today":* Kate Kelland, "Gene Maps to Transform Scientists' Work on Cancer," Reuters, December 18, 2009.

136 *"Cancer is an army of cells":* David Baltimore, "How Biology Became an Information Science," in Denning, p. 54.

138 *"Homo sapiens, the first truly free species":* Kurzweil, p. 195.

140 *"Although many genes are likely to be involved in the evolution":* Stock, p. 108.

141 *"It's as if they remember":* Jonah Lehrer, "Small, Furry . . . and Smart?" *Nature* 461 (October 2009): 864.

141 *"The obstacles to his understanding":* Ibid.

141 *In fact, scientists believe that there has to be a balance:* Jonah Lehrer, "Smart Mice," *The Frontal Cortex,* October 15, 2009, http://scienceblogs.com/cortex/2009/10/smart_mice.php.

142 *"We all know that good-looking people do well":* Sheffield et al., p. 107.

145   *"There is nothing in biology yet found that indicates the inevitability of death"*: Kurzweil, p. 320.

146   *"If something like age-1 exists in humans"*: Kaku, p. 211.

147   *Finally, in 2009, the long-awaited results came in:* Nicholas Wade, "Tests Begin on Drugs That May Slow Aging," *New York Times*, August 17, 2009, p. D4, www .nytimes.com/2009/08/18/science/18aging.html?ref=caloric_restriction.

148   *Scientists have found that sirtuin activators:* Nicholas Wade, "Quest for a Long Life Gains Scientific Respect," *New York Times*, September 29, 2009, p. D4, www .nytimes.com/2009/09/29/science/29aging.html?ref=caloric_restriction.

149   *His colleague Sinclair, in fact, admits that he:* Nicholas Wade, "Scientists Find Clues to Aging in a Red Wine Ingredient's Role in Activating a Protein," *New York Times*, November 26, 2008, p. A30, www.nytimes.com/2008/11/27/health/27aging.html?scp =6&sq=sinclair%20resveratrol&st=cse.

150   *"In five or six or seven years"*: Wade, "Quest for a Long Life," *New York Times*, September 28, 2009, p. D4, www.nytimes.com/2009/09/29/science/29aging.html ?ref=caloric_restriction.

152   *"Such interventions may become commonplace"*: Kurzweil, p. 253.

153   *"Gradually, our agonizing"*: Stock, p. 88.

153   *In 2002, with the best demographic data:* Ciara Curtin, "Fact or Fiction?: Living People Outnumber the Dead," *Scientific American*, March 2007.

154   *Every year, 79 million:* Brown, p. 5.

157   *"I believe that by 2050"*: Richard Dawkins, *A Devil's Chaplain: Reflections on Hope, Lies, Science, and Love* (New York: Houghton Mifflin Mariner, 2004), p. 113.

158   *Even more interesting is the HAR1 region of the genome:* Katherine S. Pollard, "What Makes Us Human?" *Scientific American*, May 2009, p. 44.

160   *This cell would then be reprogrammed to revert:* Nicholas Wade, "Scientists in Germany Draft Neanderthal Genome," *New York Times*, February 12, 2009, p. A12, www.nytimes.com/2009/02/13/science/13neanderthal.html?scp=3&sq=neanderthal &st=cse.

160   *"Are you going to put them in Harvard"*: Ibid.

160   *"will doubtless raise"*: Dawkins, p. 114.

160   *"A year ago, I would have said"*: Kate Wong, "Scientists Sequence Half the Woolly Mammoth's Genome," *Scientific American*, January 2009, p. 26, www .scientificamerican.com/article.cfm?id=woolly-mammoth-genome-sequenced.

167   *"Traditional Darwinian evolution now produces"*: Stock, p. 183.

4. NANOTECHNOLOGY: EVERYTHING FROM NOTHING?

173   *"The grandest dream of nanotechnology"*: Carl T. Hall, "Brave New Nano-World Lies Ahead, "*San Francisco Chronicle*, July 19, 1999, http://articles.sfgate.com/ 1999–07–19/news/17694442_1_atom-molecules-nanotech.

173   *"Eventually, the goal is not just to make computers"*: Ibid.

173   *"Nanotechnology has the potential"*: quoted in Kurzweil, p. 226.

181   *The key to these nanoparticles is their size:* James R. Heath, Mark E. Davis, and Leroy Hood, "Nanomedicine—Revolutionizing the Fight Against Cancer," *Scientific American*, February 2009, p. 44.

182   *"Because the self-assembly doesn't require"*: Emily Singer, "Stealthy Nanoparticles Attack Cancer Cells," *Technology Review*, November 4, 2009, www .technologyreview.com/business/23855/.

183   *"It's basically like putting"*: "Special Gold Nanoparticles Show Promise for 'Cooking' Cancer Cells," www.eurekalert.org/pub_releases/2009–03/acs -sgn030909.php.

184   *Yet another way to steer a molecular machine*: Thomas E. Mallouk and Ayusman Sen, "How to Build Nanotech Motors," *Scientific American,* May 2009, p. 72.

186   *"Today, it takes a room filled with computers"*: Katherine Harmon, "Could a Microchip Help to Diagnose Cancer in Minutes," *Scientific American* blog post, September 28, 2009, http://www.scientificamerican.com/blog/post.cfm?id= could-a-microchip-help-to-diagnose-2009–09–28.

189   *The question—When will Moore's law collapse?—sends shudders*: Electronic News, September 18, 2007, www.edn.com/article/CA647968.

189   *"We see that for at least the next fifteen to twenty"*: Electronic News, July 13, 2004. See also Kurzweil, p. 112, and www.nanotech-now.com/news.cgi?story_id=04803.

192   *"From the point of view of physics"*: Alexis Madrigal, "Scientist Builds World's Smallest Transistor, Gordon Moore Sighs with Relief," *Wired,* www.wired.com/ wiredscience/2008/04/scientists-buil/.

192   *"It's about the smallest"*: Ibid.

194   *"By 2050, we will surely have found ways to achieve"*: Vint Cerf, "One Is Glad to Be of Service," in Denning, p. 229.

197   *"Think of a mobile device"*: Sharon Gaudin, "Intel Sees Future with Shape-shifting Robots, Wireless Power," *Computerworld,* August 22, 2008, www.computerworld .com/s/article/9113301/Intel_sees_future_with_shape_shifting_robots_wireless _power?taxonomyId=12&pageNumber=2.

197   *"Sometime over the next forty years"*: Ibid.

200   *"Why not?"*: Ibid.

201   *"Much like you can't make a boy and a girl fall in love"*: Rudy Baum, "Nanotechnology: Drexler and Smalley Make the Case for and Against 'Molecular Assemblers,'" *Chemical & Engineering News* 81, December 1, 2003, pp. 37–42, http://pubs.acs.org/cen/coverstory/8148/8148counterpoint.html.

205   *"If a self-assembler ever does become possible"*: BBC/Discovery Channel, *Visions of the Future,* Part II, 2007.

208   *"Nanotechnology will thrive, much as photolithography thrives"*: Rodney A. Brooks, "Flesh and Machines," in Denning, p. 63.

5. FUTURE OF ENERGY: ENERGY FROM THE STARS

211   *the world consumes about 14 trillion watts of power*: Kurzweil, p. 242.

211   *U.S. oil reserves were being depleted so rapidly*: www.mkinghubbert.com/speech/ prediction.

212   *"Food and pollution are not"*: Sheffield, p. 179.

214   *China will soon surpass the United States in wind power*: www.gwec.net/index.php ?id=125.

217   *"All the geniuses here at General Motors"*: Tad Friend, "Plugged In," *The New Yorker,* August 24, 2009, pp. 50–59.

218   *"You put your hand over the exhaust pipe"*: "GM Convinced the Future Is in Fuel Cells," CBS News, September 11, 2009, www.cbsnews.com/stories/2009/09/11/tech/ main5302610.shtml?tag=mncol;lst;6.

224   *The plant will occupy 200 acres*: Business Wire, www.businesswire.com/portal/ge/ index. See also www.swampfox.ws/node/26502.

226   *Greenland's ice shelves shrank by twenty-four square miles:* Brown, p. 63.

226   *Large chunks of Antarctica's ice, which have been stable:* Brown, p. 64.

226   *According to scientists at the University of Colorado:* Brown, p. 65

227   *In 1900, the world consumed 150 million :* Brown, pp. 56–57.

230   *"Envision Pakistan, India, and China":* Peter Schwartz and Doug Randall, "An Abrupt Climate Change Scenario and Its Implications for United States National Security," Global Business Network, October 2003, p. 18. PDF available at www .gbn.com/search.php?topnavSearch=envision+pakistan%2C+india&x=0&y=0.

232   *countries bound by the London Convention:* Cornelia Dean, "Experts Ponder the Hazards of Using Technology to Save the Planet," *New York Times,* August 12, 2008, p. F4, www.nytimes.com/2008/08/12/health/12iht-ethics.3.15212327.html?_r=1&scp =10&sq=planktos&st=cse.

233   *The liquefied gas will be injected:* Matthew L. Wald, "Refitted to Bury Emissions, Plant Draws Attention," *New York Times,* September 29, 2009, p. A19, www .nytimes.com/2009/09/22/science/earth/22coal.html?ref=american_electric _power_company.

233–34   *"We view the genome as the software . . . There are already thousands . . . We think this field":* J. Craig Venter, quoted in *Oil and the Future of Energy: Climate Repair, Hydrogen, Nuclear Fuel, Renewable and Green Sources, Energy Efficiency,* editors of Scientific American (Guilford, Conn.: Lyons Press, 2007), pp. 220–21. From Venter's presentation "Synthetic Genomics" at the Conference on Synthetic Biology (SB2.0), Berkeley, California, May 20, 2006. Audio available at http:// webcast.berkeley.edu/event_details.php?webcastid=15766.

234   *"carbon bank":* Freeman J. Dyson, "Can We Control the Carbon Dioxide in the Atmosphere?" *Energy* 2 (1977): pp. 287–91.

235   *An 8-ounce glass of water is equal to:* Sheffield, p. 158.

236   *"I know what the other material is":* Ralph Lapp, quoted in "Perón's Atom," *Time,* April 2, 1951, www.time.com/time/magazine/article/0,9171,814503,00.html.

236   *"Less than that":* Seife, p. 76.

238   *"Even if the plant were flattened":* W. Wayt Gibbs, "Plan B for Energy: 8 Revolutionary Energy Sources," *Scientific American,* September 2006; reprinted April 2, 2009, www.scientificamerican.com/article.cfm?id=plan-b-for-energy-8 -ideas.

238   *"A decade ago":* Ibid.

240   *If the pellet is irregular by more than 50 nanometers:* Seife, p. 211.

242–43   *it will weigh 23,000 tons . . . ten times the amount of energy:* ITER, www.iter.org/ factsfigures.

243   *The ITER is still just a science project:* Gibbs, "Plan B," *Scientific American,* September 2006.

253   *"SSP offers a truly sustainable":* Editors of Scientific American, *Oil and the Future of Energy,* p. 217.

253   *Ben Bova, writing in the* Washington Post: Ben Bova, "To the Next President" (originally titled "An Energy Fix Written in the Stars," guest editorial, *Washington Post,* October 12, 2008), www.nss.org/settlement/ssp/bova.htm.

253   *"It sounds like a science fiction cartoon":* International Herald Tribune, September 2, 2009, p. 14. Also see Shigeru Sato and Yuji Okada, "Mitsubishi, IHI to Join $21 Bln Space Solar Project," August 31, 2009; www.bloomberg.com/apps/news?pid= newsarchive&sid=aJ529lsdk9HI.

254   *"These expenses"*: Shigeru Sato and Yuji Okada, "Mitsubishi, IHI to Join $21 Bln Space Solar Project," Bloomberg, August 31, 2009, www.bloomberg.com/apps/ news?pid=newsarchive&sid=aJ529lsdk9HI.

6. FUTURE OF SPACE TRAVEL: TO THE STARS

258   *One possibility is the Europa Ice Clipper Mission:* http://science.nasa.gov/science -news/science-at-nasa/1999/ast02feb99_1/.

268   *One game changer has been the discovery of ancient ice:* http://lcross.arc.nasa.gov.

274   *"This is an uncertain market": New York Times,* September 16, 2010, p. A3.

275   *Physicist Freeman Dyson has narrowed down some experimental technologies:* Dyson, pp. 88–99.

280   *"For transmission lines"*: Katherine Bourzac, "Making Carbon Nanotubes into Long Fibers," *Technology Review,* November 10, 2009, www.technologyreview .com/energy/23921/.

281   *Initially, the task was so difficult that no one won the prize:* BBC-TV, November 5, 2009.

283   *But finally, in May 2010, the Japan Aerospace Exploration Agency:* http://en .wikipedia.org/wiki/Ikaros.

284   *"For me, Orion . . . 2,000 bombs"*: Nicholas Dawidoff, "The Civil Heretic," *New York Times,* March 25, 2009, www.nytimes.com/2009/03/29/magazine/29Dyson-t.html ?pagewanted=7&_r=1.

289   *"The exploration of the solar system"*: Vint Cerf, "One Is Glad to Be of Service," in Denning, pp. 229–30.

290   *In 2007 and 2009, the Air Force released position papers detailing:* Scott A. Dickson, "Enabling Battlespace Persistent Surveillance: The Form, Function and Future of Smart Dust," April 2007 (Blue Horizon Paper, Center for Strategy and Technology, Air War College).

7. FUTURE OF WEALTH: WINNERS AND LOSERS

297   *"The great Islamic civilization went into decline when Muslim scholars"*: Umi Hani Sharani, "Muslims Almost Totally Dependent on Others, Says Mahathir," Muslim Institute, April 15, 2006, www.musliminstitute.com/article.php?id=499.

301   *"Heavens, no. It will be a hundred years"*: William J. Holstein, "To Gauge the Internet, Listen to the Steam Engine," *New York Times,* August 26, 2001, http:// www.nytimes.com/2001/08/26/business/26SVAL.html?scp=1&sq=%22to%20gauge %20the%20internet%22&st=cse.

302   *"attribute 90 percent of income growth in England and the United States"*: Virginia Postrel, "Avoiding Previous Blunders," *New York Times,* January 1, 2004, www .nytimes.com/2004/01/01/business/01scene.html.

303   *"A century ago, railroad companies"*: Ibid.

303   *"In the 19th century"*: Thomas L. Friedman, "Green the Bailout," *New York Times,* September 28, 2008, p. WK11, www.nytimes.com/2008/09/28/opinion/28friedman .html.

303   *From 1900 to 1925, the number of automobile start-up companies:* Steve Lohr, "New Economy; Despite Its Epochal Name, the Clicks-and-Mortar Age May Be Quietly Assimilated," *New York Times,* October 8, 2001, www.nytimes.com/2001/10/08/

business/new-economy-despite-its-epochal-name-clicks-mortar-age-may-be
-quietly.html?scp=30&sq=automobile&st=nyt.

304 *"The early 21st century saw a boom":* Ibid.

306 *"The do-it-yourself model":* Charles Gasparino, "Merrill Lynch to Offer Online
Trading," ZDNet News, June 1, 1999, www.zdnet.com/news/merrill-lynch-to-offer
-online-trading/95883.

306 *"Rarely in history has the leader in an industry":* Ibid.

309 *"In practice, the vast bulk of this 'information' ":* McRae, p. 175.

315 *"Today, knowledge and skills":* Thurow, p. 68.

315 *"With everything else dropping out of the competitive equation":* Thurow, p. 74.

315 *"in 1991 Britain became":* McRae, p. 12.

316 *"After correcting for general inflation":* Thurow, p. 67.

317 *"The prolonged migration":* James Grant, "Sometimes the Economy Needs a
Setback," *New York Times,* September 9, 2001, www.nytimes.com/2001/09/09/
opinion/sometimes-the-economy-needs-a-setback.html.

318 *"Success or failure depends":* Thurow, p. 72.

319 *"The old motors of growth":* McRae, pp. 12–13.

320 *"we are told by the World Bank that nearly 2.8 billion people":* Toffler, p. 288.

8. FUTURE OF HUMANITY: PLANETARY CIVILIZATION

333 *"People will inevitably start to look around them":* Kenichi Ohmae, *The End of the
Nation State: The Rise of Regional Economies* (New York: Free Press, 1995), p. 45.

348 *"It is impossible to imagine the height to which may be carried":* Benjamin Franklin,
letter to Joseph Priestley, quoted in Cornish, p. 173.

350 *"Science is organized knowledge. Wisdom is organized life":* http://www.brainyquote
.com/quotes/authors/i/immanuel_kant_2.html.

350 *"The saddest aspect of society right now":* http://www.brainyquote.com/quotes/
topics/topic_science4.html.

350 *"Democracy is the worst form of government":* http://www.brainyquote.com/
quotes/keywords/democracy.html.

350 *"Democracy is a device that ensures we shall be governed":* http://www.brainyquote
.com/quotes/authors/g/george_bernard_shaw_2.htm.

351 *"History is more or less bunk":* quoted in Rhodes, p. 61.

9. A DAY IN THE LIFE IN 2100

368 *"The Roots of Violence":* http://thinkexist.com/quotation/the_roots_of_violence
-wealth_without_work/191301.html.

RECOMMENDED READING

Archer, David. *The Long Thaw: How Humans Are Changing the Next 100,000 Years of Earth's Climate*. Princeton, NJ: Princeton University Press, 2009.

Bezold, Clement, ed. *2020 Visions: Health Care Information Standards and Technologies*. Rockville, MD: United States Pharmacopeial Convention, 1993.

Brockman, Max, ed. *What's Next? Dispatches on the Future of Science*. New York: Vintage, 2009.

Broderick, Damien. *The Spike: How Our Lives Are Being Transformed by Rapidly Advancing Technologies*. New York: Forge, 2001.

Broderick, Damien, ed. *Year Million: Science at the Far Edge of Knowledge*. New York: Atlas, 2008,

Brooks, Rodney A. *Flesh and Machines: How Robots Will Change Us*. New York: Vintage, 2003.

Brown, Lester. *Plan B 4.0: Mobilizing to Save Civilization*. New York: Norton, 2009.

Canton, James. *The Extreme Future: The Top Trends That Will Reshape the World for the Next 5, 10, and 20 Years*. New York: Dutton, 2006.

Coates, Joseph F., John B. Mahaffie, and Andy Hines. *2025: Scenarios of U.S. and Global Society Reshaped by Science and Technology*. Greensboro, NC: Oakhill Press, 1997.

Cornish, Edward, ed. *Futuring: The Exploration of the Future*. Bethesda, MD: World Future Society, 2004.

Crevier, Daniel. *AI: The Tumultuous History of the Search for Artificial Intelligence*. New York: Basic Books, 1993.

Davies, Paul. *The Eerie Silence: Renewing Our Search for Alien Intelligence*. Boston: Houghton Mifflin Harcourt, 2010.

Denning, Peter J., ed. *The Invisible Future: The Seamless Integration of Technology into Everyday Life*. New York: McGraw Hill, 2002.

Denning, Peter J., and Robert M. Metcalfe. *Beyond Calculation: The Next Fifty Years of Computing*. New York: Copernicus, 1997.

Dertouzos, Michael. *What Will Be: How the New World of Information Will Change Our Lives*. New York: HarperCollins, 1997.

Didsbury, Howard F., Jr., ed. *Frontiers of the 21st Century: Prelude to the New Millennium*. Bethesda, MD: World Future Society, 1999.

———. *21st Century Opportunities and Challenges: An Age of Destruction or an Age of Transformation*. Bethesda, MD: World Future Society, 2003.

Dyson, Freeman J. *The Sun, the Genome, and the Internet: Tools of Scientific Revolutions*. New York: Oxford University Press, 1999.

Foundation for the Future. *Future of Planet Earth: Seminar Proceedings*. Bellevue, WA: Foundation for the Future, 2009; www.futurefoundation.org/publications/index.htm.

———. *The Next Thousand Years.* Bellevue, WA: Foundation for the Future, 2004.

Friedman, George. *The Next 100 Years: A Forecast for the 21st Century.* New York: Doubleday, 2009.

Hanson, William. *The Edge of Medicine: The Technology That Will Change Our Lives.* New York: Palgrave Macmillan, 2008.

Kaku, Michio. *Visions: How Science Will Revolutionize the 21st Century.* New York: Anchor, 1998.

Kurzweil, Ray. *The Singularity Is Near: When Humans Transcend Biology.* New York: Viking, 2005.

McElheny, Victor K. *Drawing the Map of Life: Inside the Human Genome Project.* New York: Basic Books, 2010.

McRae, Hamish. *The World in 2020: Power, Culture, and Prosperity.* Cambridge, MA: Harvard Business School, 1995.

Mulhall, Douglas. *Our Molecular Future: How Nanotechnology, Robotics, Genetics, and Artificial Intelligence Will Transform Our World.* Amherst, NY: Prometheus, 2002.

Petersen, John L. *The Road to 2015: Profiles of the Future.* Corte Madera, CA: Waite Group, 1994.

Pickover, Clifford A., ed. *Visions of the Future: Art, Technology and Computing in the Twenty-first Century.* New York: St. Martin's Press, 1994.

Rhodes, Richard, ed. *Visions of Technology: A Century of Vital Debate About Machines, Systems, and the Human World.* New York: Simon & Schuster, 1999.

Ridley, Matt. *The Rational Optimist: How Prosperity Evolves.* New York: HarperCollins, 2010.

Rose, Steven. *The Future of the Brain: The Promise and Perils of Tomorrow's Neuroscience.* New York: Oxford University Press, 2005.

Seife, Charles. *Sun in a Bottle: The Strange History of Fusion and the Science of Wishful Thinking.* New York: Viking Penguin, 2008.

Sheffield, Charles, Marcelo Alonso, and Morton A. Kaplan, eds. *The World of 2044: Technological Development and the Future of Society.* St. Paul, MN: Paragon House, 1994.

Stock, Gregory. *Redesigning Humans: Choosing Our Genes, Changing Our Future.* Boston: Houghton Mifflin, 2003.

Thurow, Lester C. *The Future of Capitalism: How Today's Economic Forces Shape Tomorrow's World.* New York: William Morrow, 1996.

Toffler, Alvin, and Heidi Toffler. *Revolutionary Wealth.* New York: Knopf, 2006.

van der Duin, Patrick. *Knowing Tomorrow? How Science Deals with the Future.* Delft, Netherlands: Eburon, 2007.

Vinge, Vernor. *Rainbows End.* New York: Tor, 2006.

Watson, Richard. *Future Files: The 5 Trends That Will Shape the Next 50 Years.* London: Nicholas Brealey, 2008.

Weiner, Jonathan. *Long for This World: The Strange Science of Immortality.* New York: HarperCollins, 2010.

INDEX

ILLUSTRATION CREDITS

PENGUIN SCIENCE

PHYSICS OF THE IMPOSSIBLE
MICHIO KAKU

'A brilliant, provocative, freewheeling tour around the exotic shores of physics'
Independent

'A rich compendium of jaw-dropping reality checks' *The Times*

From cyborgs, starships, UFOs, aliens and antimatter to telepathy, invisibility, psychokinesis and precognition

Albert Einstein said, 'If at first an idea does not sound absurd, there is no hope for it.' *Physics of the Impossible* shows how our most far-fetched ideas today are destined to become tomorrow's science.

Michio Kaku, bestselling author and one of the world's most acclaimed physicists, looks at the technologies of the future and explains what is just around the corner, what we might have to wait a few millennia to get our hands on and how surprisingly little of it is truly impossible.

(And if that teleporter still feels a long way off, you should know that scientists have already succeeded in transmitting atoms across the lab...)

'One of the world's most distinguished physicists... takes the reader on a journey to the frontiers of science and beyond' *Guardian*

'After reading Kaku's boundless enthusiasm for the future, what you wouldn't give for a real-life time machine to travel forwards and see just how accurate his predictions are' *Sunday Telegraph*

PENGUIN SCIENCE

SIX EASY PIECES
RICHARD P. FEYNMAN

An outstanding communicator, Richard P. Feynman (1918-88) was one of this century's most brilliant and original thinkers, inspiring a generation of students with his energetic, unorthodox style of teaching. Feynman taught at Cornell, as well as the California Institute of Technology, and received the Nobel Prize in Physics for his work in quantum electrodynamics.

Drawn from his celebrated and landmark text Lectures on Physics, *Six Easy Pieces* reveals Feynman's distinctive style while introducing the essentials of physics to the general reader. The topics explored include atoms, the fundamentals of physics and its relation to other sciences, the theory of gravitation and quantum behaviour.

'If one book was all that could be passed on to the next generation of scientists it would undoubtedly have to be *Six Easy Pieces*' - John Gribbin, *New Scientist*

RICHARD P. FEYNMAN

Nobel laureate Richard Feynman was also a man who fell, often jumped, into adventure. An artist, safe-cracker, practical joker and storyteller, his life was a series of combustible combinations made possible by his unique mixture of high intelligence, unquenchable curiosity and eternal scepticism.

'His immense intelligence, irrepressible wit and touching optimism radiate from every page' *Independent on Sunday*

QED: THE STRANGE THEORY OF LIGHT AND MATTER

Quantum electrodynamics - or QED for short - is the 'strange theory' that explains how light and electrons interact. Thanks to Richard Feynman, it is also one of the rare parts of physics that is known for sure. In this lucid set of lectures, Feynman provides the definitive introduction to QED.

THE CHARACTER OF PHYSICAL LAW

This series of classic lectures, first delivered in 1960, present Feynman's unique take on the problems and puzzles that lie at the heart of physical theory - with Newton's Law of Gravitation; on whether time can ever go backwards; on maths as the supreme language of nature.

THE PLEASURE OF FINDING THINGS OUT

This collection of the Feynman's best short works shows his passion for knowledge and sense of fun at their most infectious. The revealing and inspiring pieces here span a lifetime of enthusiasm for discovering what makes the world tick – including uproarious tales of early student experiments; safecracking and outwitting US censors during the Second World War; his first lecture as a graduate student (to an audience including Albert Einstein); and the memories of the father who delighted in showing him the world and sparked his insatiable curiosity.

PENGUIN SCIENCE

THE FABRIC OF THE COSMOS
BRIAN GREENE

We are living in what is often referred to as 'the golden age of cosmology'. The study of string theory has led in recent years to fast and furious progress being made towards the discovery of the long-searched-for Theory of Everything. If it comes, this theory will describe a universe very different from that which common sense suggests – a quantum universe in which time and space can be measured only relatively, and nothing is certain. In *The Fabric of the Cosmos*, Brian Greene shows why, explains how far we have come, and speculates on how far we still have to go.

'A magnificent challenge to conventional ideas' *Financial Times*

'Manages to be both challenging and entertaining: it is highly recommended' *Independent*

'Greene threatens to do for string theory what Stephen Hawking did for black holes' *New York Magazine*

'Sends the reader's imagination hurtling through the universe on an astonishing ride. As a popularizer of exquisitely abstract science, Greene is both a skilled and kindly explicator' *The New York Times*

PENGUIN SCIENCE

IN SEARCH OF THE MULTIVERSE
JOHN GRIBBIN

In Search of The Multiverse takes us on an extraordinary journey, examining the most fundamental questions in science. What are the boundaries of our Universe? Can there be different physical laws to the ones we know? Are there in fact other universes? Do we really live in a Multiverse?

This book is a search – the ultimate search – exploring the frontiers of reality. Ideas that were once science fiction have now come to dominate modern physics. And, as John Gribbin shows, there is increasing evidence that there really is more to the Universe than we can see. Gribbin guides us through the different competing theories revealing what they have in common and what we can come to expect.

Along the way Gribbin explores the very latest thinking about quantum theory, about gravity and the fundamental forces that shape our world, about time and multiple dimensions, about matter itself, and the growth and fate of the known Universe.

John Gribbin is our best, most accessible guide to the big questions of science. And there is no bigger question than our search for the Multiverse.

He just wanted a decent book to read ...

Not too much to ask, is it? It was in 1935 when Allen Lane, Managing Director of Bodley Head Publishers, stood on a platform at Exeter railway station looking for something good to read on his journey back to London. His choice was limited to popular magazines and poor-quality paperbacks – the same choice faced every day by the vast majority of readers, few of whom could afford hardbacks. Lane's disappointment and subsequent anger at the range of books generally available led him to found a company – and change the world.

'We believed in the existence in this country of a vast reading public for intelligent books at a low price, and staked everything on it'
Sir Allen Lane, 1902–1970, founder of Penguin Books

The quality paperback had arrived – and not just in bookshops. Lane was adamant that his Penguins should appear in chain stores and tobacconists, and should cost no more than a packet of cigarettes.

Reading habits (and cigarette prices) have changed since 1935, but Penguin still believes in publishing the best books for everybody to enjoy. We still believe that good design costs no more than bad design, and we still believe that quality books published passionately and responsibly make the world a better place.

So wherever you see the little bird – whether it's on a piece of prize-winning literary fiction or a celebrity autobiography, political tour de force or historical masterpiece, a serial-killer thriller, reference book, world classic or a piece of pure escapism – you can bet that it represents the very best that the genre has to offer.

Whatever you like to read – trust Penguin.